ASTRONOMICAL CALENDAR 2023

by Guy Ottewell

Universal Workshop
www.UniversalWorkshop.com
Durham, North Carolina, USA, and Isleworth, Middlesex, England

ISBN 978-0-934546-90-4

Preface and acknowledgments

The *Astronomical Calendar* was published as a printed book for the years 1974 to 2016. It started because Professor Bill Brantley, chairman of the physics department at Furman University, Greenville, South Carolina, and creator of a famous course on "Energy," got me to show the stars to his students, and for the first many years it was under the auspices of the university and the Astronomical League.

It was continued by web pages for 2017 to 2021. Then I was able to revive it as an electronic book for 2022, with the help of Daniel Cummings, an innovator in science education (starinastar.com). For 2023 it is offered both in this way and again as a printed book.

I am again grateful to John Goss, past President of the Astronomical League, and to my son Roland for proofreading. Any surviving errors will be due to subsequent changes made by me. I thank Alastair McBeath for untinting advice about meteor showers.

A tip on how to use the electronic version of this book

It has facing pages, of letter size (8.5 by 11 inches), like a paper book. The larger the screen you can view it on, the better. And it will be best with a facing-pages view.

This can be done because it is a PDF document and will open in Adobe Reader. In the "view" menu, click "page display," then "two page view" or, even better, "two page scrolling," which allows you to move up and down with the "hand" tool. In the same dropdown, make sure that "show cover page in two page view" is enabled.

When reading that for a certain date there is, e.g., a meteor shower, you can refer to the section on meteors (remembering that the list of contents is on page 3).

Explore www.universalworkshop.com
for these and other resources:

The Astronomical Companion, universal explanation!

To Know the Stars, for children and other beginners

Albedo to Zodiac, glossary of astronomical terms

Map of the Starry Sky, huge detailed poster

Zodiac Wavy Chart for 2023 (and other years), poster-size, the Sun, Moon, and planets throughout the year

Venus: the planet and the goddess

Uranus, Neptune, Pluto

The Thousand-Yard Model of the solar system, a walk

The Under-Standing of Eclipses

Berenice's Hair, historical novel—what was the real story of how her hair became a constellation?

—and free web pages on everything from global heating to imaginary islands!

Simple calendar

Darker color means less moonlight in the following night.
Julian dates are at 0h UT on day 1 of the months.

2023

Julian Date at 0 UT between months		Sun	Mon	Tue	Wed	Thu	Fri	Sat
2459945.5		1	2	3	4	5	6	7
	January	8	9	10	11	12	13	14
		15	16	17	18	19	20	21
		22	23	24	25	26	27	28
2459976.5		29	30	31	1	2	3	4
	February	5	6	7	8	9	10	11
		12	13	14	15	16	17	18
		19	20	21	22	23	24	25
2460004.5		26	27	28	1	2	3	4
	March	5	6	7	8	9	10	11
		12	13	14	15	16	17	18
		19	20	21	22	23	24	25
2460035.5		26	27	28	29	30	31	1
	April	2	3	4	5	6	7	8
		9	10	11	12	13	14	15
		16	17	18	19	20	21	22
		23	24	25	26	27	28	29
2460065.5		30	1	2	3	4	5	6
	May	7	8	9	10	11	12	13
		14	15	16	17	18	19	20
		21	22	23	24	25	26	27
2460096.5		28	29	30	31	1	2	3
	June	4	5	6	7	8	9	10
		11	12	13	14	15	16	17
		18	19	20	21	22	23	24
2460126.5		25	26	27	28	29	30	1
	July	2	3	4	5	6	7	8
		9	10	11	12	13	14	15
		16	17	18	19	20	21	22
		23	24	25	26	27	28	29
2460157.5		30	31	1	2	3	4	5
	August	6	7	8	9	10	11	12
		13	14	15	16	17	18	19
		20	21	22	23	24	25	26
2460188.5		27	28	29	30	31	1	2
	September	3	4	5	6	7	8	9
		10	11	12	13	14	15	16
		17	18	19	20	21	22	23
		24	25	26	27	28	29	30
2460218.5		1	2	3	4	5	6	7
	October	8	9	10	11	12	13	14
		15	16	17	18	19	20	21
		22	23	24	25	26	27	28
2460249.5		29	30	31	1	2	3	4
	November	5	6	7	8	9	10	11
		12	13	14	15	16	17	18
		19	20	21	22	23	24	25
2460279.5		26	27	28	29	30	1	2
	December	3	4	5	6	7	8	9
		10	11	12	13	14	15	16
		17	18	19	20	21	22	23
		24	25	26	27	28	29	30
2460310.5		31	1					

CONTENTS

This graph summarizes the observability of the planets. The vertical dimension is elongation (angular distance from the Sun). The curve for a planet is blue when it is in the morning sky, black when in the evening sky. The thickness of the curve is proportional to the planet's magnitude (brightness). Not taken into account is the planet's declination (angular distance north or south of the equator). A planet reaches the top of the graph when it is at opposition,

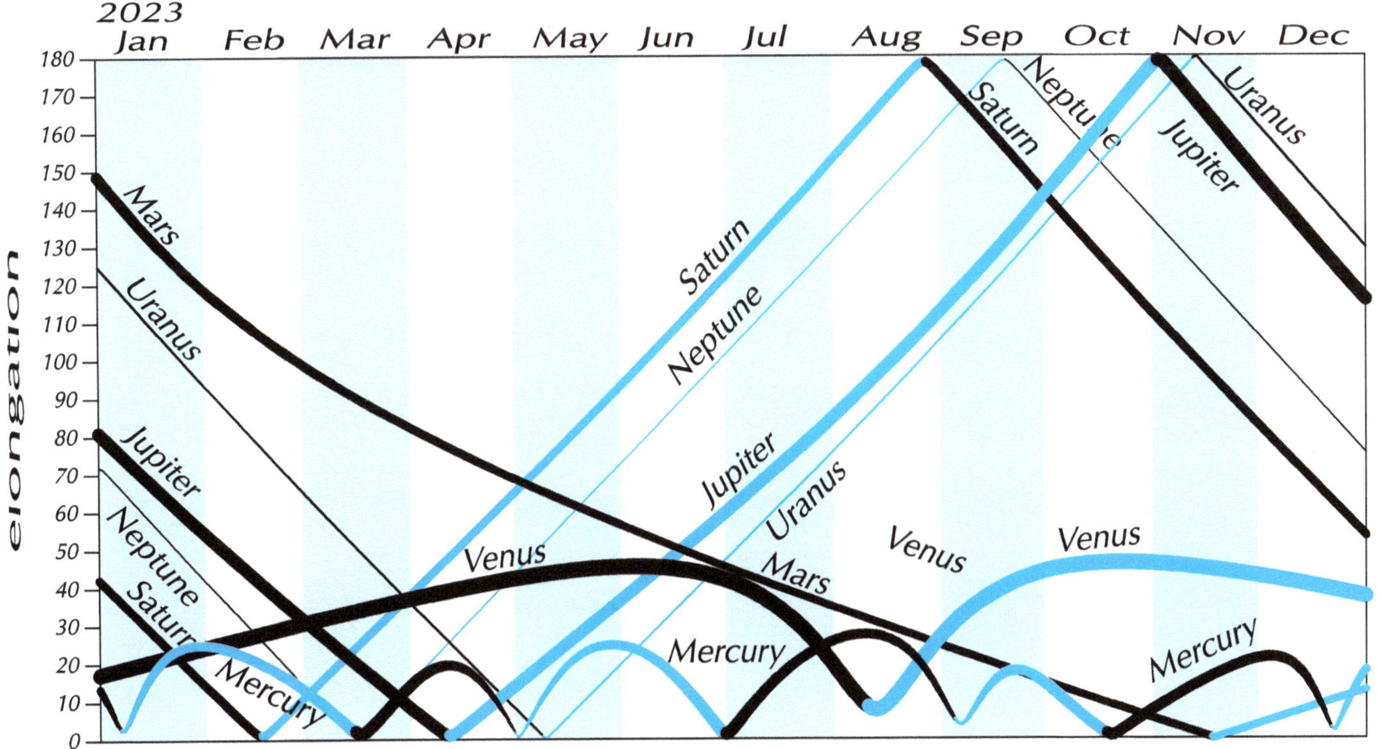

The mystique of Canopus: it allows itself to be seen at chosen times, like a seal coming up to a breathing-hole in ice. You have to know, or be lucky, to see it.

In my story of Berenice (she whose hair became a constellation), she comes before sunrise onto the deck of a ship and sees "the stars fade but one rise at the far line of the sea. What could that star be? — it must be Canopus, the great star of the south, that guides travelers in the desert but that we of the middle lands see only once a year." (But it proves to be the lighthouse of Alexandria, one of the seven wonders of the ancient world.)

And in my attempt to decipher the Navajo constellations, for the cover picture of *Astronomical Calendar 2006*, I guessed that the Monthless Star, the one that remained after Coyote blew the rest wildly into the sky, and which he dumped "in the extreme southern horizon," was Canopus. For the Navajos it shows only a couple of degrees high and is unlikely to be noticed. But it could have been borrowed from the culture of the Hopi to the south; or it could have been glimpsed by the parties sent to the four sacred peaks to fetch water, soil, and herbs for use in rituals. "Look! The star that Coyote tried to hide!"

There are more "tales of Canopus-sightings," and much else about Canopus, at https://www.universalworkshop.com/2021/03/04/the-sirius-canopus-hour-and-yemen/

The two brightest stars of our night sky, Sirius and Canopus, are both in the southern celestial hemisphere, but Canopus is much deeper south, by 36 degrees. Sirius passes overhead for Bolivia or Zimbabwe or northern Australia; Canopus for the tip of South America, which projects farther south than the other continents. We can see Sirius from North America or Europe in some part of almost every night. To see Canopus you must go south at least to latitude 37°, such as Virginia or Sicily. And there it come up over the horizon only briefly, at a time that changes through the seasons

Canopus is almost due south of Sirius. They are both in the north-south gore of the sky between the 6h and 7h lines of right ascension. They are at their highest, crossing the meridian, in what I and Fred Schaaf call the Sirius-Canopus Hour.

Every such state of the sky, or sidereal time, arrives a few minutes earlier each day. The Sirius-Canopus Hour happens around midnight at the start of each year, around sunset in March, around noon in June, around dawn in October. Such are the times when Canopus peeps above our horizon.

There is no South Pole Star. The south celestial pole, even if you are far enough south to see it, is in Octans, where there is no star brighter than magnitude 4. But Canopus has to be near the south point on the horizon, if you see it and are a cameleer on one of the ancient caravan trade routes across the Sahara.

Another feature of the north-south gore containing Canopus and Sirius is Camelopardalis (or Camelopardus), a constellation invented by Petrus Plancius about 1612 to fill a large star-poor area. The Greeks used "camel" to concoct words for African beasts they encountered later: *strouthokamêlos*, "sparrow-camel," the ostrich; *kamêlopardaislis*, "camel-pard", the giraffe. This constellation's stars, too, are not above magnitude 4. If you can see them, they suggest a camel's hump and curved neck rather than the straight neck and back of the giraffe. That may be why I forgot and painted a camel to go with the camels below. I don't want to paint a giraffe over him.

Camelopardalis is one of the far-north circumpolar constellations. It half-surrounds the north celestial pole.The animal is upside-down if you have the northern horizon behind you. In our picture, the zenith, the overhead-point, is indicated by a fictitious burst of meteors, from a radiant in Auriga. The Camel is north of it; Canopus is south of it, as are the camels plodding across a dune on their way from Egypt to a Libyan oasis, or from Algeria to Nigeria.

In our constellation system, Canopus is Alpha Carinae, chief star of Carina the keel or hull, one of three parts into which the huge ship Argo was divided. In legend, Canobus or Canopus was the helmsman of the ship of Menelaus, Helen's husband. After the fall of Troy, it was wind-driven to Egypt. Canopus died of a snake-bite, and his name was preserved in the Canopic Mouth of the Nile, and in a pleasure-city beside it, and in Alexandria's Canopic Gate leading eastward toward it; and perhaps also in the star toward which the helmsman had steered. (But *canopy* is from *kônôpeion*, from *kônôps*, "cone-face", the mosquito.)

Star Canopus, at magnitude -0.7, or 0.8 of a magnitude below Sirius, is only a quarter as bright. But, at 310 light-years, it is 36 times farther away, and its real brightness is about 700 times greater.

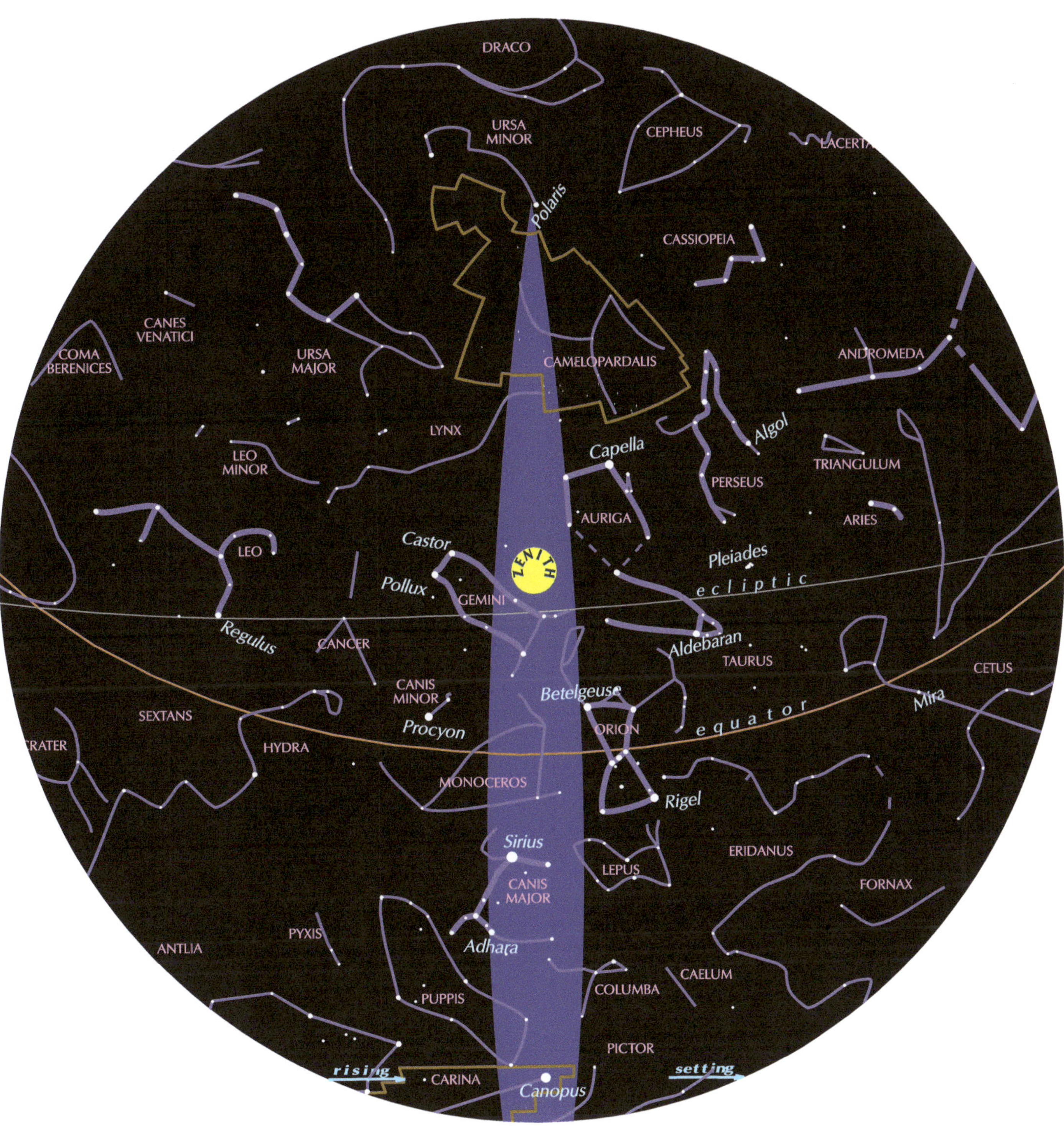
DRACO
URSA MINOR
CEPHEUS
LACERTA
Polaris
CASSIOPEIA
CANES VENATICI
COMA BERENICES
URSA MAJOR
CAMELOPARDALIS
ANDROMEDA
LYNX
Algol
Capella
LEO MINOR
TRIANGULUM
PERSEUS
AURIGA
ARIES
Castor
LEO
ZENITH
Pleiades
Pollux
ecliptic
GEMINI
Regulus
CANCER
Aldebaran
TAURUS
CETUS
CANIS MINOR
Betelgeuse
Mira
SEXTANS
Procyon
ORION
equator
CRATER
HYDRA
MONOCEROS
Rigel
Sirius
ERIDANUS
LEPUS
CANIS MAJOR
FORNAX
PYXIS
ANTLIA
Adhara
CAELUM
PUPPIS
COLUMBA
PICTOR
rising
CARINA
Canopus
setting

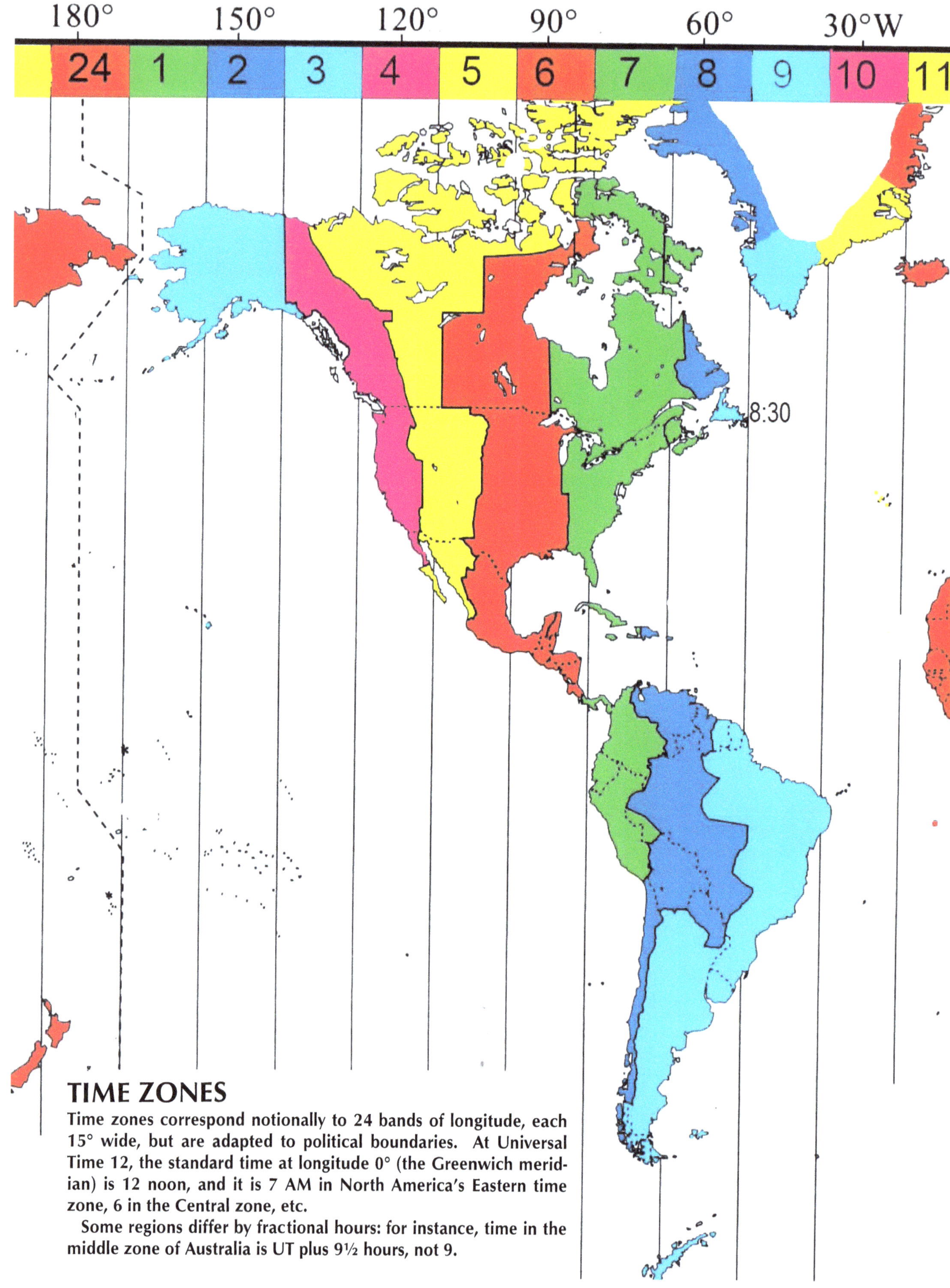

TIME ZONES

Time zones correspond notionally to 24 bands of longitude, each 15° wide, but are adapted to political boundaries. At Universal Time 12, the standard time at longitude 0° (the Greenwich meridian) is 12 noon, and it is 7 AM in North America's Eastern time zone, 6 in the Central zone, etc.

Some regions differ by fractional hours: for instance, time in the middle zone of Australia is UT plus 9½ hours, not 9.

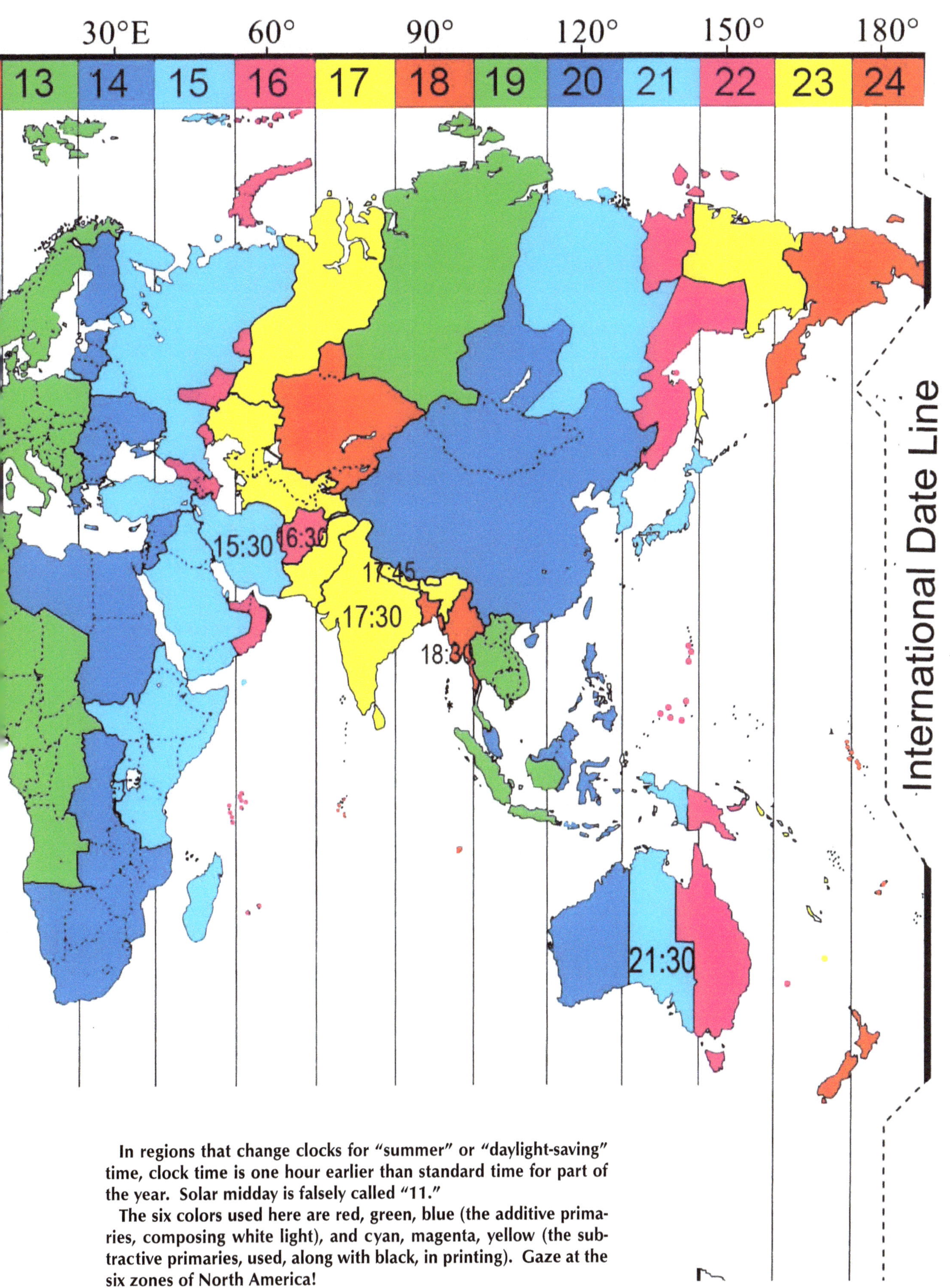

In regions that change clocks for "summer" or "daylight-saving" time, clock time is one hour earlier than standard time for part of the year. Solar midday is falsely called "11."

The six colors used here are red, green, blue (the additive primaries, composing white light), and cyan, magenta, yellow (the subtractive primaries, used, along with black, in printing). Gaze at the six zones of North America!

EXPLANATION OF THE MAIN FEATURES

The map for each month shows what is above the horizon at a convenient evening time, for latitude 40° north.

You can see the relation between map and sky by lying on your back with your feet pointing south, and hold the map over your face. The central point of the map is the overhead point or zenith. East is at the right of an Earth map, which is a view down, but at the left of a sky map, which is a view up. When instead of facing south you face east, turn the map so that the eastern horizon is at the bottom.

"Rising" and "setting" arrows show how far everything in the sky moves in one hour, rotating around the north celestial pole (near which is the Pole Star). If you look successively at the January, February, etc. maps, you will see that they form a "movie" of this rotation. The sky rotates westward (1) from hour to hour through any one night, and (2) from month to month, if viewed at the same time each month.

The approximate evening time of the maps (around 9-10 PM) are in local Sun time, disregarding summer clock-change rules.

Stars are shown down to magnitude 5.5, which is about the naked-eye limit in average conditions.

Constellations. Lines from star to star suggest the traditional pictures of the constellations. There are 88 official constellations, but in these maps we show only the more conspicuous ones. Not all the stars along the lines are bright enough to be marked. Such form-lines are usually straight, but ours curve for better expressivity. The aim is to make the constellations perceivable in the real sky (so it's no use drawing lines to stars too dim to see, or ingenious lines to non-adjacent stars to make stick figures). There is nothing official about these form-lines; you can imagine your own.

Stars are a **fixed background.** As they look on a January evening this year, so they will look next year, or indeed next century, except for small changes by *precession*, and even smaller changes by proper motion (their motion in space). In front of them is a **moving foreground**, consisting of everything inside our solar system, and inside our atmosphere, such as meteors,

Planets are shown at the 16th of the month, with symbols sized for brightness like the stars. Planets, when not too near the Sun, are visible to the naked eye, except Neptune; Uranus is just visible in good conditions.

These maps are of the mid-evening sky; there are other groupings of planets in the sky after midnight and in the twilight sky near sunset or sunrise. Mercury cannot appear except close to sunset or sunrise.

The **Moon** is shown (exaggerated 8 times in size) at 0h Universal Time on the days when it is at first quarter and full phases. This is 7 PM of the previous day by Eastern Standard Time, 6 PM by Central Time, etc. It is shown in its geocentric position, that is, without parallax; as seen from northern latitudes, it is slightly farther south.

Meteor showers listed in our calendar and described in the Meteors section are indicated by bursts of lines pointing out from their **radiant** (generally in the constellations from which the showers get their names). But some are not shown, because their radiants are not in view at map time.

The **ecliptic** is drawn as a thick curve. It marks the plane in which the Earth revolves around the Sun. The Sun appears to moves along it, at a little less than 1° a day (there are 360 degrees in the circle and 365.2425 days in the year). The planets and Moon follow it approximately, sloping across it at the ascending and descending nodes of their orbits.

The **celestial equator** curves from the east point to the west point of each map. At declination 0, it is the only line of declination shown. Ticks along it are at the 24 hours of right ascension.

Four main **lines of right ascension** are shown, at 0h, 6h, 12h, and 18h. Ticks along them are at intervals of 10° of declination.

The two points where equator and ecliptic cross are the **equinox** points, where the Sun crosses the equator in March and September. And the two points where equator and ecliptic are farthest apart are the **solstice** points, where the Sun is farthest north and south of the equator in June and December.

The March or **vernal equinox point** is the origin or zero point for measuring angles around the whole sky. (It is sometimes called the "First Point of Aries" because in ancient times it was at the western edge of that constellation. Since then it has moved most of the way through Pisces, because of the effect called precession.)

The **Milky Way** (which cannot really be seen all the way down to the horizon) is drawn as six approximate levels of brightness, from star clouds down to dark lanes and dark nebulae.

The horizon, ecliptic, celestial equator, lines of right ascension, and equator of the Milky Way are all **great circles**: they appear straight as we look out at them, but they represent planes and wrap all around us.

Map positions in the sky change slightly from year to year because of precession (see our short glossary). These maps are for the present standard **epoch** of 2000.

The difference for other latitudes on Earth. The maps show what is seen from 40° north (such as Denver, Philadelphia, Madrid, Ankara, Beijing). There is little difference for most of the U.S., Europe, the northern Middle East, central Asia, and Japan. If you travel north, stars at the south edge of the map disappear; at the north edge others spend more time above the horizon. If you stand on either pole of the Earth, stars neither rise nor set: those on the celestial equator travel around the horizon. See the small dome maps for other latitudes.

Your **longitude** makes little difference, except for the position of the Moon, which moves westward by about its own width every hour.

The **projection** used for the sky domes is **stereographic**, in which shapes remain constant but sizes are exaggerated toward the horizon.

For vastly richer detail of stars, galaxies, interesting points: see our poster *Map of the Starry Sky*.

How the sky changes for other latitudes

These smaller versions of the January evening sky dome show how different stars come into view at the northern and southern horizons.

The yellow circles represent points that pass overhead for the Arctic and Antarctic circles; the yellow line is the celestial equator

North pole

Latitude 30° north

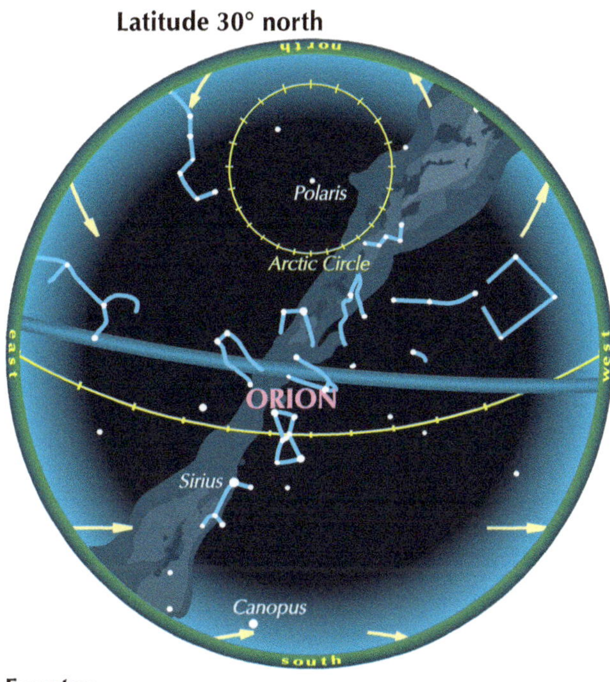

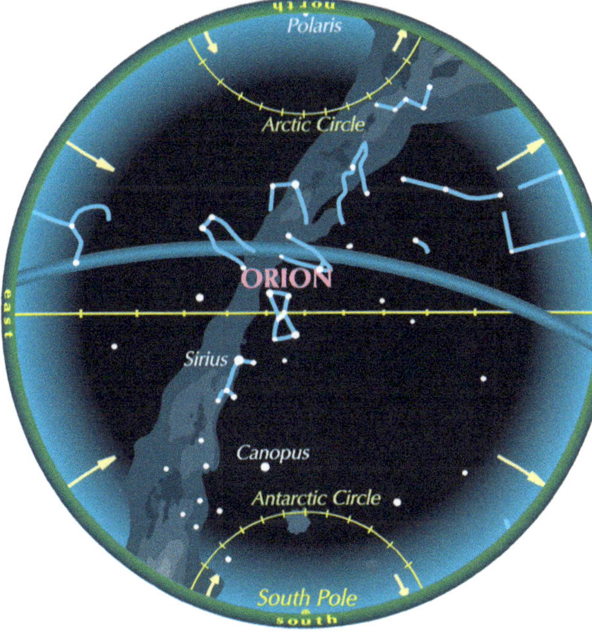

Equator

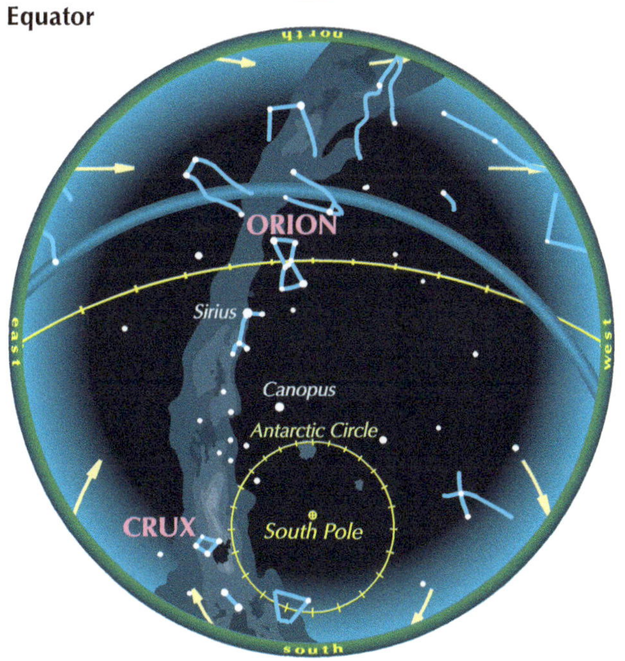

Latitude 30° south

South pole

Timetables of events

The left column gives Julian dates (number of days from 4713 BC Jan. 1 noon), useful for finding time spans between events by subtraction. The first 3 digits of the Julian date (245) are omitted, to save space.

Hours and minutes, where given, are in Universal Time. (Sometimes the hour appears as "24" or the minute as "60," because the instant is shortly before the end of the day or hour.)

Occasions such as "Moon 1.25° NNE of Venus" are **appulses**: closest apparent approaches. They are slightly different from conjunctions, when one passes north of the other as measured in right ascension or in ecliptic longitude. A quasi-conjunction is an appulse without a conjunction, and typically happens when a planet is near its stationary moment.

Occasions when three bodies are withi`in a circle of small size are "**trios**." Like appulses, they are most interesting when the bodies are bright and are not at small elongation from the Sun.

For **meteor showers**, ZHR (zenithal hourly rate) is an estimate of the number to be seen under ideal conditions at the peak time if the radiant were overhead. Actual rates may be very different. Peak times are uncertain; it's advisable to start watching the night before. Meteors are usually most abundant in the morning hours.

Paired scenes

These pages show the ten "best" days in the month for opportunities to see bright bodies (Moon, major planets, first-magnitude stars) in the morning or evening ends of the night.

That is, a ranking of the days has been calculated, based on the presence of these bodies in the scene, their brightness, and their convenient altitude above the horizon and elongation (angular distance from the Sun).

So for some days only the morning scene, or only the evening scene, is "good." Nevertheless the pair is shown, because it is interesting to see the opposite horizons at the two ends of the day.

The pink curve is the ecliptic. The blue curve is the celestial equator, and the large arrow on it shows how far the sky will rotate over the next hour.

The pictures are for latitude 40° north. They would not be much different for most of the USA or Europe; but for locations such as Australia the scenes would be very different and the "best" days different.

MINIMAL GLOSSARY

Terms used in astronomy are fully explained in our book *Albedo to Zodiac*. Here are quick-and-dirty definitions of a few essentials.

asteroids: small solar-system bodies, also called minor planets.

astronomical unit (AU): the Sun-Earth distance (about 150,000,000 kilometers or 93,000,000 miles), used for expressing distances within the solar system.

azimuth: angular distance around or parallel to the horizon, usually measured from the north counter-clockwise. For instance, east is 90°.

celestial equator: the plane in which Earth rotates, and the line around the map of the sky showing this plane.

declination: angular distance north or south of the celestial equator.

dwarf planets: a few bodies, such as Pluto and Ceres, intermediate in size between major planets (such as Earth) and minor planets (such as asteroids).

eccentricity: the measure of the shape of an orbit. A circle has eccentricity zero, a long narrow ellipse has eccentricity approaching 1.

ecliptic: the plane in which Earth revolves around the Sun, and the line around the map of the sky showing this plane.

elongation: angular distance from the Sun. When a planet's elongation is negative (westward), it is in the morning sky.

equinox: the March equinox is when the Sun, traveling on the ecliptic, crosses the celestial equator northward, and the September equinox is when it crosses southward.

inclination of an orbit: its angle to the ecliptic plane.

Julian date: a count of days since 4713 BC, simple to use in calculations instead of years-months-days-hours.

latitude: on Earth, angular distance north or south of the equator; in the sky, angular distance north or south of the ecliptic.

longitude: on Earth, angular distance around from the zero or Greenwich meridian; in the sky, angular distance around or parallel to the ecliptic, measured from the vernal equinox point.

magnitude: the astronomical way of measuring brightness; it was originally an order, from "first magnitude" down. The magnitudes of the brightest stars are around 1 or 0 or even negative; of the faintest visible to the naked eye, about 5.

node: where one plane crosses another. Ascending node: where it crosses northward.

occultation: when one body, such as the Moon, hides another, such as a planet or star.

opposition of a planet: when it is in the direction outward from the Sun, as seen from Earth, and appears approximately nearest and brightest. It may be somewhat north or south of the exact opposite point.

precession: a slow change in the map positions of everything in the sky, caused by the wobbling of Earth's rotational axis. It makes the vernal equinox point move about 1° westward along the ecliptic in 72 years.

right ascension (RA): angular distance around the sky along or parallel to the celestial equator, measured from the vernal equinox point. Can be measured in 360 degrees, but is usually measured in 24 hours.

sidereal period of a planet: the time it takes to revolve around the Sun, in relation to the "starry" (sidereal) background.

solstice: the June solstice is when the Sun appears farthest north, and the December solstice is when it appears farthest south.

stationary moment: when a planet ceases to move eastward or westward, at the beginning or end of its apparent retrograde path.

synodic period of a planet: the time it takes to travel around the sky, as seen from Earth which is itself moving; for instance the time from one opposition to the next.

universal time (UT): the time by the Sun at the zero or Greenwich meridian of longitude on Earth.

vernal equinox point: where the ecliptic crosses the celestial equator northward. It is the origin (zero point) for mapping all positions in the sky.

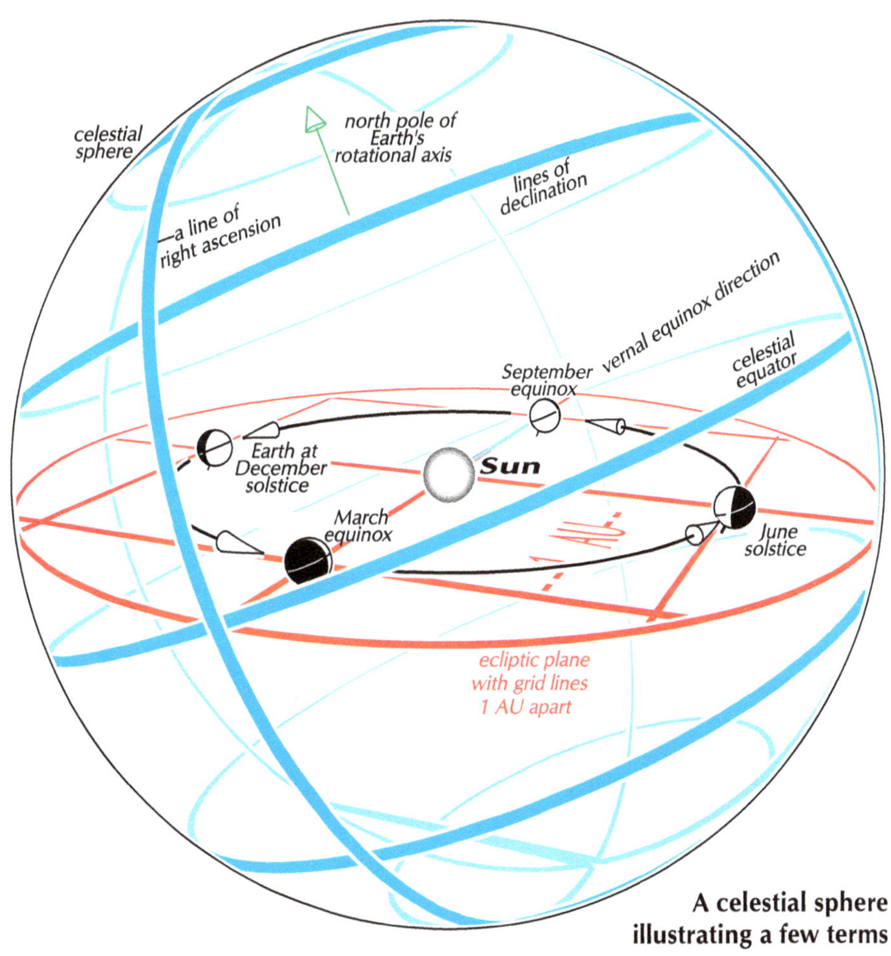

A celestial sphere illustrating a few terms

JANUARY

SKY DOME

Evening sky
for latitude 40° north

about 10 PM at the 5th,
9 PM at the 20th
of the month

sidereal time
5h

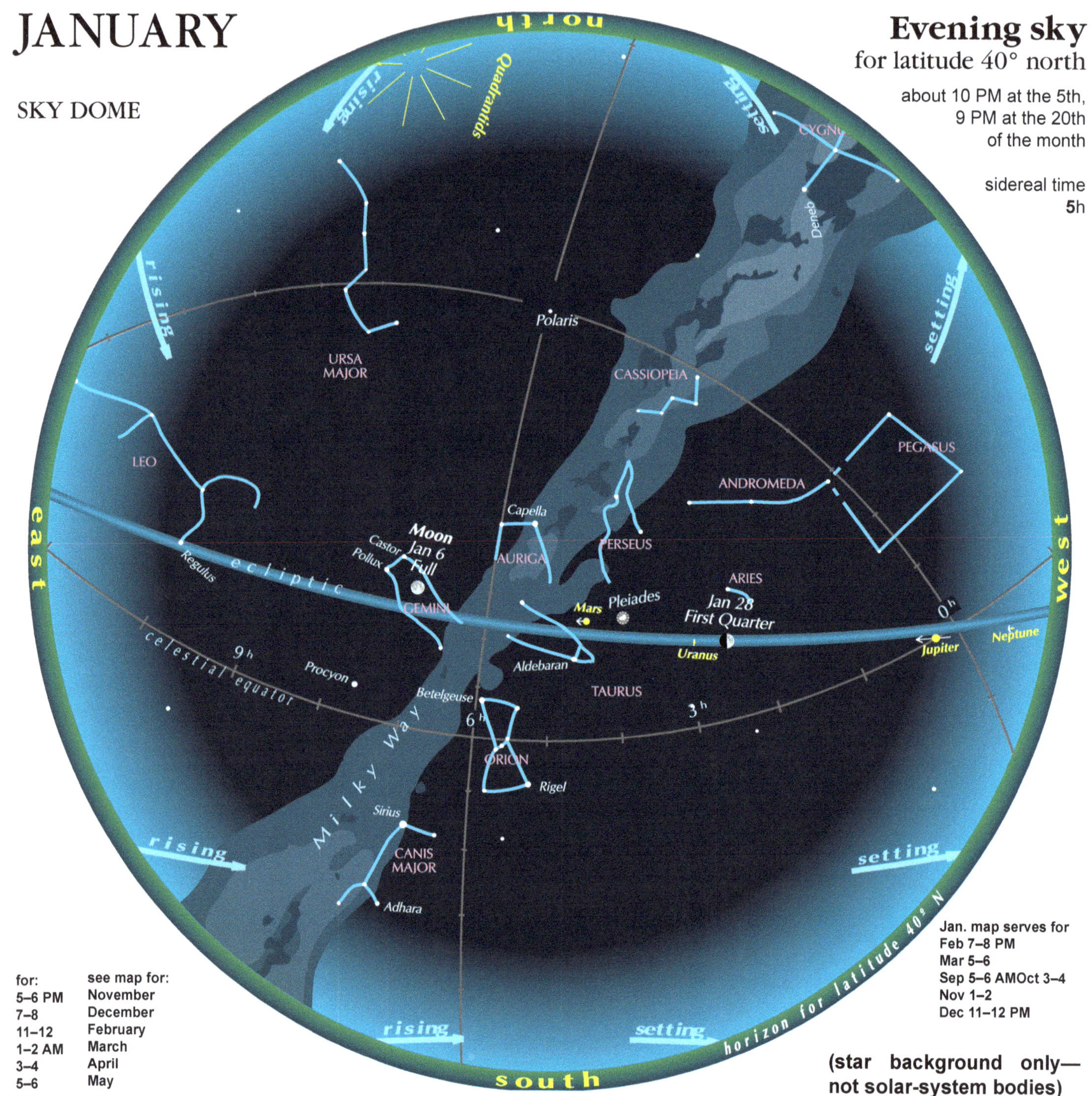

for:	see map for:
5–6 PM	November
7–8	December
11–12	February
1–2 AM	March
3–4	April
5–6	May

Jan. map serves for
Feb 7–8 PM
Mar 5–6
Sep 5–6 AMOct 3–4
Nov 1–2
Dec 11–12 PM

(star background only—not solar-system bodies)

Horizon scenes

In pictures like that on the facing page, angular distances are true from a center, which I usually set 10° below the horizon. This reminds us that we are on a spherical planet. (The center could be on the horizon, making that a straight line; or at the zenith, making the horizon a circle.)

The flying Moon is shown, twice real size, at the picture time and the same time 1 and 2 days before and after. Its position is affected by parallax: seen from farther south, it would appear farther north. The arrows between positions are without parallax (as seen from Earth's center), showing the difference parallax makes. Figures such as "+77" mean that the Moon is 77 hours "old," i.e. past its new moment.

The Sun is shown, also at twice size, even if underground. Arrows through the Sun and planets show their movement over 5 days, in relation to the starry background. Venus and Saturn are drawn at 150 times scale, so that you can see Venus's crescent shape and the current orientation of Saturn's rings; other planets have symbols sized for brightness so that they can be compared with the stars.

The curving form-lines that I use for constellations are thick for more conspicuous constellations, thin for others, which in other pictures are omitted.

A broad arrow on the celestial equator shows how far the sky will rotate in the next hour, carrying stars up from the eastern and down to the western horizon.

The Milky Way and dimmer planets and stars are unlikely to be visible in twilight or near the horizon. They are shown because they aid perception of where we are looking out into our solar system and galaxy.

2023 Jan

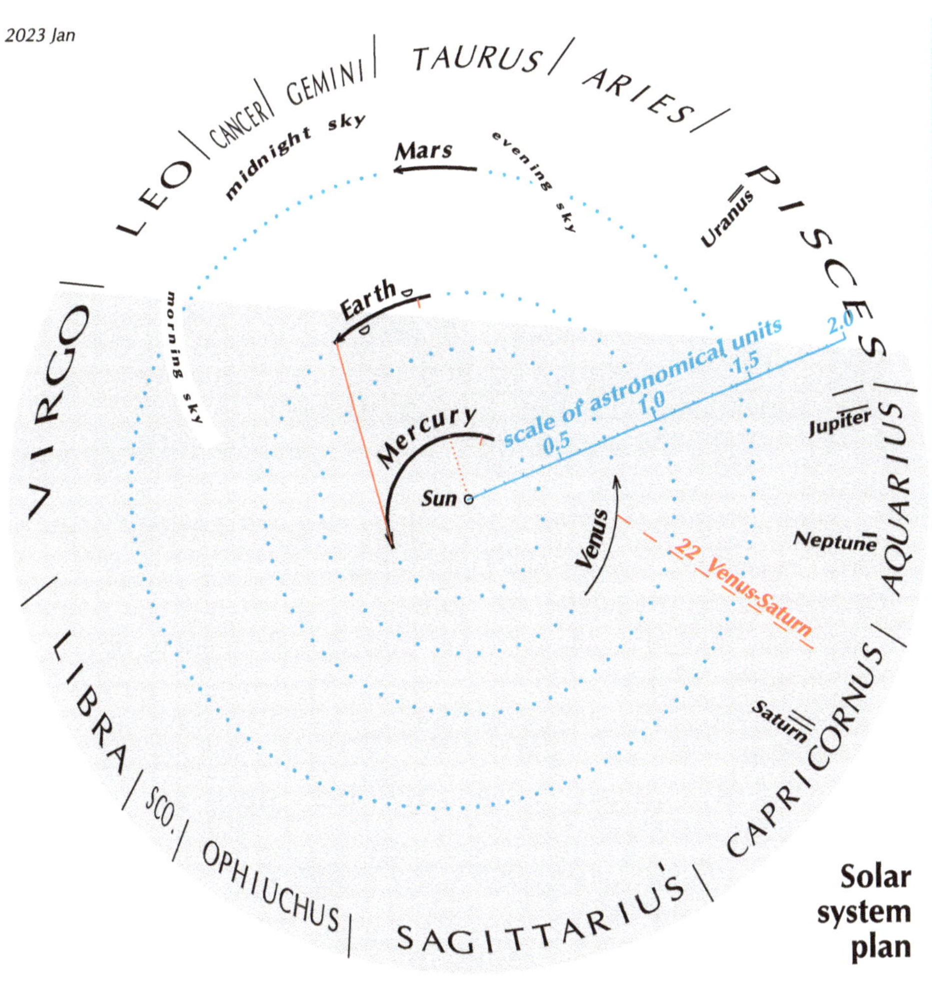

In these views from ecliptic north, arrows (thinner when south of the ecliptic plane) are the paths of the four inner planets. Dots along the rest of the orbits are 5 days apart (and are black for the part of its course that a planet has trodden since the beginning of the year).

Semicircles show the sunlit side of the new and full Moon (vastly exaggerated in size and distance).

Pairs of lines point outward to the more remote planets.

Phenomena such as perihelia (represented by ticks) and conjunctions (represented by lines between planets) are at dates that can be found in the list.

Gray covers the half of the universe below the horizon around 10 PM at mid-month (as seen from the equator).

The zodiacal constellations are in directions from the Earth at mid-month (not from the Sun).

aphelion
perihelion
node
opposition
greatest elongation
conjunction with Sun
conjunction of planets (in longitude)

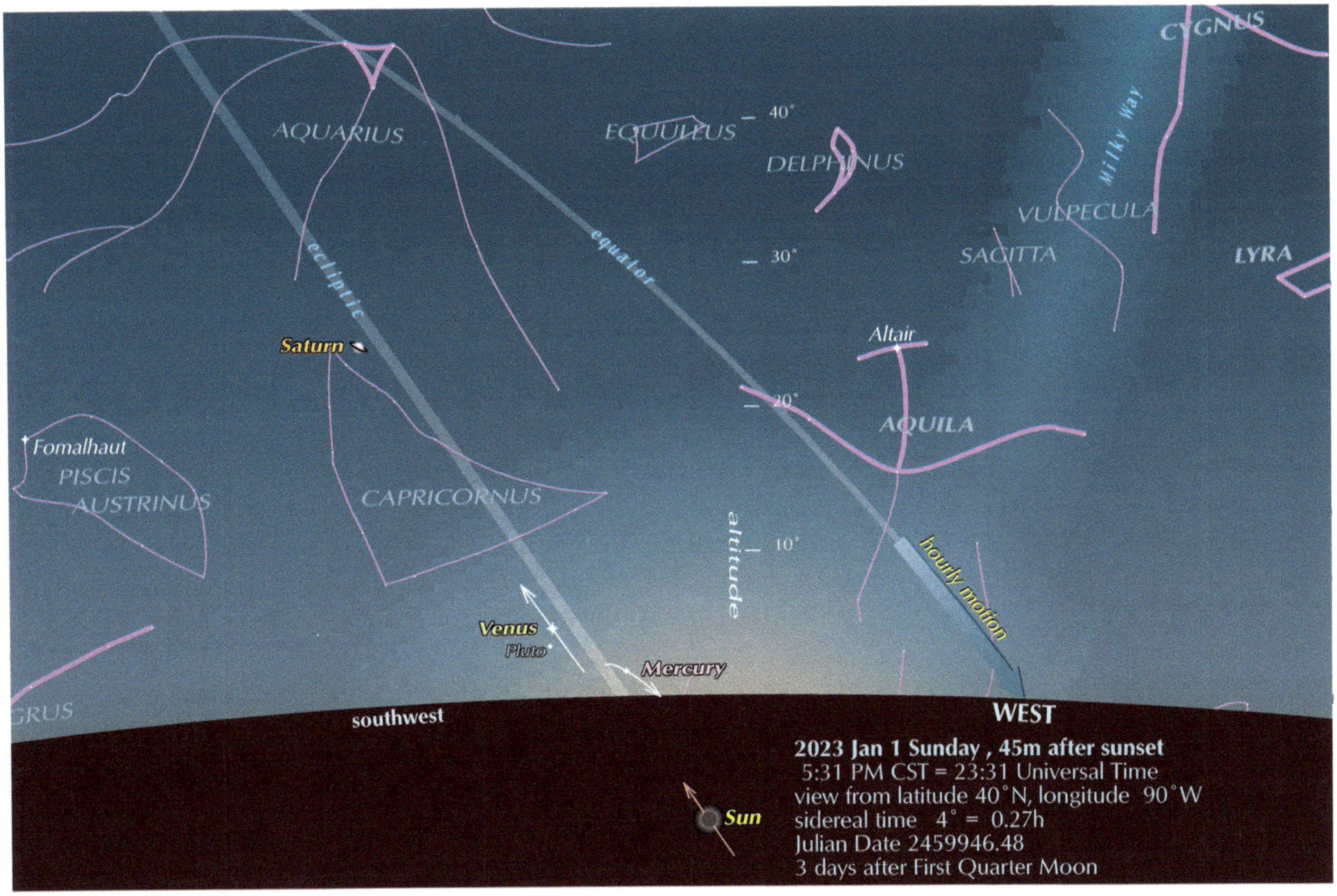

2023 Jan 1 Sunday , 45m after sunset
5:31 PM CST = 23:31 Universal Time
view from latitude 40˚N, longitude 90˚W
sidereal time 4˚ = 0.27h
Julian Date 2459946.48
3 days after First Quarter Moon

2023 JD 245-				UT	JANUARY
9945.5	Jan	1	SUN		Gregorian calendar Jan 1 = Julian calendar 2022 Dec 19
9946.143	Jan	1	SUN	15	Moon at ascending node; longitude 41.8°
9946.438	Jan	1	SUN	23	Moon 0.76° N of Uranus; 124° from Sun in evening sky; magnitudes -11.2 and 5.7; occultation
9947.345	Jan	2	Mon	20	Mercury at perihelion; 0.3075 AU from the Sun
9947.688	Jan	3	Tue	5	Moon 2.45° SE of Pleiades; 138° from Sun in evening sky
9948.354	Jan	3	Tue	21	Moon 0.61° SE of Mars; 146° and 145° from Sun in evening sky; magnitudes -11.8 and -1.1; occultation
9948.417	Jan	3	Tue	22	Moon 7.9° N of Aldebaran; 146° from Sun in evening sky; magnitudes -11.8 and 0.9
9948.5	Jan	4	Wed	0	Quadrantid meteors; ZHR 120; 3 days before full Moon
9949.217	Jan	4	Wed	17	Earth at perihelion; 0.9833 AU from the Sun
9949.807	Jan	5	Thu	7:22	Latest sunrise, at latitude 40° north
9950.292	Jan	5	Thu	19	Moon 3.0° N of M35 cluster; 167° from Sun in evening sky; magnitudes -12.3 and 5.3
9951.464	Jan	6	Fri	23:09	Full Moon
9951.854	Jan	7	SAT	9	Moon 5.4° S of Castor; 174° and 169° from Sun in morning sky; magnitudes -12.5 and 1.5
9952.036	Jan	7	SAT	13	Mercury at inferior conjunction with the Sun; 0.671 AU from Earth; latitude 5.97°
9952.083	Jan	7	SAT	14	Moon 1.90° S of Pollux; 172° and 171° from Sun in morning midnight sky; magnitudes -12.4 and 1.2
9952.806	Jan	8	SUN	7	Asteroid 2 Pallas at opposition in longitude; magnitude 7.6
9952.890	Jan	8	SUN	9	Moon at apogee; distance 63.73 Earth-radii
9953.271	Jan	8	SUN	19	Moon 3.7° NNE of Beehive Cluster; 160° from Sun in morning sky; magnitudes -12.1 and 3.7
9955.188	Jan	10	Tue	17	Moon 4.3° NNE of Regulus; 139° and 140° from Sun in morning sky; magnitudes -11.6 and 1.4
9957.343	Jan	12	Thu	20	Mars stationary in right ascension; resumes direct motion
9957.373	Jan	12	Thu	21	Mars stationary in longitude; resumes direct motion
9957.375	Jan	12	Thu	21	Mars 7.9° E of the Pleiades; 136° and 128° from Sun in evening sky; quasi-conjunction
9957.553	Jan	13	Fri	1	Mercury at northernmost latitude from the ecliptic plane, 7.0°
9957.704	Jan	13	Fri	5	Jupiter crosses equator northward
9958.5	Jan	14	SAT		Julian calendar 2023 Jan 1
9958.953	Jan	14	SAT	11	Mars at southernmost declination, 24.46°
9959.592	Jan	15	SUN	2:12	Last quarter Moon
9959.604	Jan	15	SUN	3	Moon 3.4° NNE of Spica; 90° from Sun in morning sky; magnitudes -10.1 and 1.0
9960.774	Jan	16	Mon	7	Moon at descending node; longitude 220.1°
9961.940	Jan	17	Tue	11	Venus at southernmost latitude from the ecliptic plane, -3.4°
9962.796	Jan	18	Wed	7	Pluto at conjunction with the Sun; 35.673 AU from Earth; latitude -2.34°
9962.979	Jan	18	Wed	12	Moon 2.08° NNE of Antares; 48° from Sun in morning sky; magnitudes -8.2 and 1.0

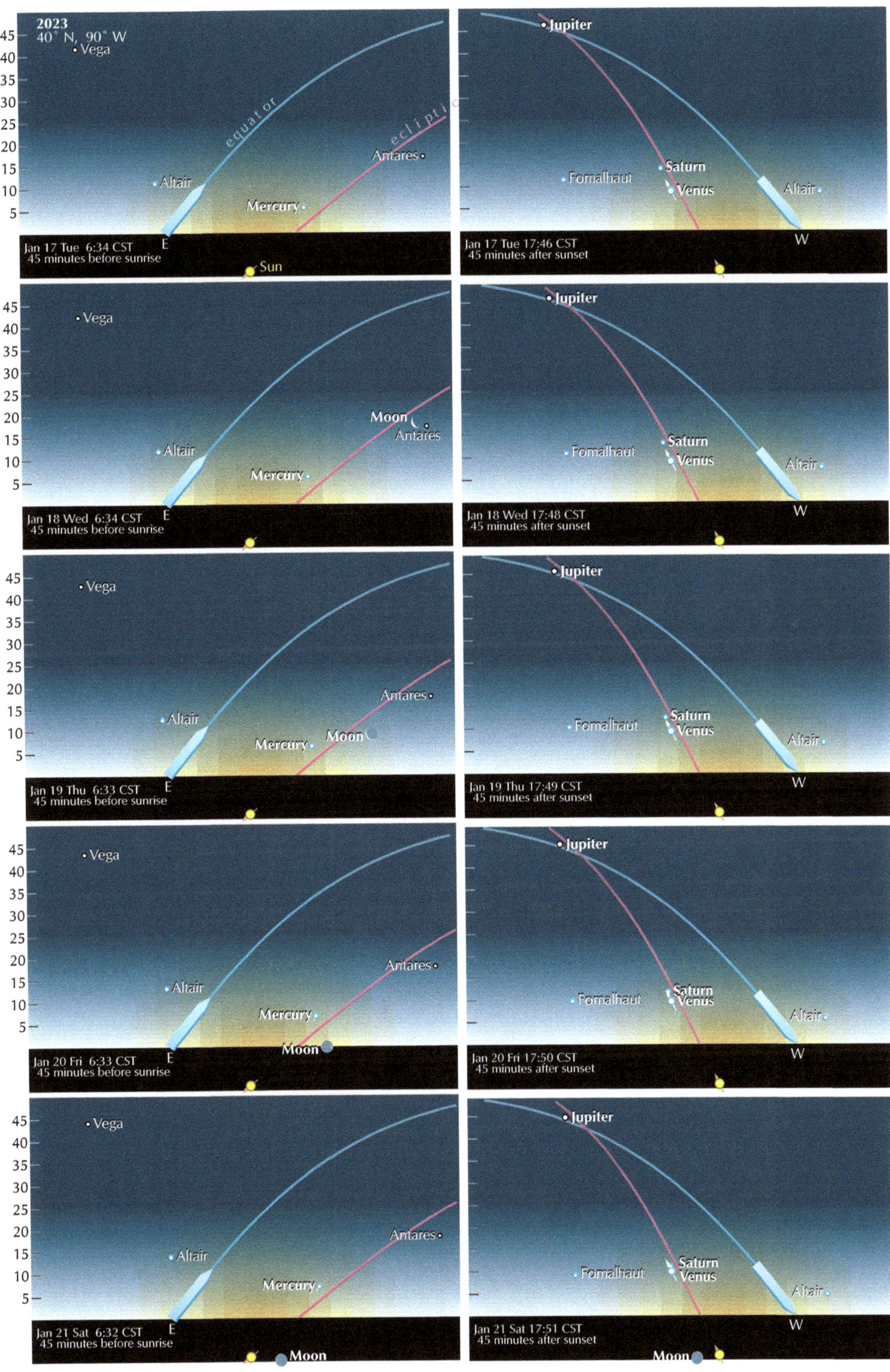

2023
40° N, 90° W
Vega
equator
ecliptic
Antares
Altair
Mercury
E
Jan 17 Tue 6:34 CST
45 minutes before sunrise
Sun
Jupiter
Saturn
Fomalhaut
Venus
Altair
W
Jan 17 Tue 17:46 CST
45 minutes after sunset
Vega
Moon
Antares
Altair
Mercury
E
Jan 18 Wed 6:34 CST
45 minutes before sunrise
Jupiter
Saturn
Fomalhaut
Venus
Altair
W
Jan 18 Wed 17:48 CST
45 minutes after sunset
Vega
Antares
Altair
Mercury
Moon
E
Jan 19 Thu 6:33 CST
45 minutes before sunrise
Jupiter
Saturn
Fomalhaut
Venus
Altair
W
Jan 19 Thu 17:49 CST
45 minutes after sunset
Vega
Antares
Altair
Mercury
Moon
E
Jan 20 Fri 6:33 CST
45 minutes before sunrise
Jupiter
Saturn
Fomalhaut
Venus
Altair
W
Jan 20 Fri 17:50 CST
45 minutes after sunset
Vega
Antares
Altair
Mercury
E
Jan 21 Sat 6:32 CST
45 minutes before sunrise
Moon
Jupiter
Saturn
Fomalhaut
Venus
Altair
W
Jan 21 Sat 17:51 CST
45 minutes after sunset
Moon

9962.989	Jan	18	Wed	12	Mercury stationary in right ascension; resumes direct motion
9963.047	Jan	18	Wed	13	Mercury stationary in longitude; resumes direct motion
9964.627	Jan	20	Fri	3	Sun enters Capricornus, at longitude 299.77° on the ecliptic
9964.853	Jan	20	Fri	8	Sun enters the astrological sign Aquarius, i.e. its longitude is 300°
9964.854	Jan	20	Fri	9	Moon 6.9° S of Mercury; 22° from Sun in morning sky; magnitudes -6.3 and 0.3
9964.895	Jan	20	Fri	9	Jupiter at perihelion; 4.9510 AU from the Sun
9966.371	Jan	21	SAT	20:55	Moon at perigee; distance 55.91 Earth-radii; nearest in year; only 0.0 hours after new Moon
9966.371	Jan	21	SAT	20:54	New Moon; beginning of lunation 1238
9967.363	Jan	22	SUN	21	Uranus stationary in longitude; resumes direct motion
9967.438	Jan	22	SUN	23	Venus 0.34° SE of Saturn; 22° from Sun in evening sky; magnitudes -3.9 and 0.9
9967.515	Jan	23	Mon	0	Uranus stationary in right ascension; resumes direct motion
9967.900	Jan	23	Mon	10	Moon, Venus, and Saturn within circle of diameter 3.57°; about 22° from the Sun in the evening sky; magnitudes -6, -4, 1
9967.917	Jan	23	Mon	10	Moon 3.6° SE of Saturn; 23° and 22° from Sun in evening sky; magnitudes -6.4 and 0.9
9967.958	Jan	23	Mon	11	Moon 3.2° SE of Venus; 23° and 22° from Sun in evening sky; magnitudes -6.4 and -3.9
9969.854	Jan	25	Wed	9	Moon 2.42° SE of Neptune; 49° and 48° from Sun in evening sky; magnitudes -8.3 and 7.9
9970.667	Jan	26	Thu	4	Moon 1.63° SE of Jupiter; 60° and 59° from Sun in evening sky; magnitudes -8.9 and -2.2
9973.138	Jan	28	SAT	15:19	First quarter Moon
9973.170	Jan	28	SAT	16	Moon at ascending node; longitude 38.8°
9973.667	Jan	29	SUN	4	Moon 0.92° N of Uranus; 96° from Sun in evening sky; magnitudes -10.3 and 5.7; occultation
9974.738	Jan	30	Mon	6	Mercury at westernmost elongation; 25.0° from Sun in morning sky; magnitude -0.1
9974.917	Jan	30	Mon	10	Moon 2.21° SE of Pleiades; 110° from Sun in evening sky
9975.708	Jan	31	Tue	5	Moon 0.28° E of Mars; 119° from Sun in evening sky; magnitudes -11.0 and -0.3; occultation

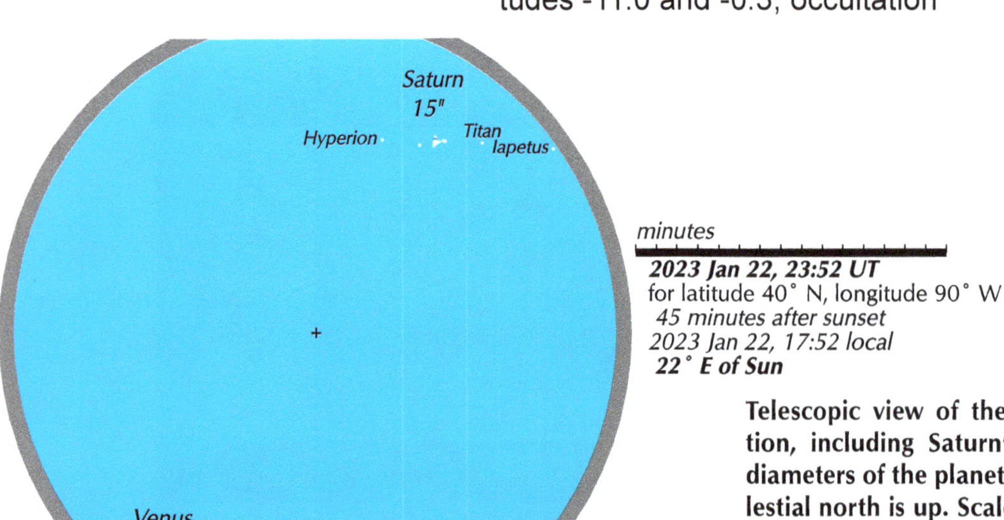

Telescopic view of the Jan. 22 Venus-Saturn conjunction, including Saturn's major satellites. The angular diameters of the planets are given in seconds of arc. Celestial north is up. Scale is shown by a line with ticks at intervals of 1 minute of arc. Also noted are Universal Time, location, local time, and elongation from the Sun.

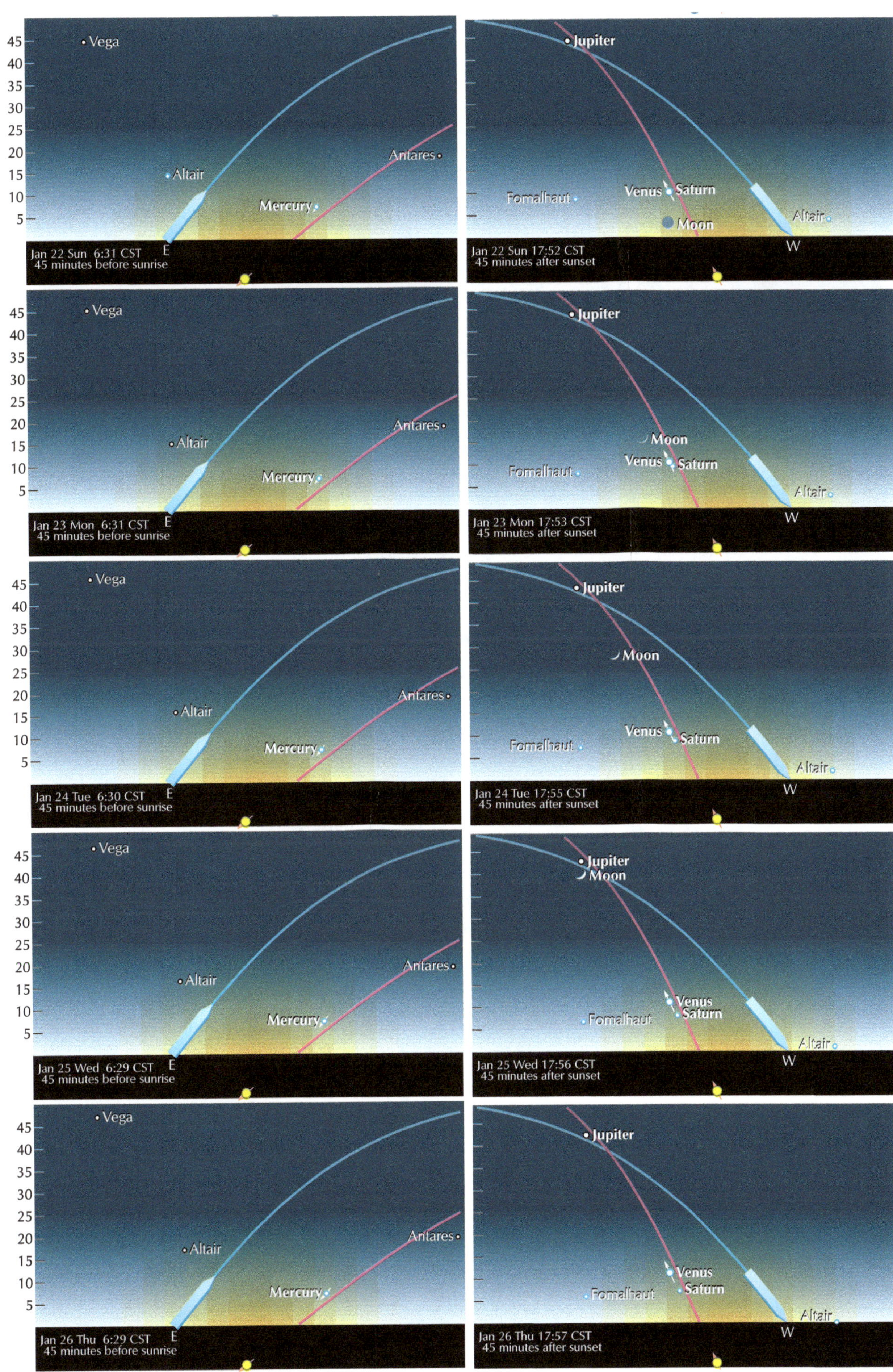
Vega
Altair
Antares
Mercury
E
Jan 22 Sun 6:31 CST
45 minutes before sunrise
Jupiter
Fomalhaut
Venus
Saturn
Moon
Altair
W
Jan 22 Sun 17:52 CST
45 minutes after sunset
Jan 23 Mon 6:31 CST
45 minutes before sunrise
Jan 23 Mon 17:53 CST
45 minutes after sunset
Jan 24 Tue 6:30 CST
45 minutes before sunrise
Jan 24 Tue 17:55 CST
45 minutes after sunset
Jan 25 Wed 6:29 CST
45 minutes before sunrise
Jan 25 Wed 17:56 CST
45 minutes after sunset
Jan 26 Thu 6:29 CST
45 minutes before sunrise
Jan 26 Thu 17:57 CST
45 minutes after sunset

FEBRUARY

SKY DOME

Evening sky

for latitude 40° north

about 10 PM at the 5th,
9 PM at the 20th
of the month

sidereal time
7h

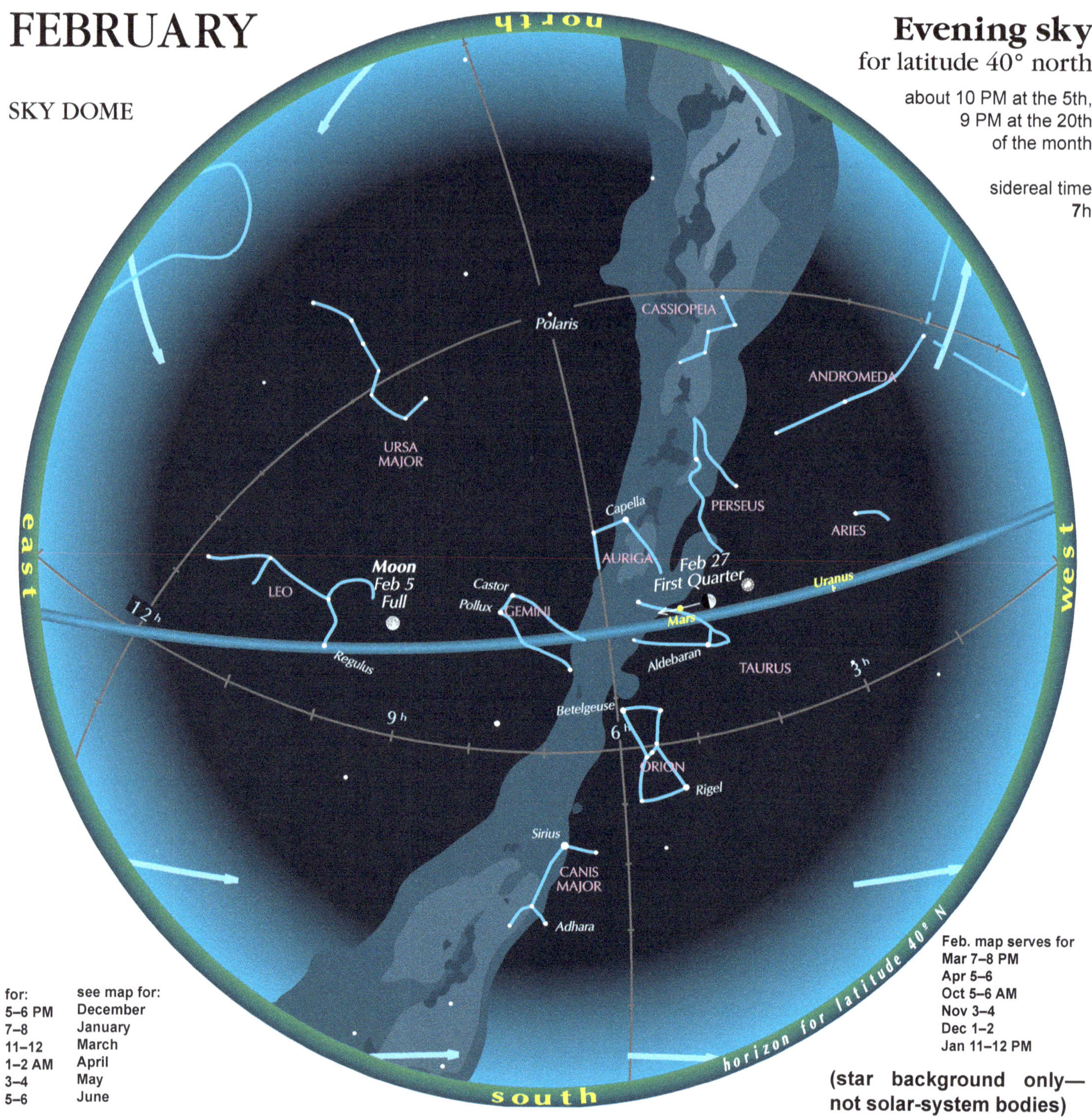

for:	see map for:
5–6 PM	December
7–8	January
11–12	March
1–2 AM	April
3–4	May
5–6	June

Feb. map serves for
Mar 7–8 PM
Apr 5–6
Oct 5–6 AM
Nov 3–4
Dec 1–2
Jan 11–12 PM

(star background only—not solar-system bodies)

2023 Feb

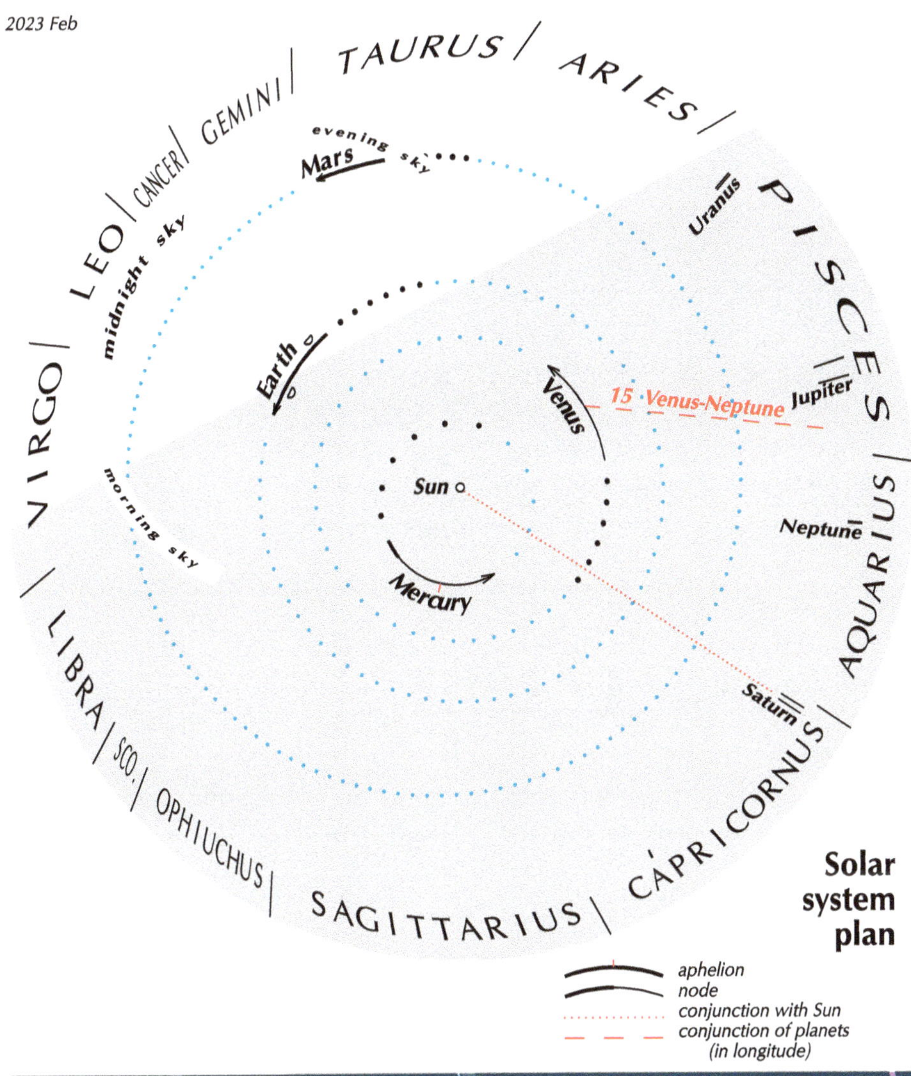

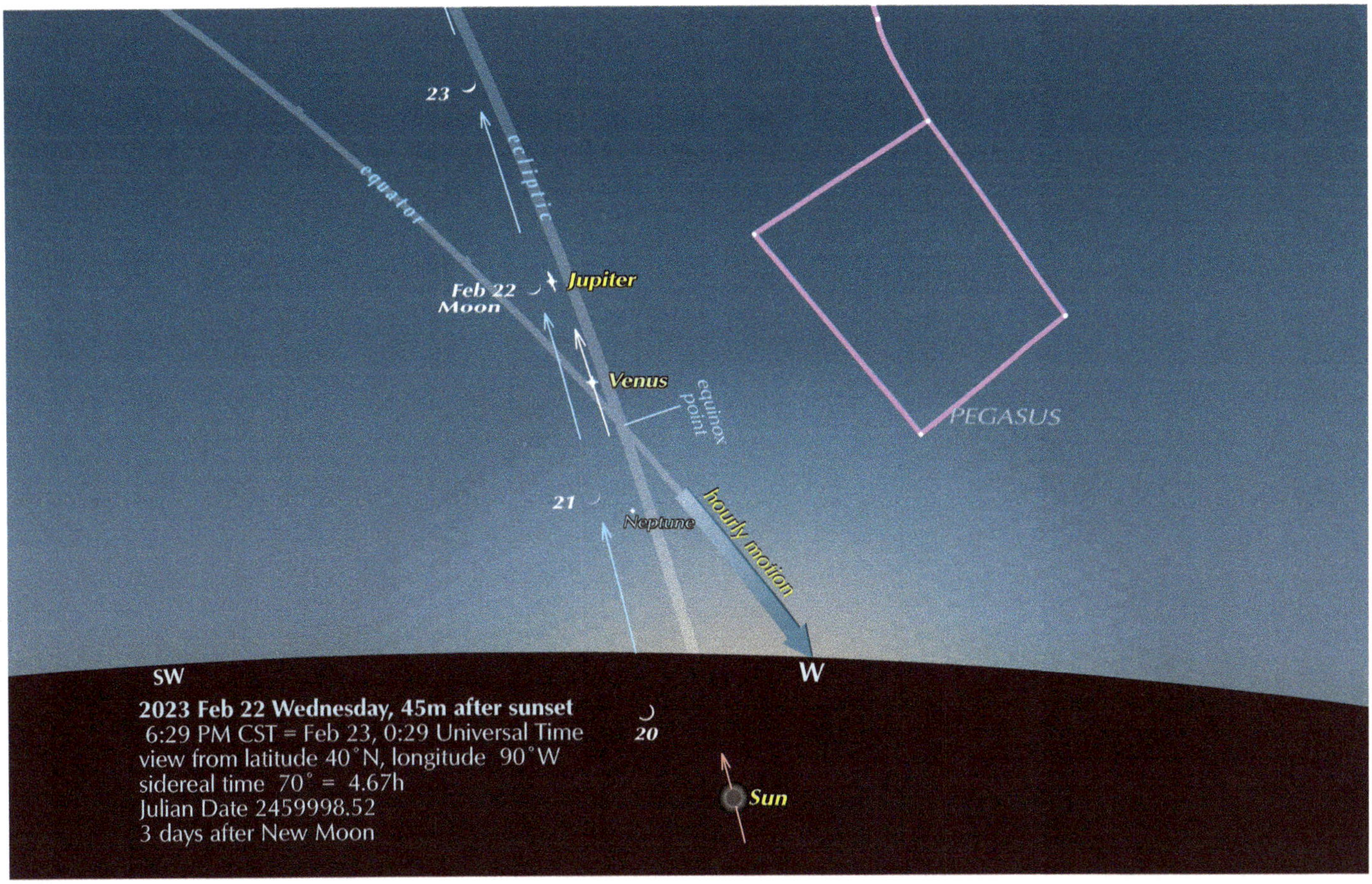

2023 Feb 22 Wednesday, 45m after sunset
6:29 PM CST = Feb 23, 0:29 Universal Time
view from latitude 40°N, longitude 90°W
sidereal time 70° = 4.67h
Julian Date 2459998.52
3 days after New Moon

FEBRUARY

9977.5	Feb	2	Thu		Groundhog Day
9977.542	Feb	2	Thu	1	Moon 3.1° N of M35 cluster; 139° from Sun in evening sky; magnitudes -11.6 and 5.3
9978.039	Feb	2	Thu	13	Mercury at southernmost declination, -21.69°
9979.104	Feb	3	Fri	15	Moon 5.3° S of Castor; 156° and 154° from Sun in evening sky; magnitudes -12.0 and 1.5
9979.354	Feb	3	Fri	21	Moon 1.89° S of Pollux; 159° and 158° from Sun in evening sky; magnitudes -12.1 and 1.2
9979.613	Feb	4	SAT	3	Uranus at east quadrature, 90° from the Sun
9979.867	Feb	4	SAT	9	Moon at apogee; distance 63.73 Earth-radii
9980.542	Feb	5	SUN	1	Moon 3.7° NNE of Beehive Cluster; 171° and 172° from Sun in evening midnight sky; magnitudes -12.4 and 3.7
9980.956	Feb	5	SUN	11	Mercury at descending node through the ecliptic plane
9981.270	Feb	5	SUN	18:30	Full Moon
9982.438	Feb	6	Mon	23	Moon 4.2° NNE of Regulus; 166° and 168° from Sun in morning sky; magnitudes -12.3 and 1.4
9986.854	Feb	11	SAT	9	Moon 3.2° NNE of Spica; 118° from Sun in morning sky; magnitudes -11.0 and 1.0
9987.131	Feb	11	SAT	15	The equation of time is at a minimum of -14.23 minutes
9987.815	Feb	12	SUN	8	Moon at descending node; longitude 216.9°
9989.168	Feb	13	Mon	16:02	Last quarter Moon
9989.5	Feb	14	Tue		St. Valentine's Day
9990.333	Feb	14	Tue	20	Moon 1.83° NNE of Antares; 75° and 76° from Sun in morning sky; magnitudes -9.6 and 1.0
9991.042	Feb	15	Wed	13	Venus 0.03° E of Neptune; 28° from Sun in evening sky; magnitudes -4.0 and 8.0; quasi-conjunction
9991.330	Feb	15	Wed	20	Mercury at aphelion; 0.4667 AU from the Sun
9992.202	Feb	16	Thu	17	Saturn at conjunction with the Sun; 10.812 AU from Earth; latitude -1.39°
9992.403	Feb	16	Thu	22	Sun enters Aquarius, at longitude 327.94° on the ecliptic
9994.439	Feb	18	SAT	23	Sun enters the astrological sign Pisces, i.e. its longitude is 330°
9994.458	Feb	18	SAT	23	Moon 3.5° SE of Mercury; 20° and 19° from Sun in morning sky; magnitudes -6.1 and -0.2
9994.873	Feb	19	SUN	8:57	Moon at perigee; distance 56.17 Earth-radii; only 22.2 hours before new Moon
9995.604	Feb	20	Mon	3	Moon 3.4° SE of Saturn; 5° and 3° from Sun in morning sky; magnitudes -4.6 and 0.8
9995.797	Feb	20	Mon	7:08	New Moon; beginning of lunation 1239
9997.354	Feb	21	Tue	21	Moon 2.21° SE of Neptune; 22° and 21° from Sun in evening sky; magnitudes -6.3 and 8.0
9997.5	Feb	22	Wed		Ash Wednesday
9997.917	Feb	22	Wed	10	Moon 1.85° SE of Venus; 29° from Sun in evening sky; magnitudes -6.9 and -4.0
9998.479	Feb	22	Wed	24	Moon 1.09° SE of Jupiter; 37° and 36° from Sun in evening sky; magnitudes -7.4 and -2.1; occultation
0000.290	Feb	24	Fri	19	Moon at ascending node; longitude 35.9°

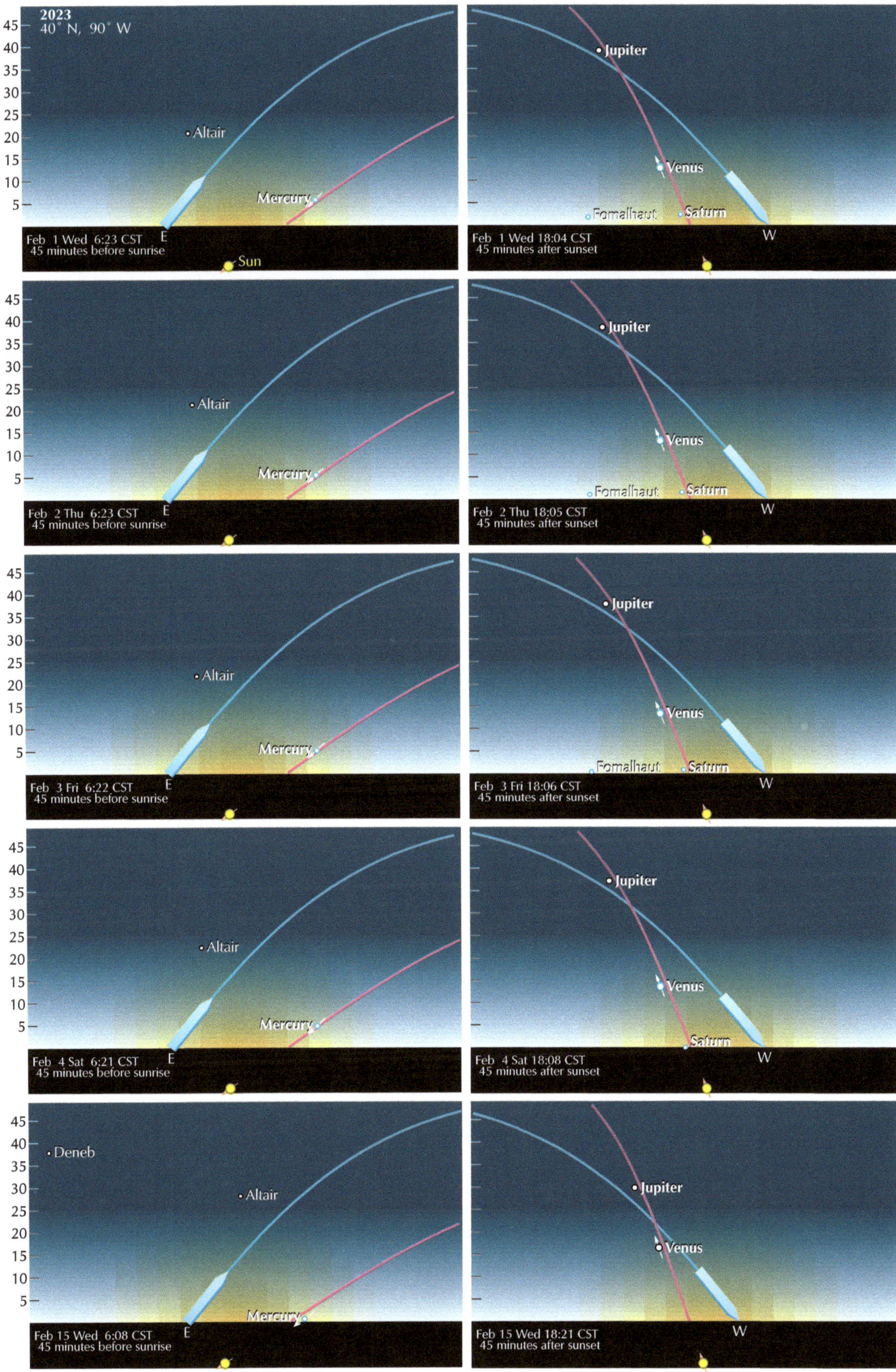
2023
40° N, 90° W
45
40
35
30
25
20
15
10
5
Altair
Mercury
E
Feb 1 Wed 6:23 CST
45 minutes before sunrise
Sun
Jupiter
Venus
Fomalhaut
Saturn
W
Feb 1 Wed 18:04 CST
45 minutes after sunset
Feb 2 Thu 6:23 CST
45 minutes before sunrise
Feb 2 Thu 18:05 CST
45 minutes after sunset
Feb 3 Fri 6:22 CST
45 minutes before sunrise
Feb 3 Fri 18:06 CST
45 minutes after sunset
Feb 4 Sat 6:21 CST
45 minutes before sunrise
Feb 4 Sat 18:08 CST
45 minutes after sunset
Deneb
Feb 15 Wed 6:08 CST
45 minutes before sunrise
Feb 15 Wed 18:21 CST
45 minutes after sunset

0001.021	Feb	25	SAT	13	Moon 1.19° NNW of Uranus; 69° from Sun in evening sky; magnitudes -9.2 and 5.8; occultation
0002.208	Feb	26	SUN	17	Moon 1.94° SE of Pleiades; 83° from Sun in evening sky
0002.837	Feb	27	Mon	8:05	First quarter Moon
0003.688	Feb	28	Tue	5	Moon 1.07° N of Mars; 100° and 99° from Sun in evening sky; magnitudes -10.4 and 0.4; occultation

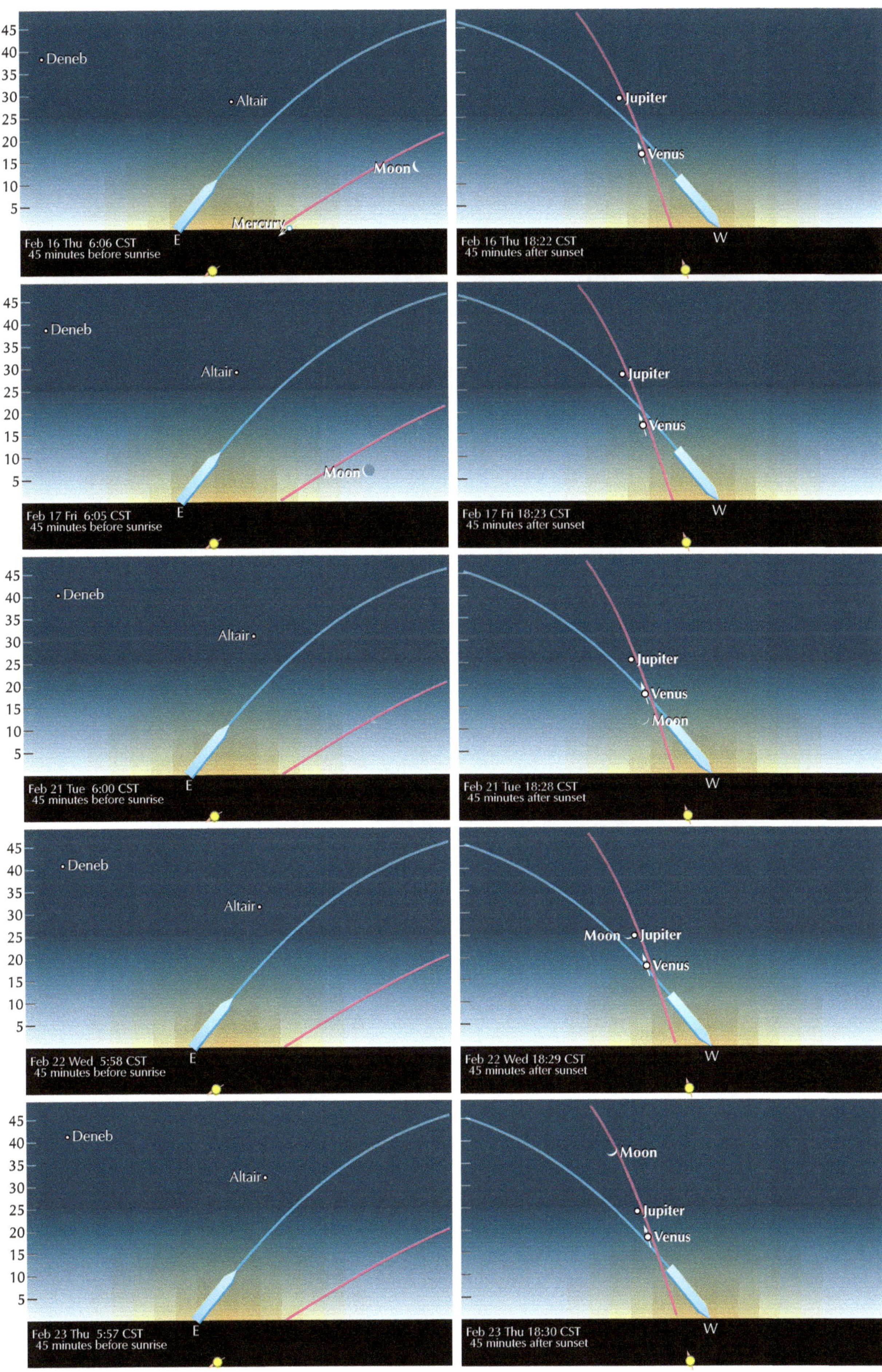

Deneb
Altair
Moon
Mercury
E
Feb 16 Thu 6:06 CST
45 minutes before sunrise
Jupiter
Venus
W
Feb 16 Thu 18:22 CST
45 minutes after sunset
Feb 17 Fri 6:05 CST
45 minutes before sunrise
Feb 17 Fri 18:23 CST
45 minutes after sunset
Feb 21 Tue 6:00 CST
45 minutes before sunrise
Feb 21 Tue 18:28 CST
45 minutes after sunset
Feb 22 Wed 5:58 CST
45 minutes before sunrise
Feb 22 Wed 18:29 CST
45 minutes after sunset
Feb 23 Thu 5:57 CST
45 minutes before sunrise
Feb 23 Thu 18:30 CST
45 minutes after sunset

MARCH

Evening sky
for latitude 40° north

about 10 PM at the 5th,
9 PM at the 20th
of the month

sidereal time
9h

SKY DOME

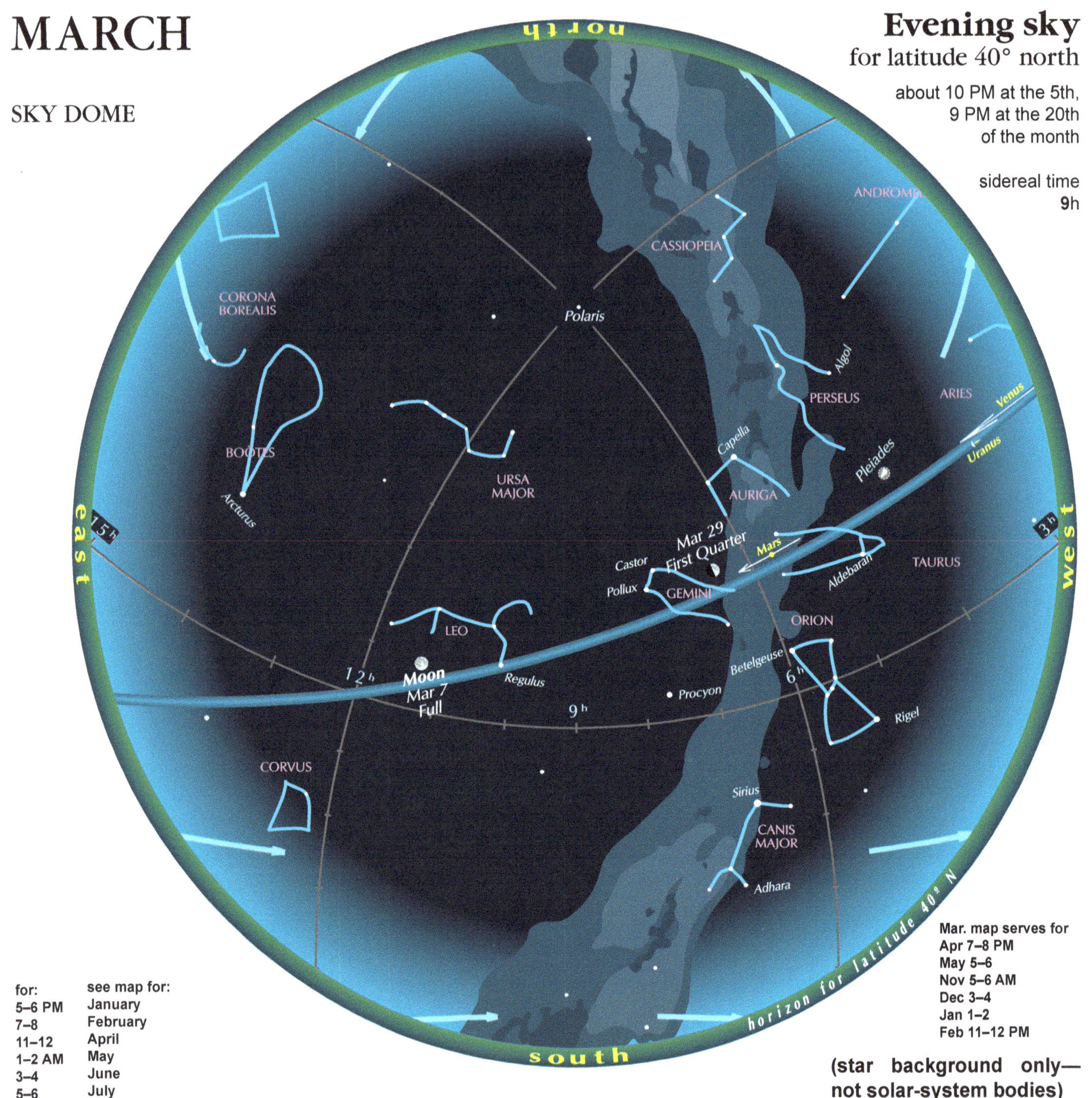

for:	see map for:
5–6 PM	January
7–8	February
11–12	April
1–2 AM	May
3–4	June
5–6	July

Mar. map serves for
Apr 7–8 PM
May 5–6
Nov 5–6 AM
Dec 3–4
Jan 1–2
Feb 11–12 PM

(star background only—not solar-system bodies)

2023 Mar

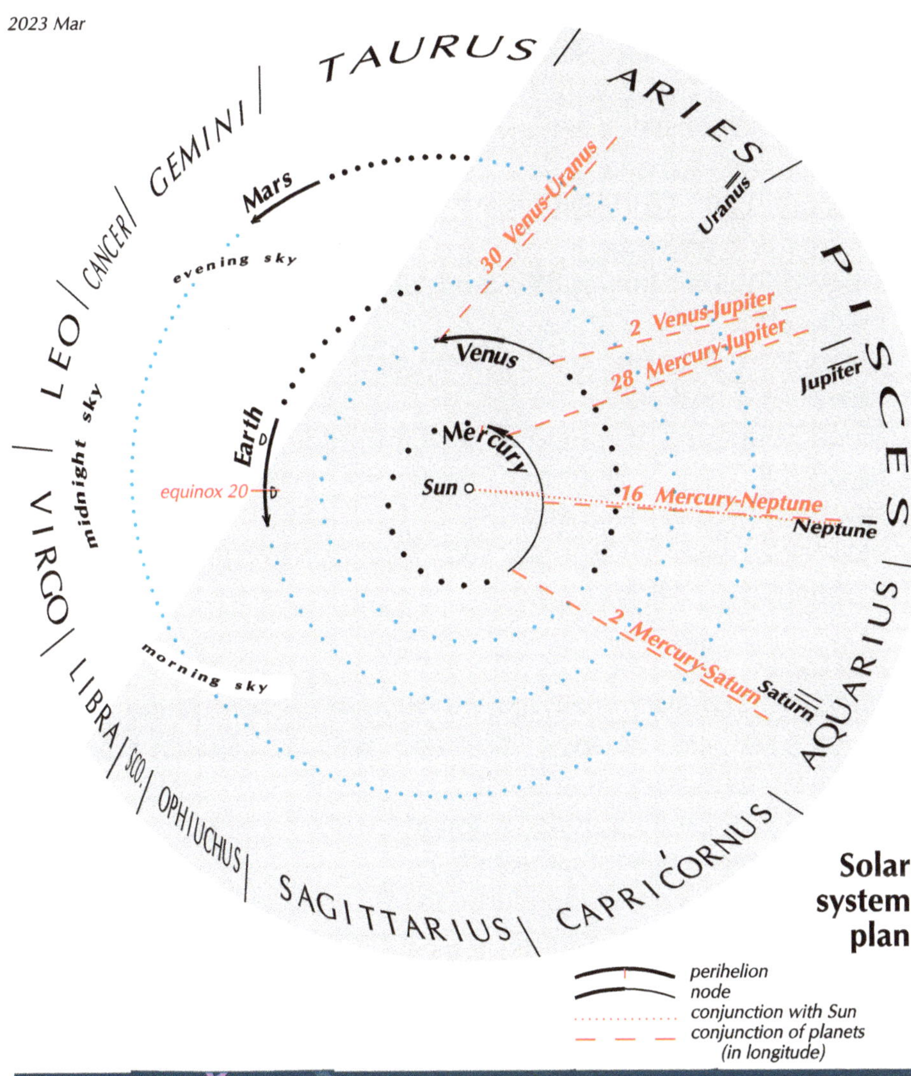

TAURUS
ARIES
GEMINI
CANCER
LEO
VIRGO
LIBRA
SCO.
OPHIUCHUS
SAGITTARIUS
CAPRICORNUS
AQUARIUS
PISCES
Mars
evening sky
midnight sky
morning sky
Uranus
30 Venus-Uranus
2 Venus-Jupiter
28 Mercury-Jupiter
Jupiter
Venus
Earth
Mercury
Sun
equinox 20
16 Mercury-Neptune
Neptune
2 Mercury-Saturn
Saturn
Solar system plan
perihelion
node
conjunction with Sun
conjunction of planets (in longitude)

Pleiades
25
24
Uranus
Venus
ARIES
CASSIOPEIA
ANDROMEDA
Moon Mar 23
equator
ecliptic
hourly motion
22
Jupiter
Mercury
21
Sun
W
NW
2023 Mar 23 Thursday, 45m after sunset
8:01 PM CDT = Mar 24, 1:01 Universal Time
view from latitude 40°N, longitude 90°W
sidereal time 106° = 7.09h
Julian Date 2460027.54
2 days after New Moon

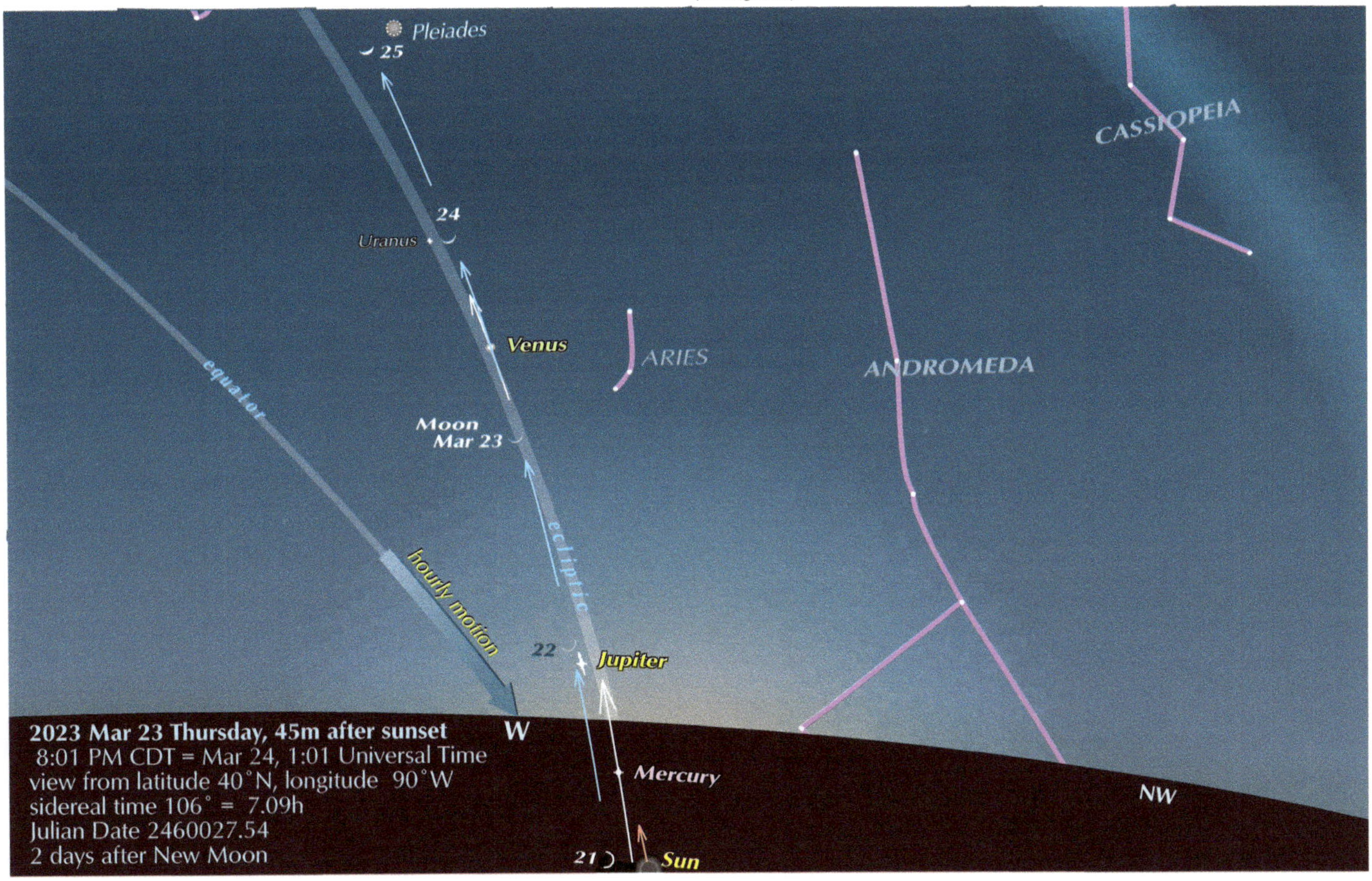

MARCH

0004.813	Mar	1	Wed	8	Moon 3.4° N of M35 cluster; 112° from Sun in evening sky; magnitudes -10.8 and 5.3
0005.729	Mar	2	Thu	6	Venus 0.49° NNW of Jupiter; 31° from Sun in evening sky; magnitudes -4.0 and -2.1
0006.125	Mar	2	Thu	15	Mercury 0.88° SE of Saturn; 12° from Sun in morning sky; magnitudes -0.6 and 0.9
0006.375	Mar	2	Thu	21	Moon 5.2° S of Castor; 129° and 128° from Sun in evening sky; magnitudes -11.3 and 1.5
0006.625	Mar	3	Fri	3	Moon 1.75° S of Pollux; 131° from Sun in evening sky; magnitudes -11.3 and 1.2
0007.245	Mar	3	Fri	18	Moon at apogee; distance 63.64 Earth-radii
0007.813	Mar	4	SAT	8	Moon 3.8° NNE of Beehive Cluster; 144° from Sun in evening sky; magnitudes -11.7 and 3.7
0009.708	Mar	6	Mon	5	Moon 4.2° NNE of Regulus; 165° from Sun in evening sky; magnitudes -12.2 and 1.4
0011.029	Mar	7	Tue	12:42	Full Moon
0011.585	Mar	8	Wed	2	Mercury at southernmost latitude from the ecliptic plane, -7.0°
0014.083	Mar	10	Fri	14	Moon 3.00° NNE of Spica; 145° and 146° from Sun in morning sky; magnitudes -11.8 and 1.0
0014.872	Mar	11	SAT	9	Moon at descending node; longitude 214.7°
0015.5	Mar	12	SUN		Clocks forward 1 hour (America)
0015.990	Mar	12	SUN	12	Sun enters Pisces, at longitude 351.63° on the ecliptic
0017.583	Mar	14	Tue	2	Moon 1.59° NNE of Antares; 103° from Sun in morning sky; magnitudes -10.6 and 1.0
0018.228	Mar	14	Tue	17	Venus at ascending node through the ecliptic plane
0018.590	Mar	15	Wed	2:09	Last quarter Moon
0019.487	Mar	15	Wed	24	Neptune at conjunction with the Sun; 30.905 AU from Earth; latitude -1.21°
0020.250	Mar	16	Thu	18	Mars at east quadrature, 90° from the Sun
0020.250	Mar	16	Thu	18	Mercury 0.38° SE of Neptune; 2° and 1° from Sun in evening sky; magnitudes -1.8 and 8.0
0020.5	Mar	17	Fri		St. Patrick's Day
0020.939	Mar	17	Fri	11	Mercury at superior conjunction with the Sun; 1.353 AU from Earth; latitude -5.53°
0021.5	Mar	18	SAT	0:00	Day and night equal, at latitude 40° north
0022.517	Mar	19	SUN	0	Mars at northernmost declination, 25.61°
0023.132	Mar	19	SUN	15:10	Moon at perigee; distance 56.86 Earth-radii
0023.250	Mar	19	SUN	18	Moon 3.3° SE of Saturn; 28° and 27° from Sun in morning sky; magnitudes -6.7 and 1.0
0024.372	Mar	20	Mon	21	Dwarf planet 1 Ceres at opposition in longitude; magnitude 7.0
0024.391	Mar	20	Mon	21:23	Sun enters the astrological sign Aries, i.e. its longitude is 0°
0024.391	Mar	20	Mon	21:23	March or vernal (northern spring) equinox
0024.875	Mar	21	Tue	9	Moon 2.11° SE of Neptune; 6° and 5° from Sun in morning sky; magnitudes -4.6 and 8.0
0025.226	Mar	21	Tue	17:26	New Moon; beginning of lunation 1240

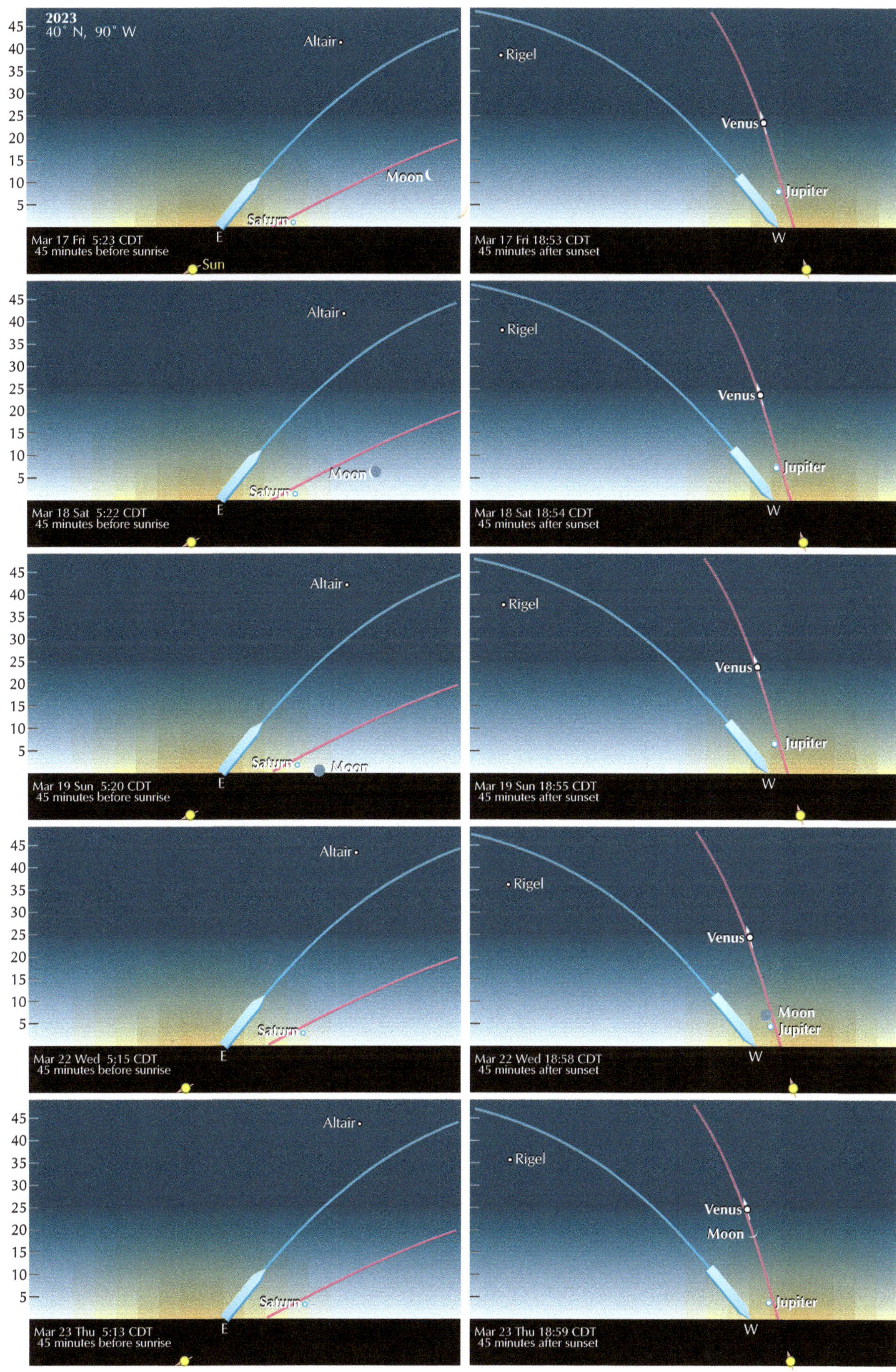

2023
40° N, 90° W
45
40
35
30
25
20
15
10
5
Altair
Moon
Saturn
E
Mar 17 Fri 5:23 CDT
45 minutes before sunrise
Sun
Rigel
Venus
Jupiter
W
Mar 17 Fri 18:53 CDT
45 minutes after sunset
Altair
Moon
Saturn
E
Mar 18 Sat 5:22 CDT
45 minutes before sunrise
Rigel
Venus
Jupiter
W
Mar 18 Sat 18:54 CDT
45 minutes after sunset
Altair
Saturn
Moon
E
Mar 19 Sun 5:20 CDT
45 minutes before sunrise
Rigel
Venus
Jupiter
W
Mar 19 Sun 18:55 CDT
45 minutes after sunset
Altair
Saturn
E
Mar 22 Wed 5:15 CDT
45 minutes before sunrise
Rigel
Venus
Moon
Jupiter
W
Mar 22 Wed 18:58 CDT
45 minutes after sunset
Altair
Saturn
E
Mar 23 Thu 5:13 CDT
45 minutes before sunrise
Rigel
Venus
Moon
Jupiter
W
Mar 23 Thu 18:59 CDT
45 minutes after sunset

0025.604	Mar	22	Wed	3	Moon 1.65° SE of Mercury; 6° and 5° from Sun in evening sky; magnitudes -4.6 and -1.7
0026.375	Mar	22	Wed	21	Moon 0.59° ESE of Jupiter; 15° from Sun in evening sky; magnitudes -5.6 and -2.1; occultation
0026.5	Mar	23	Thu		1st day of Ramadan (1444 A.H.)
0027.589	Mar	24	Fri	2	Moon at ascending node; longitude 34.3°
0027.958	Mar	24	Fri	11	Moon 0.24° E of Venus; 36° from Sun in evening sky; magnitudes -7.3 and -4.0; occultation
0028.5	Mar	25	SAT	0	Moon 1.44° NNW of Uranus; 42° from Sun in evening sky; magnitudes -7.7 and 5.8
0029.5	Mar	26	SUN		Clocks forward 1 hour (Europe)
0029.583	Mar	26	SUN	2	Moon 1.77° SE of Pleiades; 56° and 55° from Sun in evening sky
0030.643	Mar	27	Mon	3	Mercury at ascending node through the ecliptic plane
0031.729	Mar	28	Tue	6	Mercury 1.28° NNW of Jupiter; 11° from Sun in evening sky; magnitudes -1.4 and -2.1
0032.063	Mar	28	Tue	14	Moon 2.30° N of Mars; 84° from Sun in evening sky; magnitudes -9.8 and 0.9
0032.100	Mar	28	Tue	14	Moon, Mars, and M35 clu within circle of diameter 3.53°; about 84° from the Sun in the evening sky; magnitudes -10, 1, 5
0032.125	Mar	28	Tue	15	Moon 3.5° N of M35 cluster; 85° from Sun in evening sky; magnitudes -9.8 and 5.3
0032.606	Mar	29	Wed	2:32	First quarter Moon
0033.667	Mar	30	Thu	4	Moon 5.0° S of Castor; 102° and 101° from Sun in evening sky; magnitudes -10.4 and 1.5
0033.792	Mar	30	Thu	7	Mars 1.18° N of M35 cluster; 83° from Sun in evening sky; magnitudes 0.9 and 5.3
0033.917	Mar	30	Thu	10	Moon 1.57° S of Pollux; 104° from Sun in evening sky; magnitudes -10.5 and 1.2
0034.396	Mar	30	Thu	22	Venus 1.22° NNW of Uranus; 37° from Sun in evening sky; magnitudes -4.0 and 5.8
0034.974	Mar	31	Fri	11	Moon at apogee; distance 63.49 Earth-radii
0035.104	Mar	31	Fri	15	Moon 3.9° NNE of Beehive Cluster; 117° from Sun in evening sky; magnitudes -10.9 and 3.7
0035.314	Mar	31	Fri	20	Mercury at perihelion; 0.3075 AU from the Sun

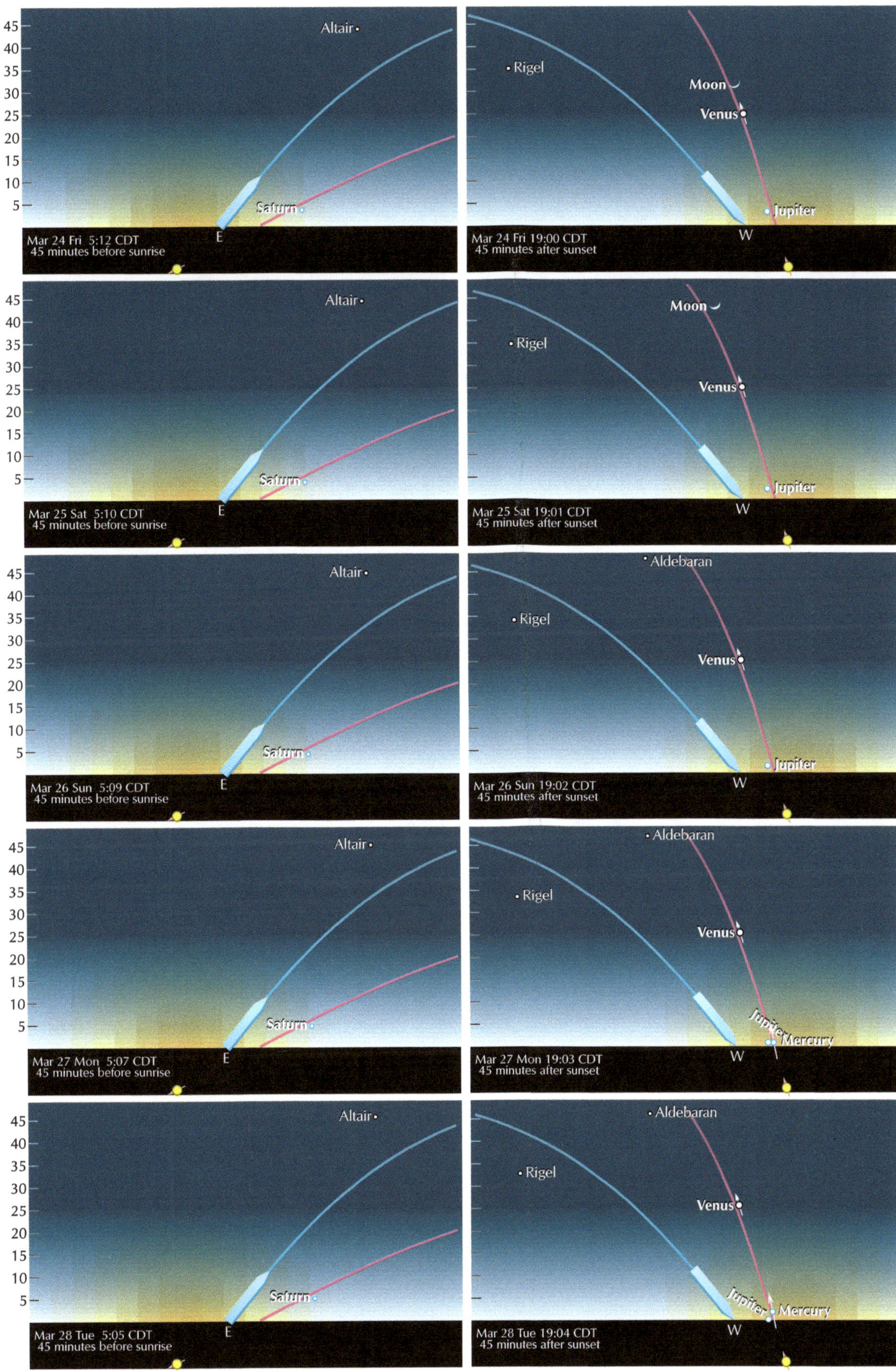

45
40
35
30
25
20
15
10
5
Altair
Saturn
E
Mar 24 Fri 5:12 CDT
45 minutes before sunrise
Rigel
Moon
Venus
Jupiter
W
Mar 24 Fri 19:00 CDT
45 minutes after sunset
Altair
Saturn
E
Mar 25 Sat 5:10 CDT
45 minutes before sunrise
Moon
Rigel
Venus
Jupiter
W
Mar 25 Sat 19:01 CDT
45 minutes after sunset
Altair
Saturn
E
Mar 26 Sun 5:09 CDT
45 minutes before sunrise
Aldebaran
Rigel
Venus
Jupiter
W
Mar 26 Sun 19:02 CDT
45 minutes after sunset
Altair
Saturn
E
Mar 27 Mon 5:07 CDT
45 minutes before sunrise
Aldebaran
Rigel
Venus
Jupiter
Mercury
W
Mar 27 Mon 19:03 CDT
45 minutes after sunset
Altair
Saturn
E
Mar 28 Tue 5:05 CDT
45 minutes before sunrise
Aldebaran
Rigel
Venus
Jupiter
Mercury
W
Mar 28 Tue 19:04 CDT
45 minutes after sunset

APRIL

SKY DOME

Evening sky

for latitude 40° north

about 10 PM at the 5th,
9 PM at the 20th
of the month

sidereal time
11h

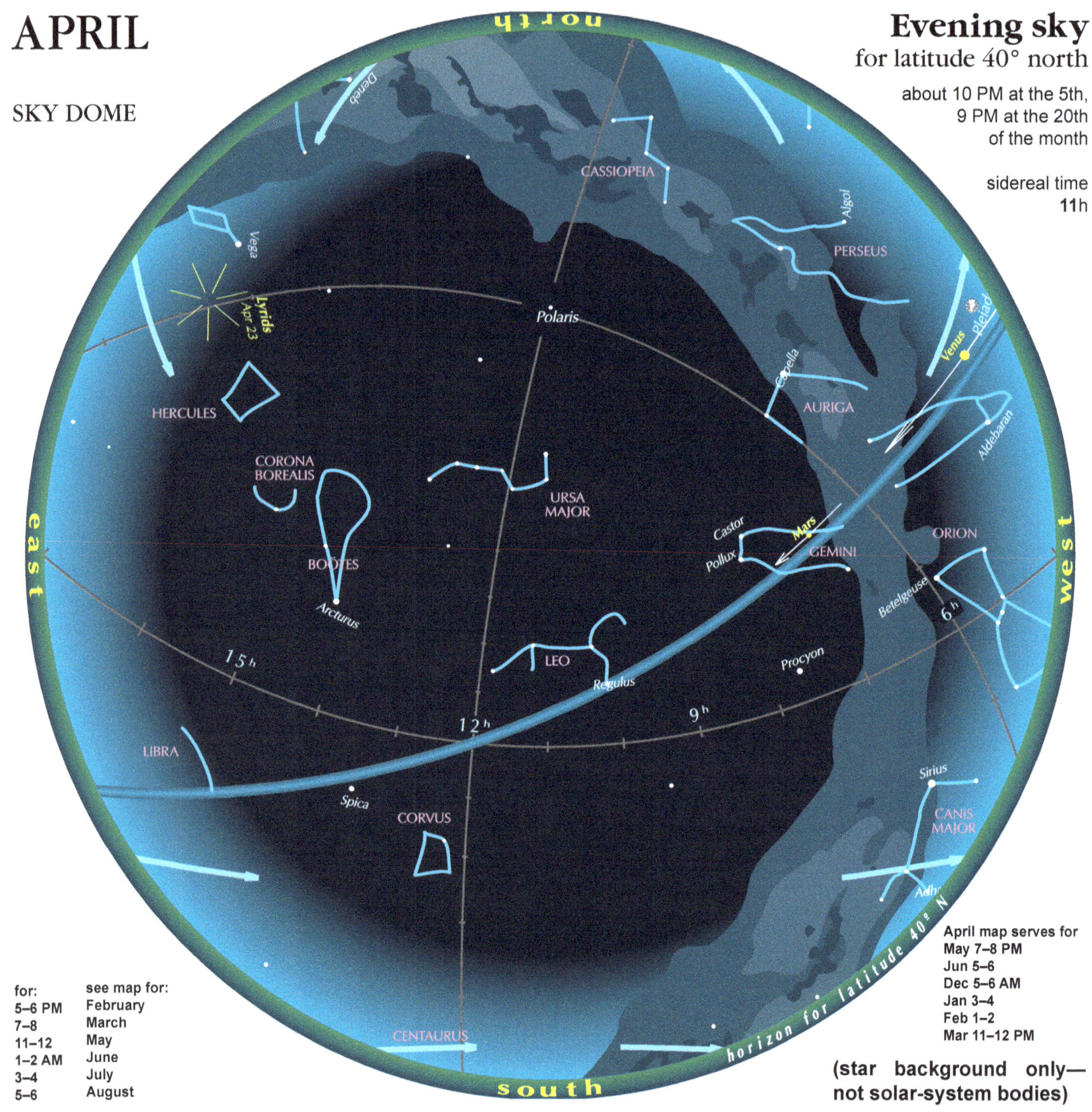

for:	see map for:
5–6 PM	February
7–8	March
11–12	May
1–2 AM	June
3–4	July
5–6	August

April map serves for
May 7–8 PM
Jun 5–6
Dec 5–6 AM
Jan 3–4
Feb 1–2
Mar 11–12 PM

(star background only—not solar-system bodies)

2023 Apr

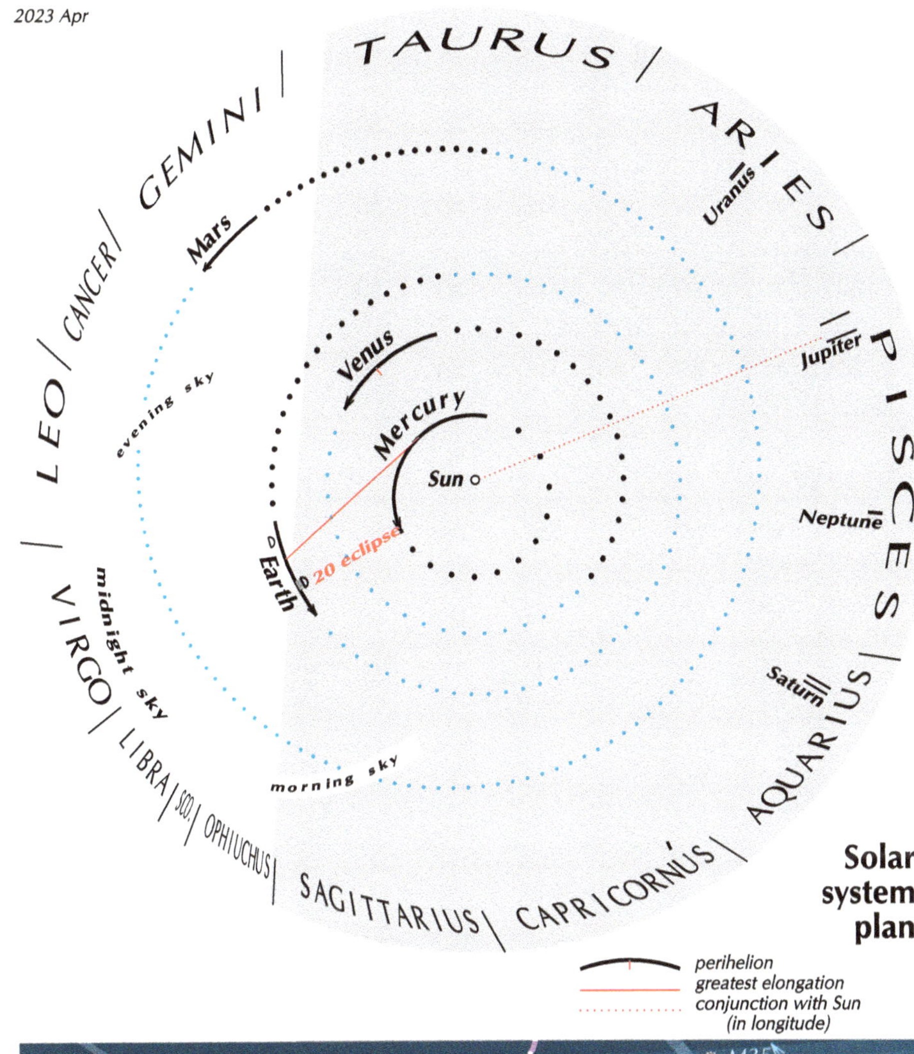

2023 Apr 22 Saturday, 45m after sunset
8:31 PM CDT = Apr 23, 1:31 Universal Time
view from latitude 40°N, longitude 90°W
sidereal time 144° = 9.57h
Julian Date 2460057.56
3 days after New Moon

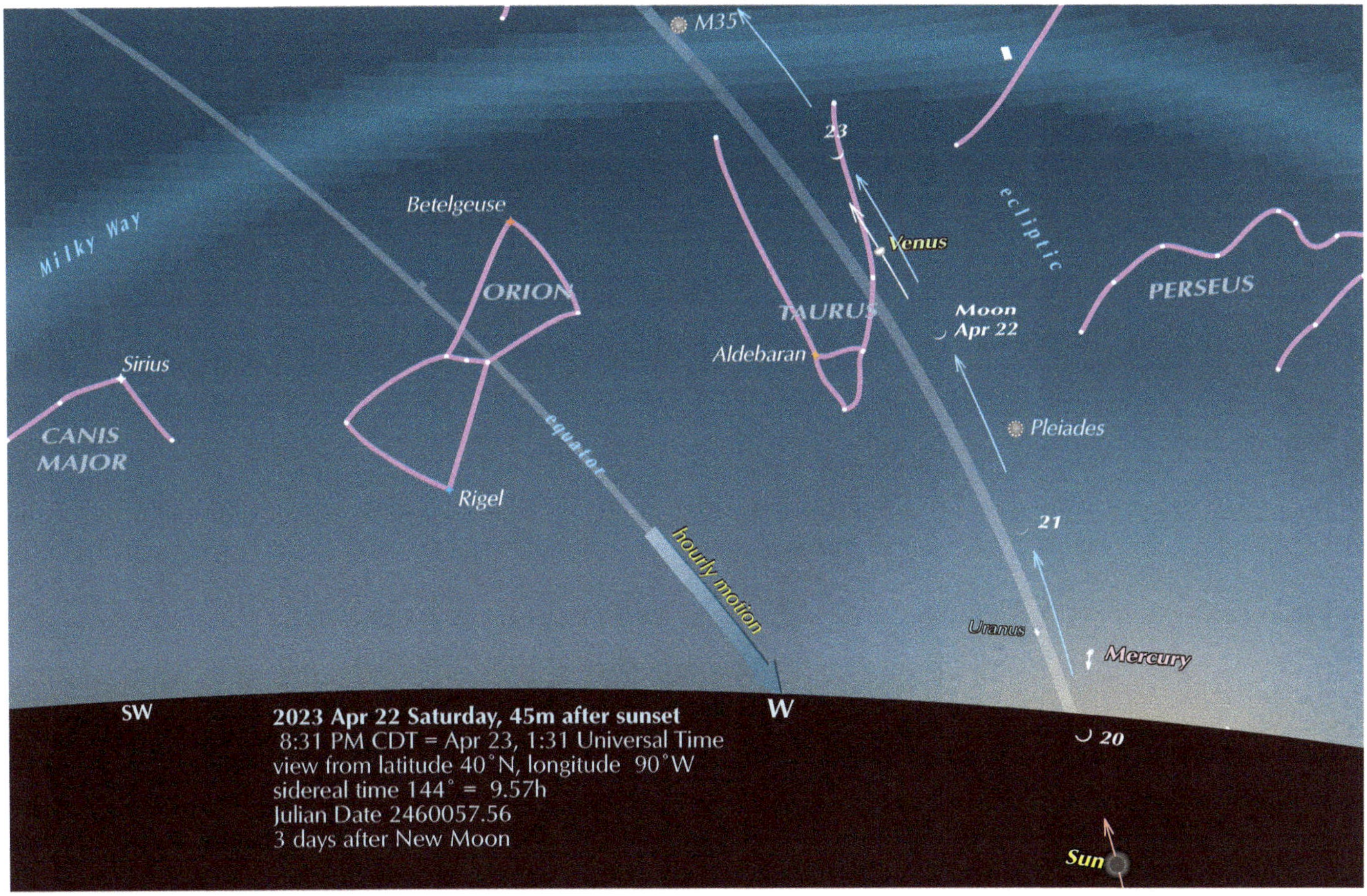

APRIL

0035.5	Apr	1	SAT		All Fools' Day
0036.5	Apr	2	SUN		Palm Sunday
0037.000	Apr	2	SUN	12	Moon 4.2° NNE of Regulus; 138° from Sun in evening sky; magnitudes -11.5 and 1.4
0040.692	Apr	6	Thu	4:36	Full Moon
0041.095	Apr	6	Thu	14	Pluto at northernmost declination, -22.48°
0041.354	Apr	6	Thu	21	Moon 2.95° NNE of Spica; 172° from Sun in midnight sky; magnitudes -12.5 and 1.0
0041.5	Apr	7	Fri		Good Friday
0042.080	Apr	7	Fri	14	Moon at descending node; longitude 214.0°
0043.5	Apr	9	SUN		Easter
0044.813	Apr	10	Mon	8	Moon 1.49° NNE of Antares; 130° from Sun in morning sky; magnitudes -11.4 and 1.0
0045.522	Apr	11	Tue	1	Mercury at northernmost latitude from the ecliptic plane, 7.0°
0046.083	Apr	11	Tue	14	Venus 2.52° SE of Pleiades; 39° from Sun in evening sky
0046.418	Apr	11	Tue	22	Mercury at easternmost elongation; 19.5° from Sun in evening sky; magnitude 0.1
0046.425	Apr	11	Tue	22	Jupiter at conjunction with the Sun; 5.955 AU from Earth; latitude -1.28°
0047.883	Apr	13	Thu	9:12	Last quarter Moon
0050.374	Apr	15	SAT	21	The equation of time is 0
0050.605	Apr	16	SUN	2:31	Moon at perigee; distance 57.69 Earth-radii
0050.771	Apr	16	SUN	7	Moon 3.2° SE of Saturn; 52° from Sun in morning sky; magnitudes -8.4 and 1.1
0052.047	Apr	17	Mon	13	Venus at perihelion; 0.7184 AU from the Sun
0052.313	Apr	17	Mon	20	Moon 2.05° SE of Neptune; 31° from Sun in morning sky; magnitudes -7.0 and 7.9
0053.967	Apr	19	Wed	11	Sun enters Aries, at longitude 29.14° on the ecliptic
0054.188	Apr	19	Wed	17	Venus 7.4° N of Aldebaran; 41° from Sun in evening sky; magnitudes -4.1 and 0.9
0054.250	Apr	19	Wed	18	Moon 0.33° NE of Jupiter; 6° from Sun in morning sky; magnitudes -4.5 and -2.0
0054.677	Apr	20	Thu	4:15	New Moon; beginning of lunation 1241; annular-total eclipse of the Sun
0054.843	Apr	20	Thu	8	Sun enters the astrological sign Taurus, i.e. its longitude is 30°
0054.981	Apr	20	Thu	12	Moon at ascending node; longitude 34.0°
0055.854	Apr	21	Fri	8	Mercury stationary in longitude; starts retrograde motion
0055.875	Apr	21	Fri	9	Moon 1.78° SE of Mercury; 15° from Sun in evening sky; magnitudes -5.5 and 2.1
0056.021	Apr	21	Fri	13	Moon 1.61° N of Uranus; 17° from Sun in evening sky; magnitudes -5.7 and 5.8
0056.042	Apr	21	Fri	13	Moon, Mercury, and Uranus within circle of diameter 3.86°; about 16° from the Sun in the evening sky; magnitudes -6, 2, 6
0056.163	Apr	21	Fri	16	Mercury stationary in right ascension; starts retrograde motion
0056.792	Apr	22	SAT	7	Mercury 3.8° NW of Uranus; 14° and 16° from Sun in evening sky; magnitudes 2.4 and 5.8; quasi-conjunction

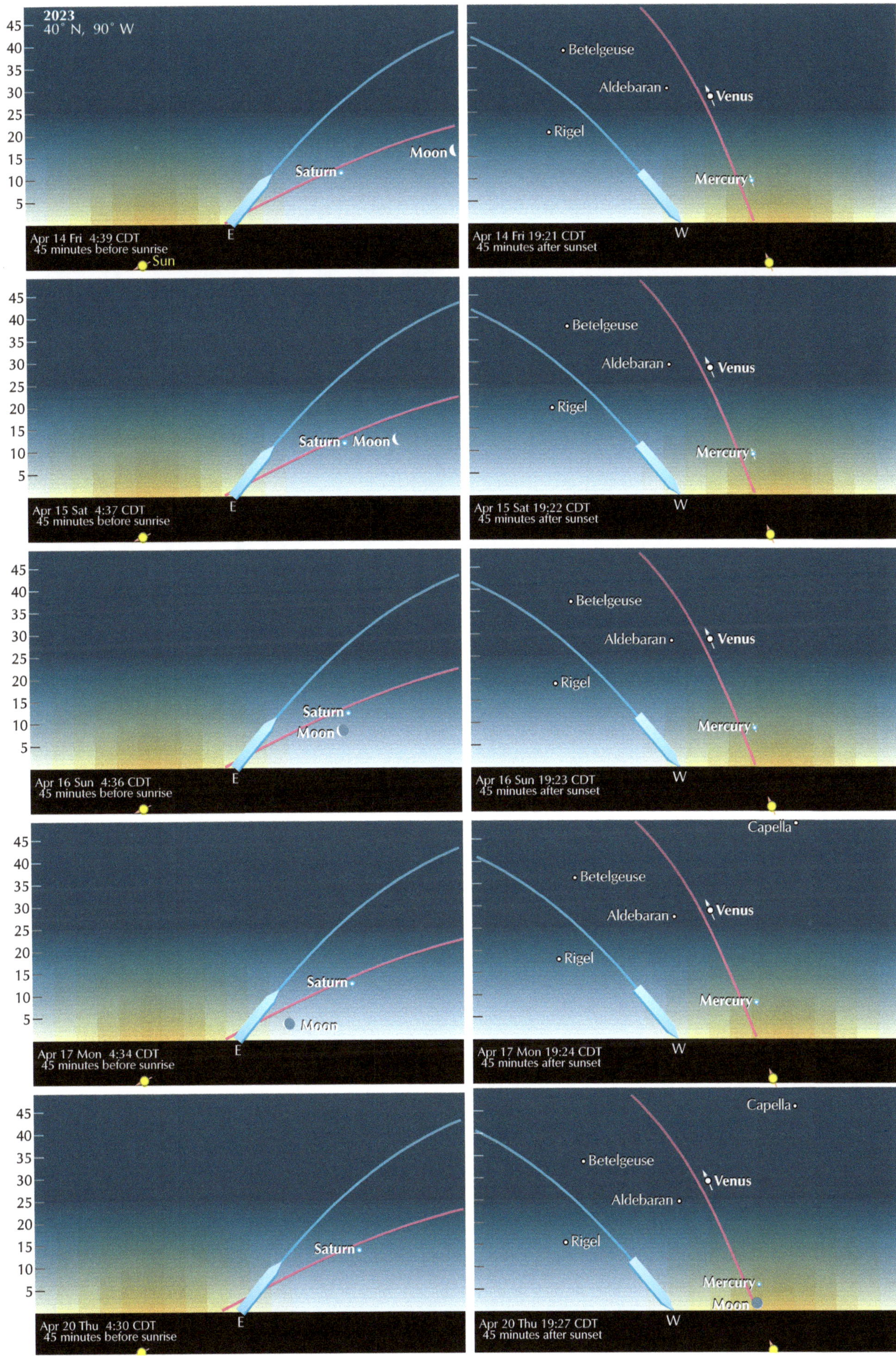
2023
40° N, 90° W
45
40
35
30
25
20
15
10
5
Moon
Saturn
E
Apr 14 Fri 4:39 CDT
45 minutes before sunrise
Sun
Betelgeuse
Aldebaran
Venus
Rigel
Mercury
W
Apr 14 Fri 19:21 CDT
45 minutes after sunset
Saturn
Moon
Apr 15 Sat 4:37 CDT
45 minutes before sunrise
Apr 15 Sat 19:22 CDT
45 minutes after sunset
Apr 16 Sun 4:36 CDT
45 minutes before sunrise
Apr 16 Sun 19:23 CDT
45 minutes after sunset
Capella
Apr 17 Mon 4:34 CDT
45 minutes before sunrise
Apr 17 Mon 19:24 CDT
45 minutes after sunset
Apr 20 Thu 4:30 CDT
45 minutes before sunrise
Apr 20 Thu 19:27 CDT
45 minutes after sunset

0056.979	Apr	22	SAT	12	Moon 1.72° SE of Pleiades; 29° and 28° from Sun in evening sky
0057.193	Apr	22	SAT	17	Mars at northernmost latitude from the ecliptic plane, 1.8°
0057.5	Apr	23	SUN	0	Lyrid meteors; ZHR 18; 3 days after new Moon
0058.042	Apr	23	SUN	13	Moon 1.31° N of Venus; 41° from Sun in evening sky; magnitudes -7.6 and -4.1
0059.126	Apr	24	Mon	15	Middle of eclipse season: Sun is at same longitude as Moon's ascending node, 34.2°
0059.479	Apr	24	Mon	24	Moon 3.6° N of M35 cluster; 58° from Sun in evening sky; magnitudes -8.6 and 5.3
0060.646	Apr	26	Wed	4	Moon 3.2° N of Mars; 71° from Sun in evening sky; magnitudes -9.2 and 1.3
0061.021	Apr	26	Wed	13	Moon 5.0° S of Castor; 75° from Sun in evening sky; magnitudes -9.4 and 1.5
0061.250	Apr	26	Wed	18	Moon 1.51° S of Pollux; 78° and 77° from Sun in evening sky; magnitudes -9.5 and 1.2
0062.389	Apr	27	Thu	21:20	First quarter Moon
0062.438	Apr	27	Thu	23	Moon 4.0° NNE of Beehive Cluster; 91° and 90° from Sun in evening sky; magnitudes -10.0 and 3.7
0062.786	Apr	28	Fri	7	Moon at apogee; distance 63.39 Earth-radii
0064.333	Apr	29	SAT	20	Moon 4.3° NNE of Regulus; 111° from Sun in evening sky; magnitudes -10.7 and 1.4

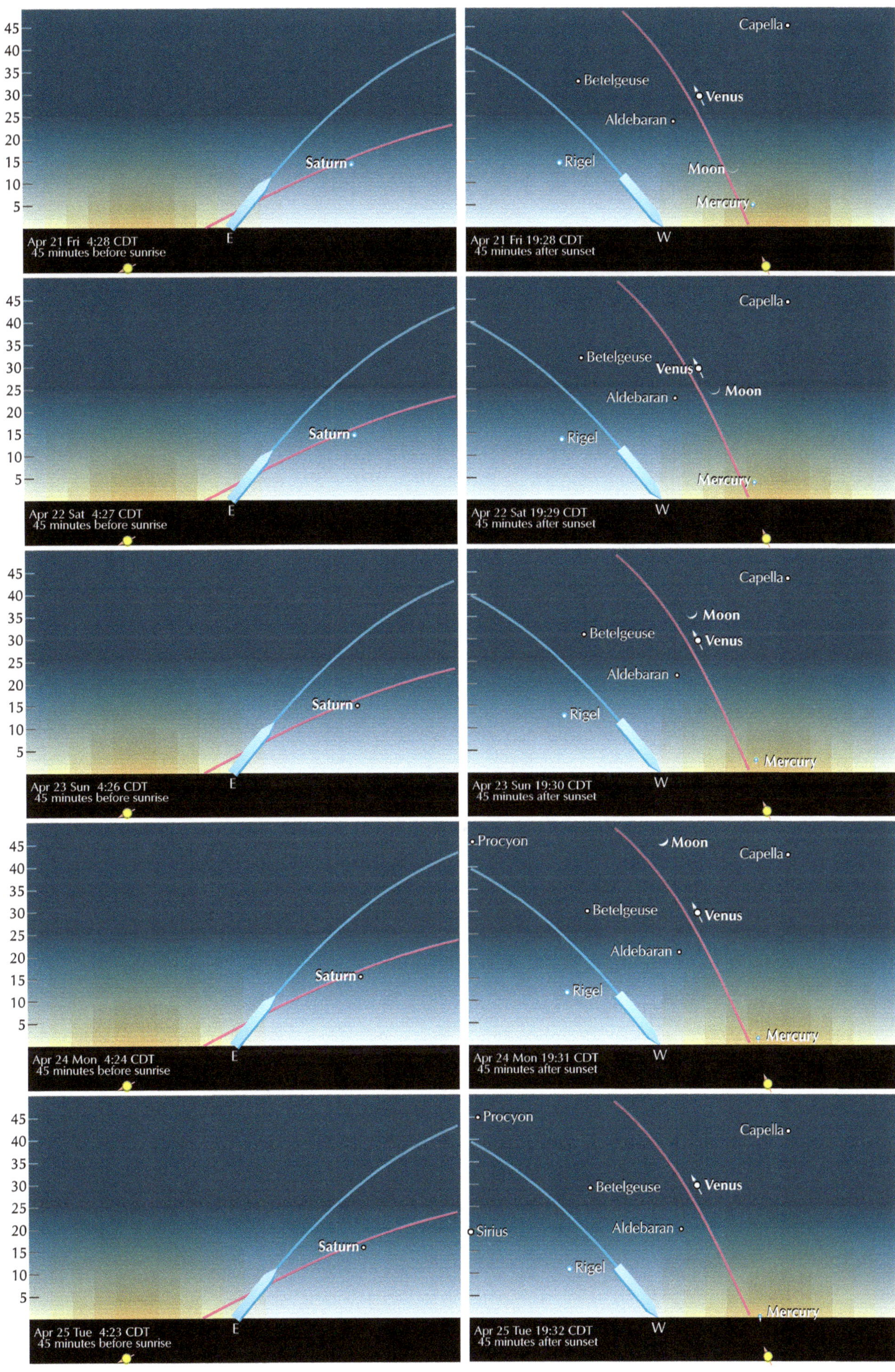
45
40
35
30
25
20
15
10
5
Saturn
E
Apr 21 Fri 4:28 CDT
45 minutes before sunrise
Capella
Betelgeuse
Venus
Aldebaran
Rigel
Moon
Mercury
W
Apr 21 Fri 19:28 CDT
45 minutes after sunset
Apr 22 Sat 4:27 CDT
45 minutes before sunrise
Apr 22 Sat 19:29 CDT
45 minutes after sunset
Apr 23 Sun 4:26 CDT
45 minutes before sunrise
Apr 23 Sun 19:30 CDT
45 minutes after sunset
Procyon
Apr 24 Mon 4:24 CDT
45 minutes before sunrise
Apr 24 Mon 19:31 CDT
45 minutes after sunset
Sirius
Apr 25 Tue 4:23 CDT
45 minutes before sunrise
Apr 25 Tue 19:32 CDT
45 minutes after sunset

MAY

SKY DOME

Evening sky
for latitude 40° north

about 10 PM at the 5th,
9 PM at the 20th
of the month

sidereal time
13h

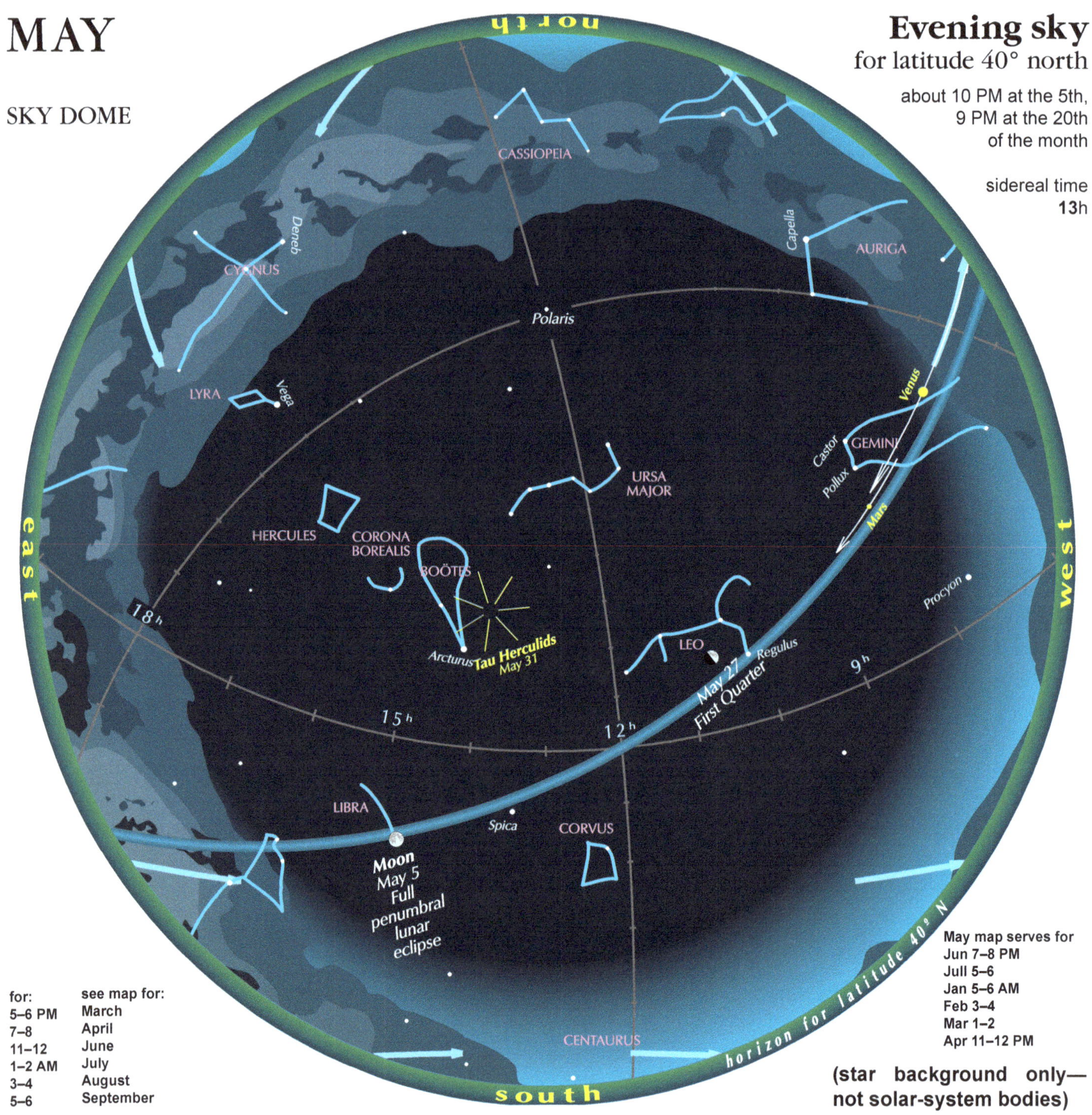

for:	see map for:
5–6 PM	March
7–8	April
11–12	June
1–2 AM	July
3–4	August
5–6	September

May map serves for
Jun 7–8 PM
Jull 5–6
Jan 5–6 AM
Feb 3–4
Mar 1–2
Apr 11–12 PM

(star background only—not solar-system bodies)

2023 May

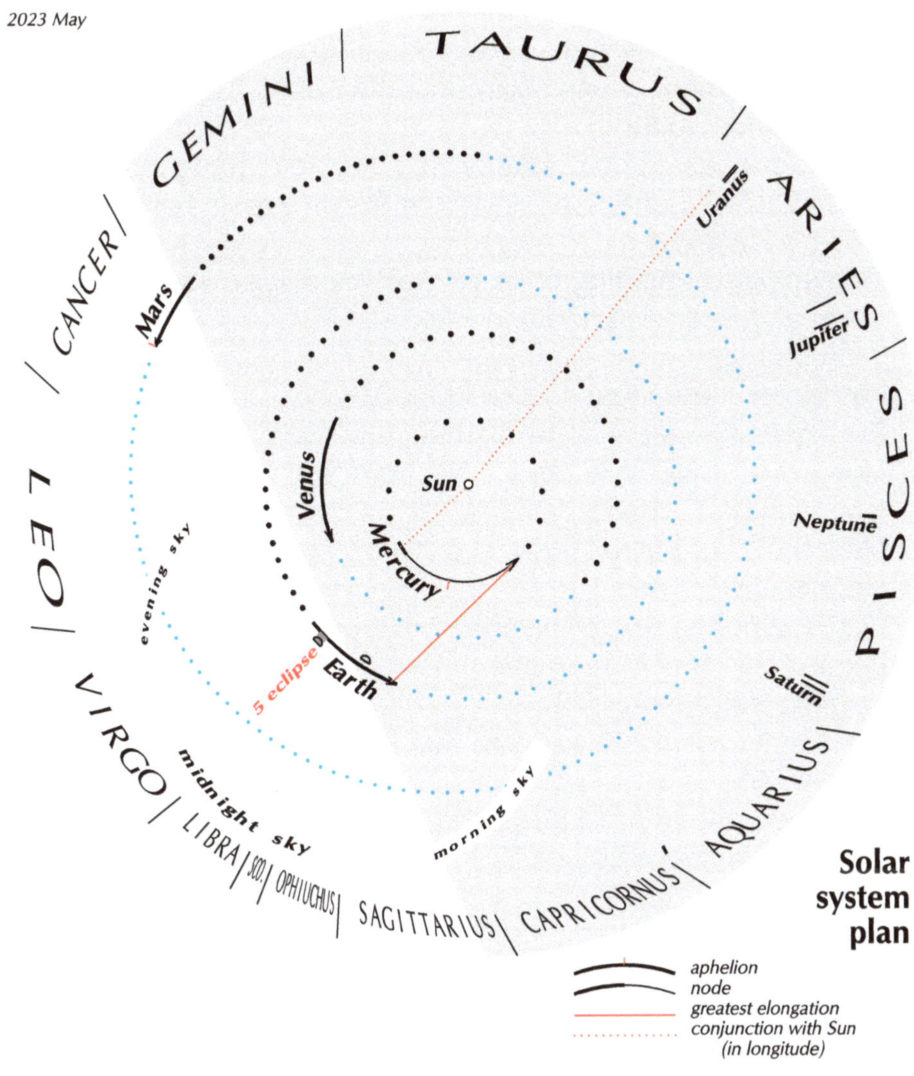
TAURUS
GEMINI
CANCER
LEO
VIRGO
LIBRA
SCO.
OPHIUCHUS
SAGITTARIUS
CAPRICORNUS
AQUARIUS
PISCES
ARIES
Uranus
Jupiter
Neptune
Saturn
Mars
Venus
Sun
Mercury
Earth
5 eclipse
evening sky
midnight sky
morning sky
Solar system plan
aphelion
node
greatest elongation
conjunction with Sun (in longitude)

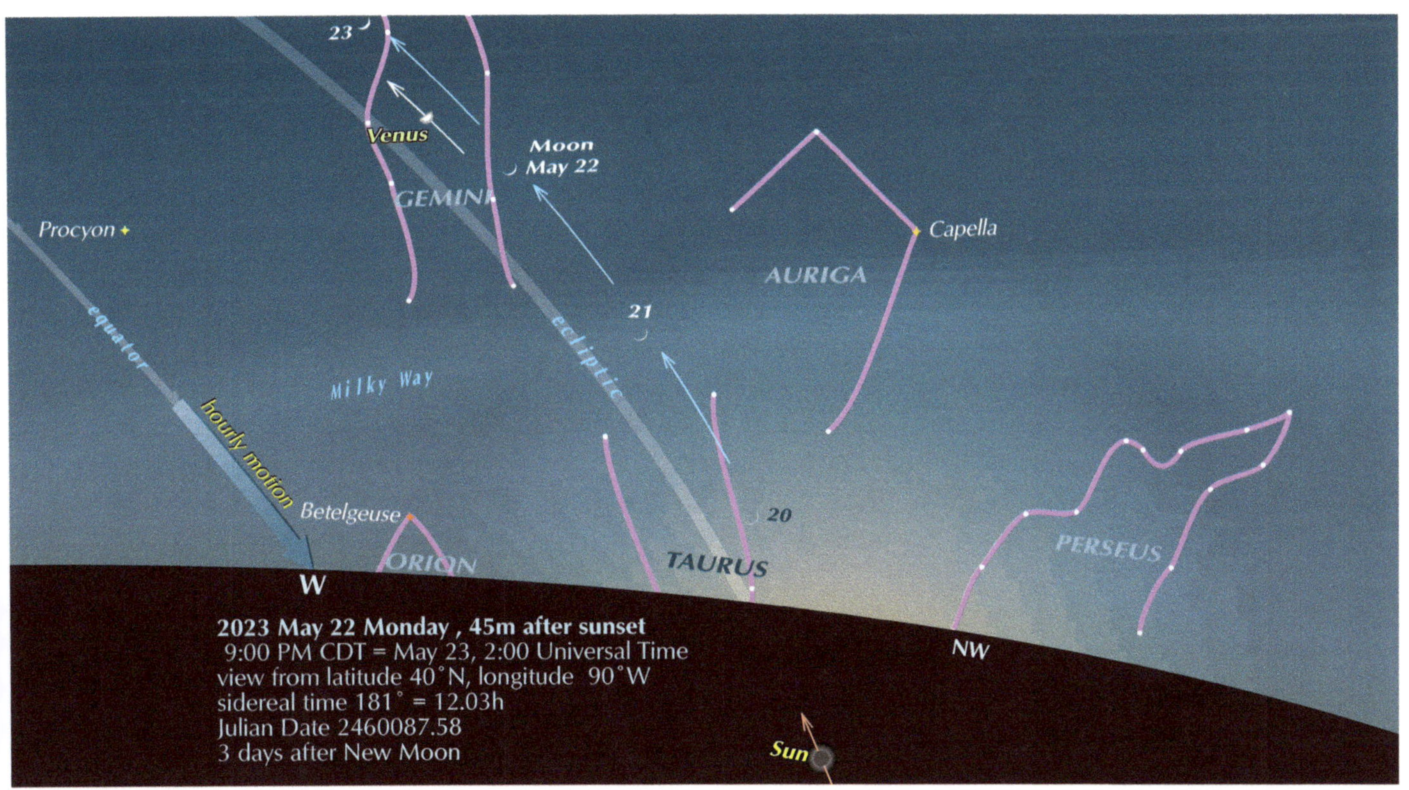
23
Venus
Moon May 22
GEMINI
Procyon
Capella
AURIGA
21
equator
ecliptic
Milky Way
hourly motion
Betelgeuse
20
ORION
TAURUS
PERSEUS
W
NW
Sun
2023 May 22 Monday , 45m after sunset
9:00 PM CDT = May 23, 2:00 Universal Time
view from latitude 40°N, longitude 90°W
sidereal time 181° = 12.03h
Julian Date 2460087.58
3 days after New Moon

MAY

0065.661	May	1	Mon	4	Pluto stationary in longitude; starts retrograde motion
0066.473	May	1	Mon	23	Mercury at inferior conjunction with the Sun; 0.564 AU from Earth; latitude 0.89°
0067.002	May	2	Tue	12	Pluto stationary in right ascension; starts retrograde motion
0068.708	May	4	Thu	5	Moon 2.98° NE of Spica; 161° from Sun in evening sky; magnitudes -12.2 and 1.0
0068.925	May	4	Thu	10	Mercury at descending node through the ecliptic plane
0069.416	May	4	Thu	22	Moon at descending node; longitude 214.1°
0070.233	May	5	Fri	17:36	Full Moon; penumbral eclipse of the Moon
0070.5	May	6	SAT	0	Eta Aquarid meteors; ZHR 50; near full Moon
0072.083	May	7	SUN	14	Moon 1.48° NNE of Antares; 156° from Sun in morning sky; magnitudes -12.1 and 1.0
0073.563	May	9	Tue	2	Mars 5.0° S of Pollux; 65° and 66° from Sun in evening sky; magnitudes 1.4 and 1.2
0073.732	May	9	Tue	6	Venus at northernmost latitude from the ecliptic plane, 3.4°
0074.208	May	9	Tue	17	Venus 1.76° N of M35 cluster; 44° from Sun in evening sky; magnitudes -4.2 and 5.3
0074.333	May	9	Tue	20	Uranus at conjunction with the Sun; 20.660 AU from Earth; latitude -0.33°
0074.428	May	9	Tue	22	Venus at northernmost declination, 26.08°
0075.713	May	11	Thu	5:07	Moon at perigee; distance 57.91 Earth-radii
0077.103	May	12	Fri	14:28	Last quarter Moon
0078.167	May	13	SAT	16	Moon 3.0° SE of Saturn; 76° from Sun in morning sky; magnitudes -9.6 and 1.1
0078.778	May	14	SUN	7	Mercury stationary in right ascension; resumes direct motion
0078.830	May	14	SUN	8	The equation of time is at a maximum of 3.64 minutes
0079.082	May	14	SUN	14	Sun enters Taurus, at longitude 53.52° on the ecliptic
0079.299	May	14	SUN	19	Mercury at aphelion; 0.4667 AU from the Sun
0079.634	May	15	Mon	3	Mercury stationary in longitude; resumes direct motion
0079.646	May	15	Mon	4	Moon 1.96° SE of Neptune; 57° from Sun in morning sky; magnitudes -8.6 and 7.9
0081.928	May	17	Wed	10	Mars and Saturn at heliocentric opposition; longitudes 150.6° and 330.6°
0082.042	May	17	Wed	13	Moon 0.74° NNW of Jupiter; 26° from Sun in morning sky; magnitudes -6.5 and -2.1; occultation
0082.042	May	17	Wed	13	Mercury 6.2° E of Jupiter; 21° and 26° from Sun in morning sky; magnitudes 1.7 and -2.1; quasi-conjunction
0082.318	May	17	Wed	20	Moon at ascending node; longitude 34.0°
0082.479	May	17	Wed	24	Moon 3.3° NNW of Mercury; 21° from Sun in morning sky; magnitudes -6.0 and 1.7
0083.479	May	18	Thu	24	Moon 1.71° NNW of Uranus; 8° from Sun in morning sky; magnitudes -4.8 and 5.9
0084.163	May	19	Fri	15:55	New Moon; beginning of lunation 1242
0084.333	May	19	Fri	20	Moon 1.74° SE of Pleiades; 3° and 4° from Sun in evening sky
0085.799	May	21	SUN	7	Sun enters the astrological sign Gemini, i.e. its longitude is 60°

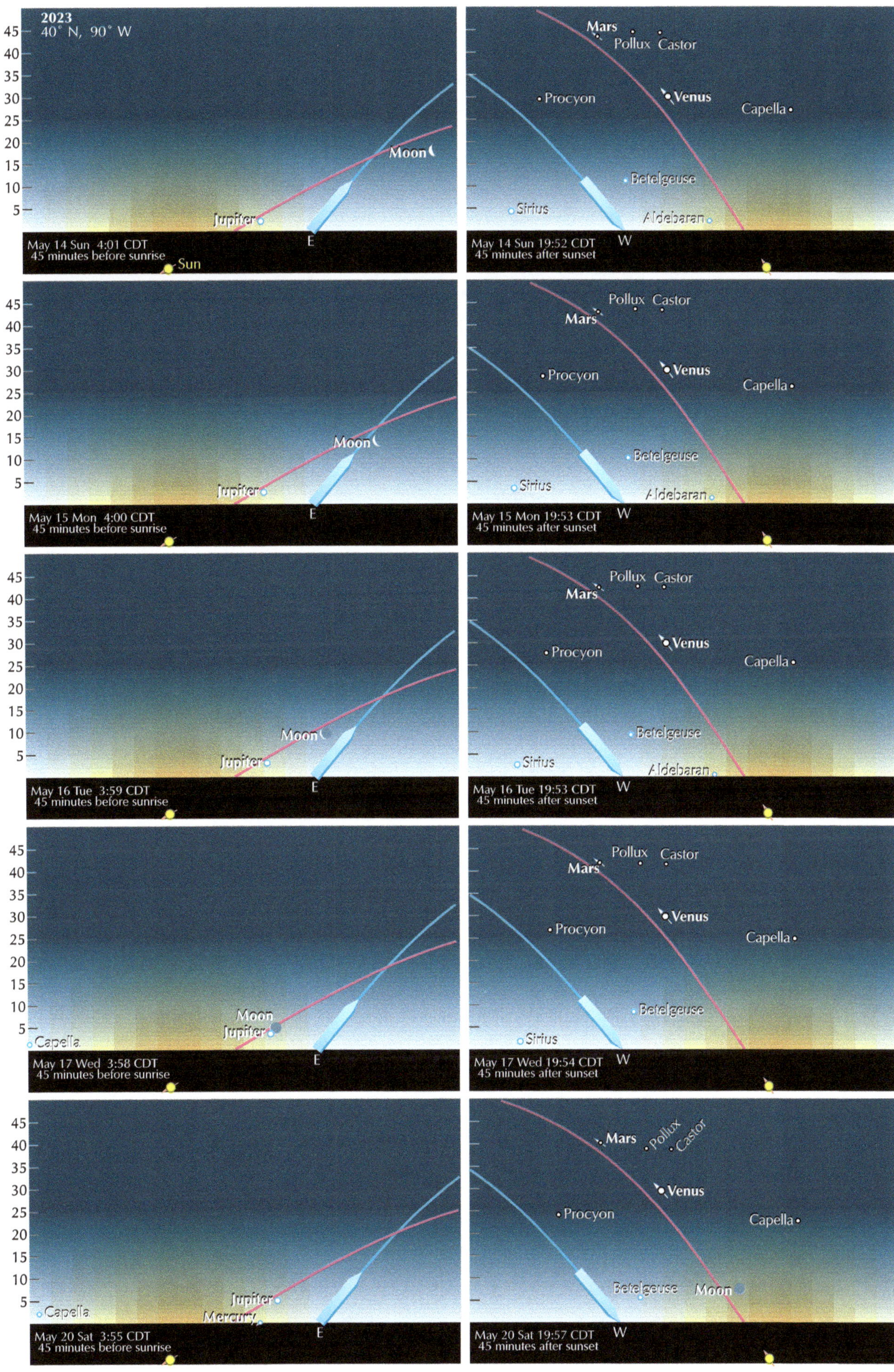
2023
40° N, 90° W
45
40
35
30
25
20
15
10
5
Moon
Jupiter
E
May 14 Sun 4:01 CDT
45 minutes before sunrise
Sun
Mars
Pollux
Castor
Procyon
Venus
Capella
Betelgeuse
Sirius
Aldebaran
W
May 14 Sun 19:52 CDT
45 minutes after sunset
May 15 Mon 4:00 CDT
45 minutes before sunrise
May 15 Mon 19:53 CDT
45 minutes after sunset
May 16 Tue 3:59 CDT
45 minutes before sunrise
May 16 Tue 19:53 CDT
45 minutes after sunset
Capella
May 17 Wed 3:58 CDT
45 minutes before sunrise
May 17 Wed 19:54 CDT
45 minutes after sunset
Mercury
May 20 Sat 3:55 CDT
45 minutes before sunrise
May 20 Sat 19:57 CDT
45 minutes after sunset

0086.833	May	22	Mon	8	Moon 3.5° N of M35 cluster; 32° and 31° from Sun in evening sky; magnitudes -6.8 and 5.3
0088.042	May	23	Tue	13	Moon 2.20° N of Venus; 45° from Sun in evening sky; magnitudes -7.8 and -4.2
0088.354	May	23	Tue	21	Moon 5.0° S of Castor; 49° from Sun in evening sky; magnitudes -8.0 and 1.5
0088.583	May	24	Wed	2	Moon 1.57° S of Pollux; 51° from Sun in evening sky; magnitudes -8.1 and 1.2
0089.333	May	24	Wed	20	Moon 3.7° NNE of Mars; 59° from Sun in evening sky; magnitudes -8.6 and 1.5
0089.771	May	25	Thu	7	Moon 3.9° NNE of Beehive Cluster; 64° from Sun in evening sky; magnitudes -8.9 and 3.7
0090.566	May	26	Fri	2	Moon at apogee; distance 63.42 Earth-radii
0091.396	May	26	Fri	22	Venus 7.3° S of Castor; 45° and 46° from Sun in evening sky; magnitudes -4.3 and 1.5
0091.688	May	27	SAT	5	Moon 4.2° NNE of Regulus; 85° and 84° from Sun in evening sky; magnitudes -9.8 and 1.4
0092.141	May	27	SAT	15:23	First quarter Moon
0092.5	May	28	SUN		Whit Sunday
0092.947	May	28	SUN	11	Saturn at west quadrature, 90° from the Sun
0093.726	May	29	Mon	5	Mercury at westernmost elongation; 24.9° from Sun in morning sky; magnitude 0.6
0094.375	May	29	Mon	21	Venus 4.0° SSW of Pollux; 45° and 46° from Sun in evening sky; magnitudes -4.3 and 1.2
0095.359	May	30	Tue	21	Mars at aphelion; 1.6659 AU from the Sun
0096.083	May	31	Wed	14	Moon 2.91° NNE of Spica; 135° and 134° from Sun in evening sky; magnitudes -11.5 and 1.0

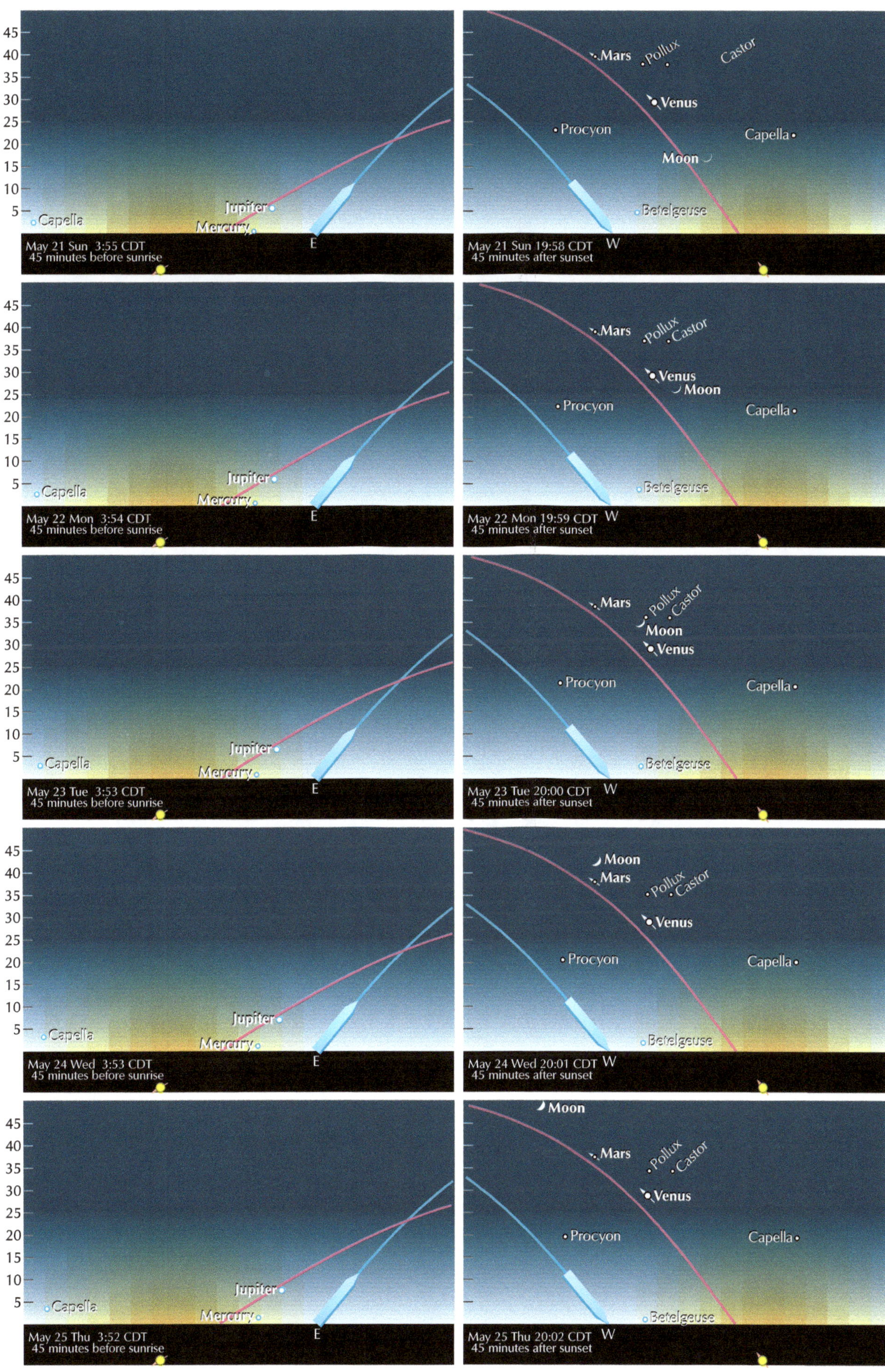
May 21 Sun 3:55 CDT
45 minutes before sunrise
Capella
Jupiter
Mercury
E
May 21 Sun 19:58 CDT
45 minutes after sunset
W
Mars
Pollux
Castor
Venus
Procyon
Capella
Moon
Betelgeuse
May 22 Mon 3:54 CDT
45 minutes before sunrise
May 22 Mon 19:59 CDT
45 minutes after sunset
May 23 Tue 3:53 CDT
45 minutes before sunrise
May 23 Tue 20:00 CDT
45 minutes after sunset
May 24 Wed 3:53 CDT
45 minutes before sunrise
May 24 Wed 20:01 CDT
45 minutes after sunset
May 25 Thu 3:52 CDT
45 minutes before sunrise
May 25 Thu 20:02 CDT
45 minutes after sunset

JUNE

SKY DOME

Evening sky
for latitude 40° north

about 10 PM at the 5th,
9 PM at the 20th
of the month

sidereal time
15h

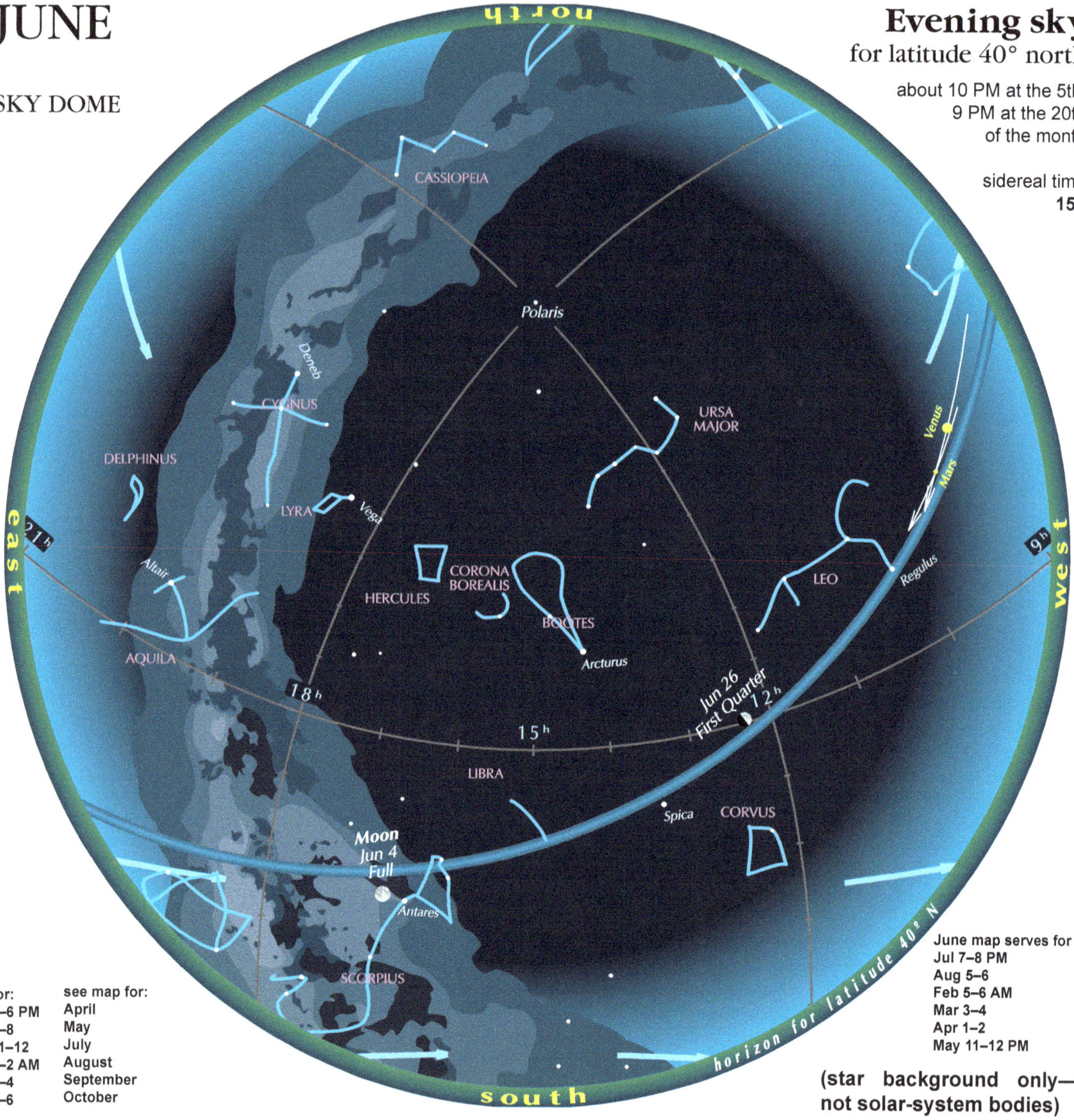

for:	see map for:
5–6 PM	April
7–8	May
11–12	July
1–2 AM	August
3–4	September
5–6	October

June map serves for
Jul 7–8 PM
Aug 5–6
Feb 5–6 AM
Mar 3–4
Apr 1–2
May 11–12 PM

(star background only—not solar-system bodies)

2023 Jun

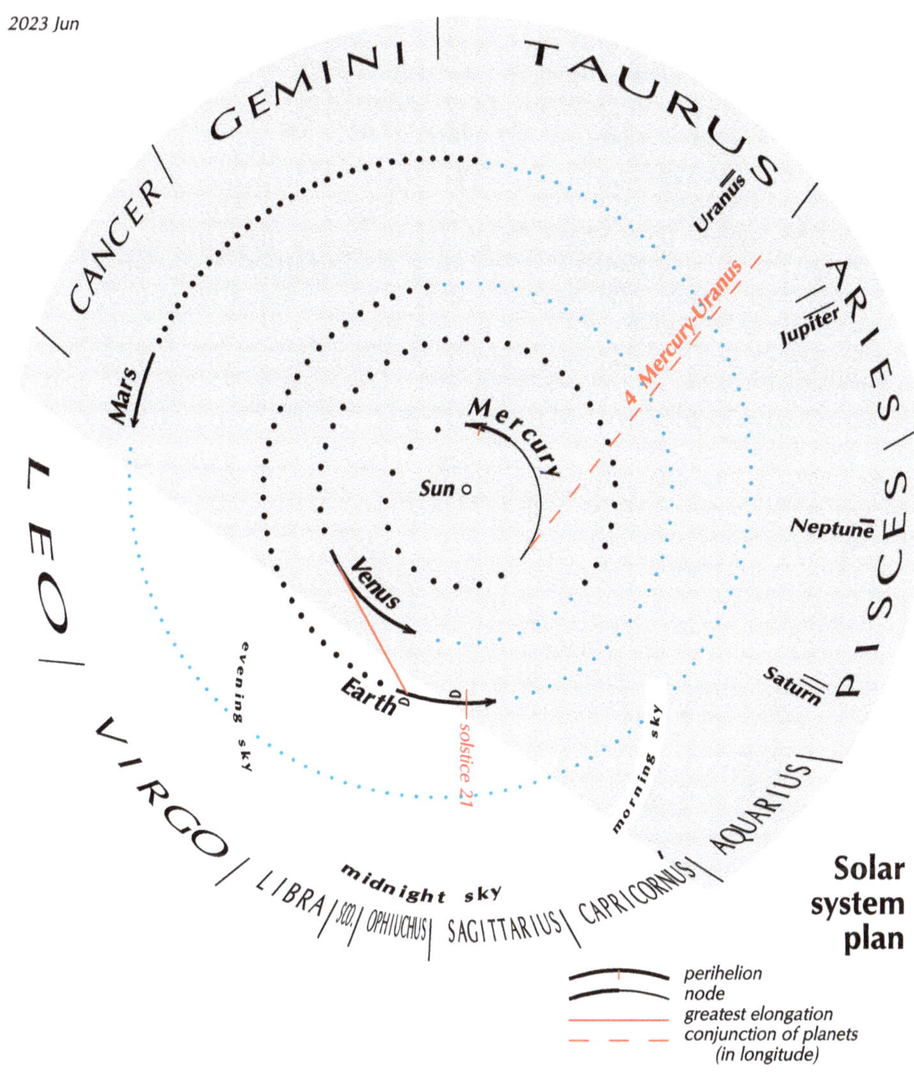

Solar system plan

perihelion
node
greatest elongation
conjunction of planets (in longitude)

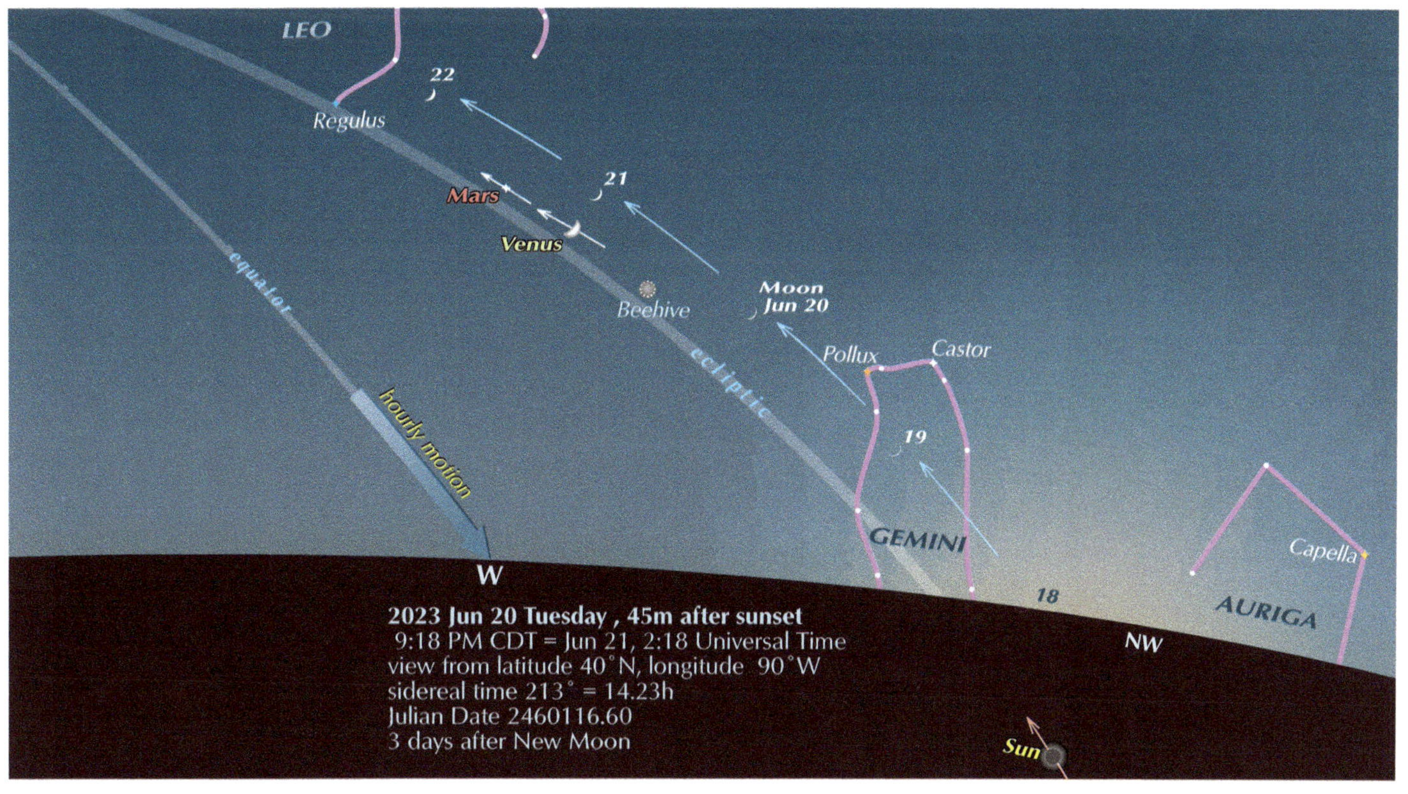

JUNE

0096.767	Jun	1	Thu	6	Moon at descending node; longitude 213.6°
0098.458	Jun	2	Fri	23	Mars 0.16° NNE of Beehive Cluster; 56° from Sun in evening sky; magnitudes 1.6 and 3.7
0099.458	Jun	3	SAT	23	Moon 1.52° NNE of Antares; 176° and 175° from Sun in evening midnight sky; magnitudes -12.7 and 1.0
0099.554	Jun	4	SUN	1	Mercury at southernmost latitude from the ecliptic plane, -7.0°
0099.616	Jun	4	SUN	3	Venus dichotomy (D-shape)
0099.655	Jun	4	SUN	3:43	Full Moon
0099.944	Jun	4	SUN	11	Venus at easternmost elongation; 45.4° from Sun in evening sky; magnitude -4.3
0100.542	Jun	5	Mon	1	Mercury 2.72° SE of Uranus; 24° from Sun in morning sky; magnitudes 0.1 and 5.8
0102.468	Jun	6	Tue	23:14	Moon at perigee; distance 57.21 Earth-radii
0102.5	Jun	7	Wed	0	Daytime Arietid meteors; ZHR 30; 3 days after full Moon
0105.458	Jun	9	Fri	23	Moon 2.72° SE of Saturn; 101° and 102° from Sun in morning sky; magnitudes -10.6 and 1.0
0106.313	Jun	10	SAT	19:31	Last quarter Moon
0106.917	Jun	11	SUN	10	Moon 1.80° SE of Neptune; 82° and 83° from Sun in morning sky; magnitudes -9.8 and 7.9
0107.521	Jun	12	Mon	1	Mercury 6.1° SE of the Pleiades; 20° and 21° from Sun in morning sky
0108.825	Jun	13	Tue	8	The equation of time is 0
0109.504	Jun	14	Wed	0	Moon at ascending node; longitude 33.1°
0109.521	Jun	14	Wed	1	Venus 0.78° NNE of Beehive Cluster; 45° from Sun in evening sky; magnitudes -4.4 and 3.7
0109.688	Jun	14	Wed	4:31	Earliest sunrise, at latitude 40° north
0109.750	Jun	14	Wed	6	Moon 1.41° NNW of Jupiter; 47° from Sun in morning sky; magnitudes -8.0 and -2.1
0110.875	Jun	15	Thu	9	Moon 1.89° NNW of Uranus; 33° from Sun in morning sky; magnitudes -7.0 and 5.8
0111.625	Jun	16	Fri	3	Moon 1.71° SE of Pleiades; 24° and 25° from Sun in morning sky
0112.292	Jun	16	Fri	19	Moon 4.2° N of Mercury; 17° and 16° from Sun in morning sky; magnitudes -5.5 and -0.8
0112.542	Jun	17	SAT	1	Mercury 4.3° NNW of Aldebaran; 16° from Sun in morning sky; magnitudes -0.8 and 0.9
0113.151	Jun	17	SAT	16	Saturn stationary in longitude; starts retrograde motion
0113.693	Jun	18	SUN	4:38	New Moon; beginning of lunation 1243
0114.057	Jun	18	SUN	13	Saturn stationary in right ascension; starts retrograde motion
0114.146	Jun	18	SUN	16	Moon 3.5° N of M35 cluster; 7° and 5° from Sun in evening sky; magnitudes -4.5 and 5.3
0115.667	Jun	20	Tue	4	Moon 5.1° S of Castor; 23° and 24° from Sun in evening sky; magnitudes -6.1 and 1.5
0115.917	Jun	20	Tue	10	Moon 1.72° S of Pollux; 26° from Sun in evening sky; magnitudes -6.3 and 1.2

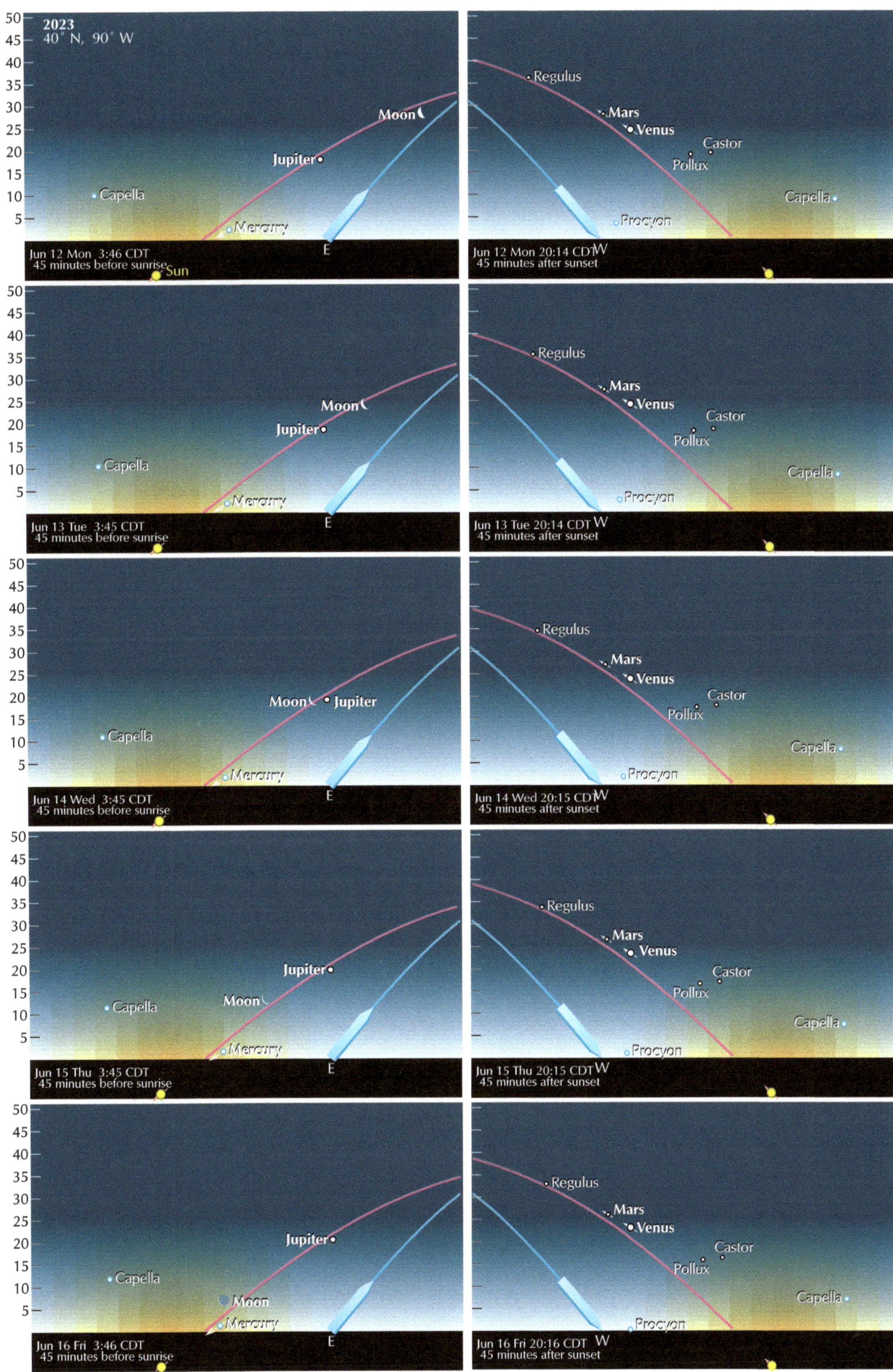

2023
40° N, 90° W
Moon
Jupiter
Capella
Mercury
E
Jun 12 Mon 3:46 CDT
45 minutes before sunrise
Sun
Regulus
Mars
Venus
Castor
Pollux
Capella
Procyon
W
Jun 12 Mon 20:14 CDT
45 minutes after sunset
Jun 13 Tue 3:45 CDT
45 minutes before sunrise
Jun 13 Tue 20:14 CDT
45 minutes after sunset
Jun 14 Wed 3:45 CDT
45 minutes before sunrise
Jun 14 Wed 20:15 CDT
45 minutes after sunset
Jun 15 Thu 3:45 CDT
45 minutes before sunrise
Jun 15 Thu 20:15 CDT
45 minutes after sunset
Jun 16 Fri 3:46 CDT
45 minutes before sunrise
Jun 16 Fri 20:16 CDT
45 minutes after sunset

0117.083	Jun	21	Wed	14	Moon 3.8° NNE of Beehive Cluster; 38° from Sun in evening sky; magnitudes -7.3 and 3.7
0117.125	Jun	21	Wed	15:01	June (northern summer) solstice
0117.125	Jun	21	Wed	15:01	Sun enters the astrological sign Cancer, i.e. its longitude is 90°
0117.636	Jun	22	Thu	3	Sun enters Gemini, at longitude 90.49° on the ecliptic
0117.646	Jun	22	Thu	4	Moon 3.5° NNE of Venus; 44° from Sun in evening sky; magnitudes -7.7 and -4.4
0117.842	Jun	22	Thu	8	Moon, Venus, and Mars within circle of diameter 4.95°; about 46° from the Sun in the evening sky; magnitudes -8, -4, 2
0118.063	Jun	22	Thu	14	Moon 3.6° NNE of Mars; 49° and 48° from Sun in evening sky; magnitudes -8.0 and 1.7
0118.272	Jun	22	Thu	19	Moon at apogee; distance 63.56 Earth-radii
0118.613	Jun	23	Fri	3	Mercury at ascending node through the ecliptic plane
0118.979	Jun	23	Fri	12	Moon 4.0° NNE of Regulus; 59° and 58° from Sun in evening sky; magnitudes -8.6 and 1.4
0121.826	Jun	26	Mon	7:50	First quarter Moon
0123.284	Jun	27	Tue	19	Mercury at perihelion; 0.3075 AU from the Sun
0123.458	Jun	27	Tue	23	Moon 2.75° NNE of Spica; 109° and 108° from Sun in evening sky; magnitudes -10.7 and 1.0
0123.604	Jun	28	Wed	3	Mercury 0.08° SE of M35 cluster; 4° from Sun in morning sky; magnitudes -1.9 and 5.3
0124.016	Jun	28	Wed	12	Moon at descending node; longitude 211.9°
0124.315	Jun	28	Wed	19:33	Latest sunset, at latitude 40° north
0126.011	Jun	30	Fri	12	Mercury at northernmost declination, 24.40°
0126.144	Jun	30	Fri	15	Neptune stationary in longitude; starts retrograde motion

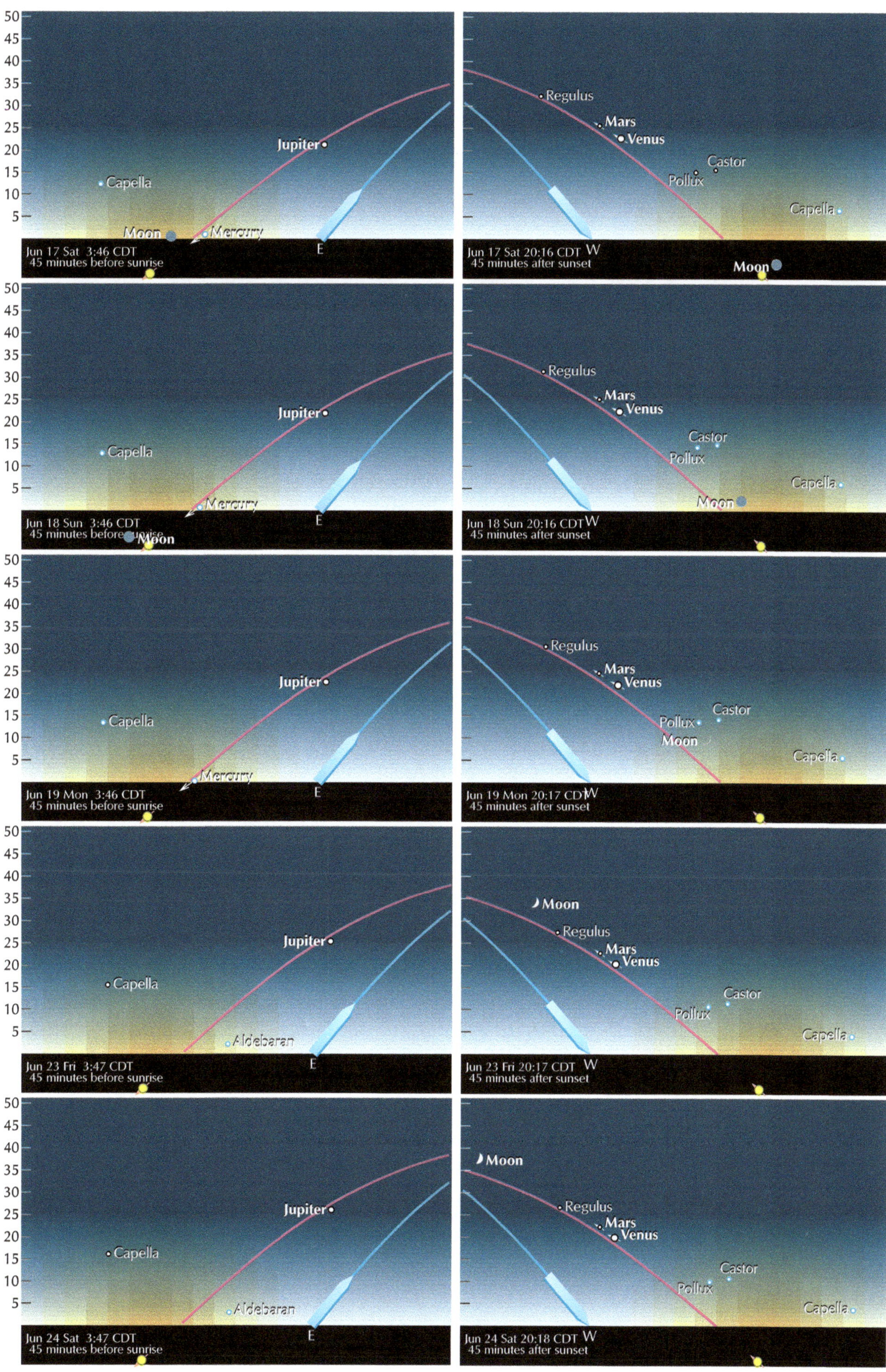

Jupiter
Capella
Moon
Mercury
E
Jun 17 Sat 3:46 CDT
45 minutes before sunrise
Regulus
Mars
Venus
Castor
Pollux
Capella
W
Moon
Jun 17 Sat 20:16 CDT
45 minutes after sunset
Jun 18 Sun 3:46 CDT
45 minutes before sunrise
Jun 18 Sun 20:16 CDT
45 minutes after sunset
Jun 19 Mon 3:46 CDT
45 minutes before sunrise
Jun 19 Mon 20:17 CDT
45 minutes after sunset
Aldebaran
Jun 23 Fri 3:47 CDT
45 minutes before sunrise
Jun 23 Fri 20:17 CDT
45 minutes after sunset
Jun 24 Sat 3:47 CDT
45 minutes before sunrise
Jun 24 Sat 20:18 CDT
45 minutes after sunset

JULY

SKY DOME

Evening sky
for latitude 40° north

about 10 PM at the 5th,
9 PM at the 20th
of the month

sidereal time
17h

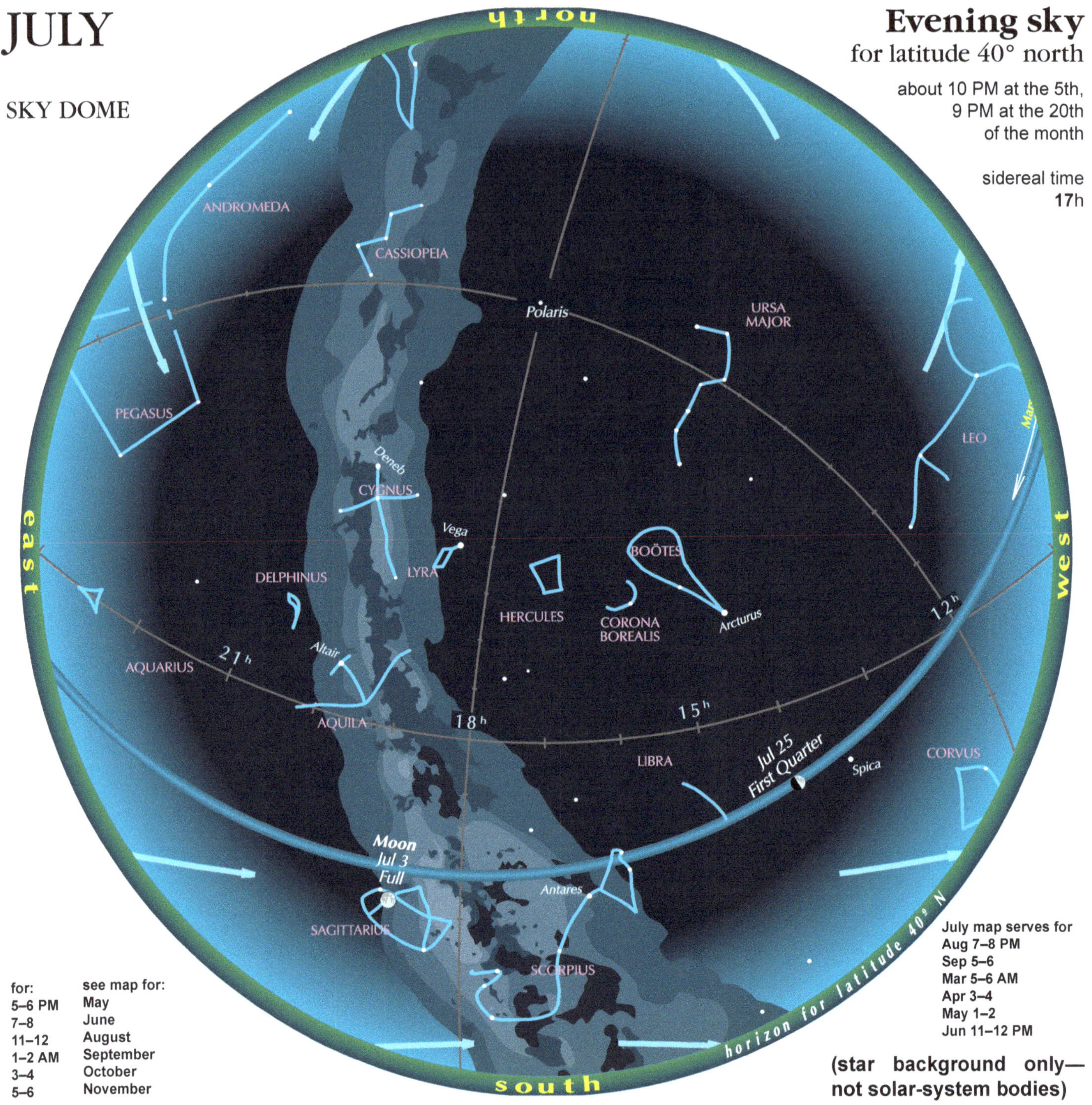

for:	see map for:
5–6 PM	May
7–8	June
11–12	August
1–2 AM	September
3–4	October
5–6	November

July map serves for
Aug 7–8 PM
Sep 5–6
Mar 5–6 AM
Apr 3–4
May 1–2
Jun 11–12 PM

(star background only—not solar-system bodies)

2023 Jul

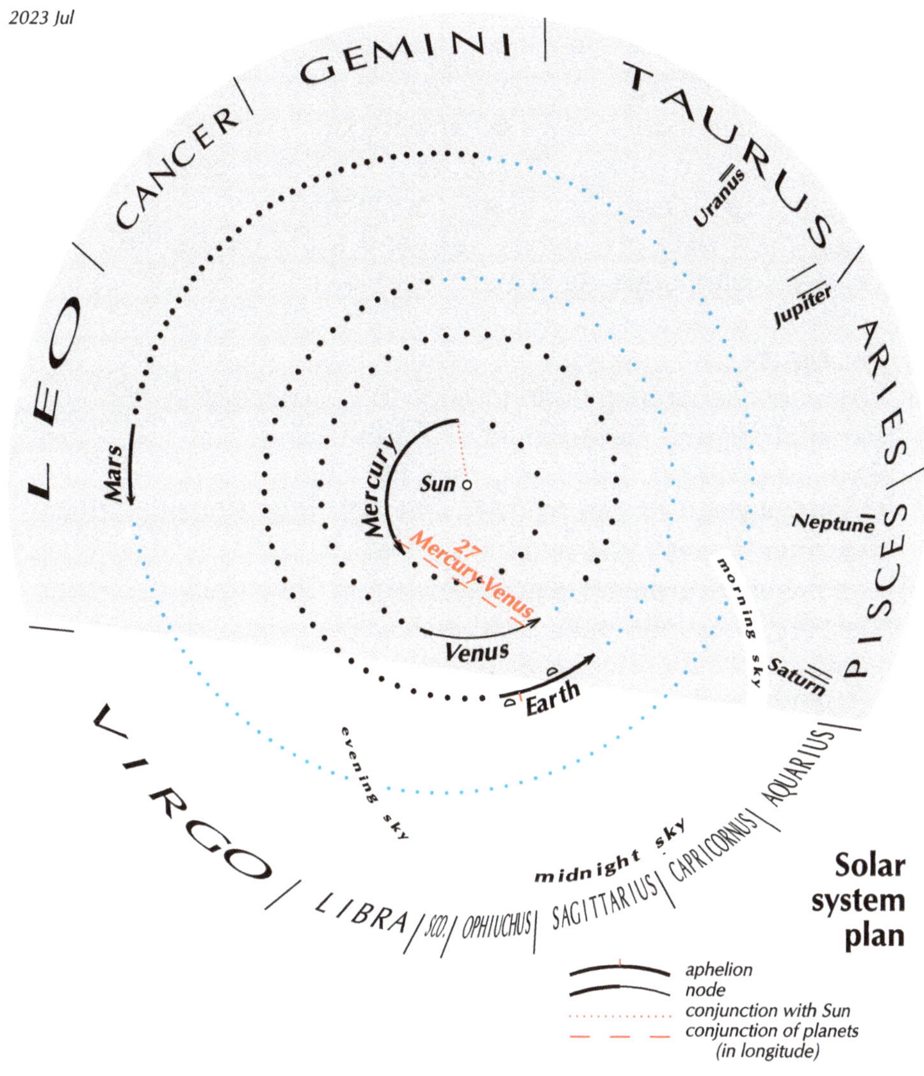
GEMINI
TAURUS
CANCER
LEO
ARIES
PISCES
VIRGO
LIBRA
SCO.
OPHIUCHUS
SAGITTARIUS
CAPRICORNUS
AQUARIUS
Uranus
Jupiter
Neptune
Saturn
Mars
Mercury
Sun
27
Mercury-Venus
Venus
Earth
morning sky
evening sky
midnight sky
Solar system plan
aphelion
node
conjunction with Sun
conjunction of planets
(in longitude)

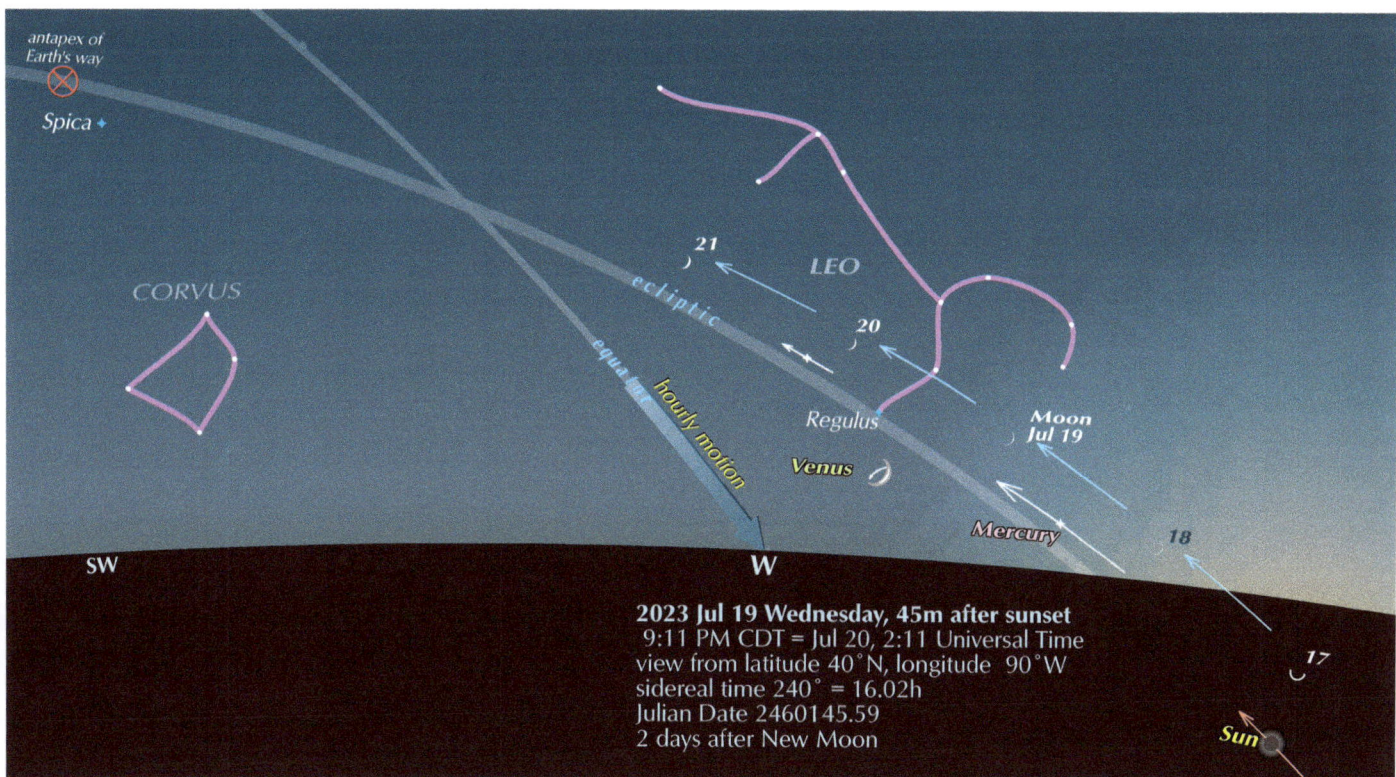

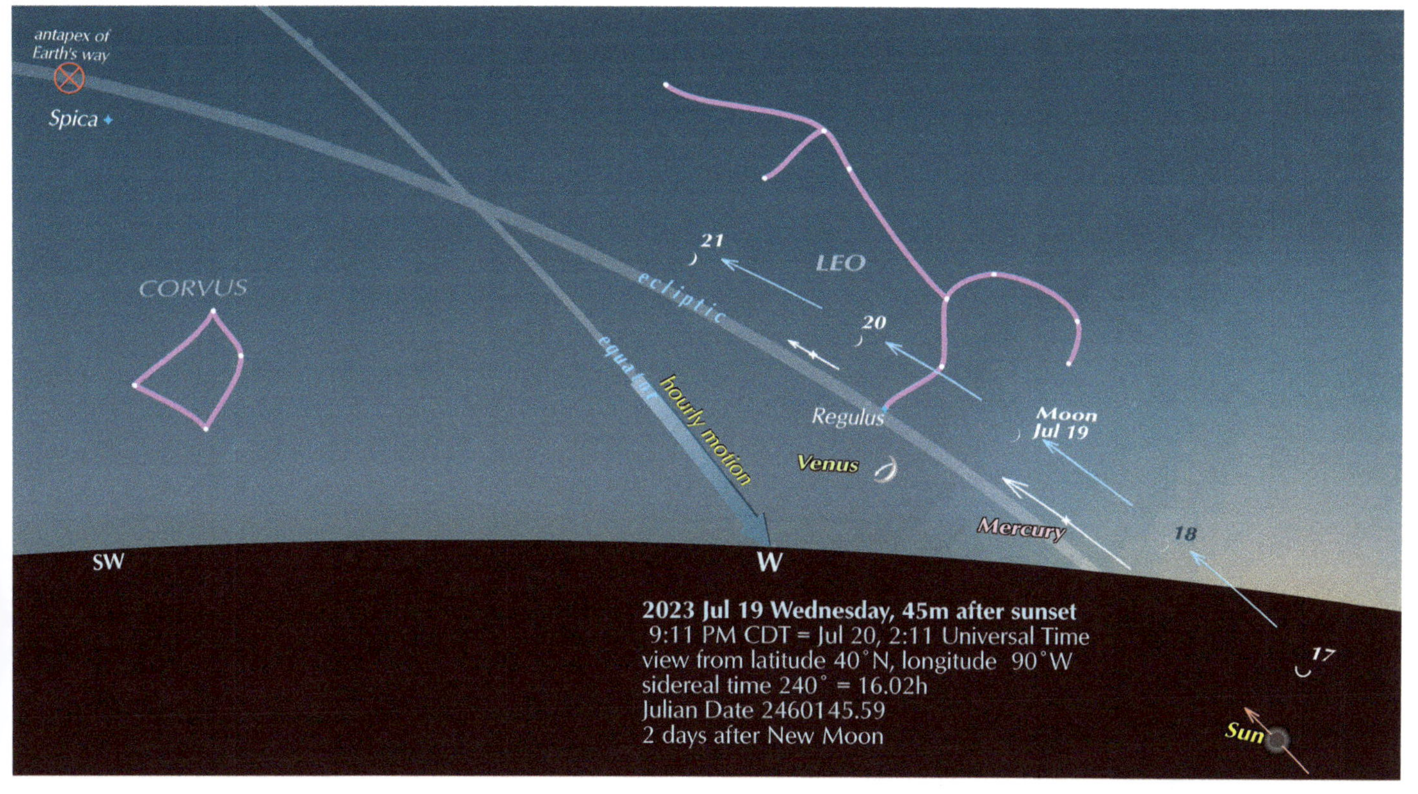
antapex of Earth's way
Spica
CORVUS
LEO
21
20
ecliptic
equator
hourly motion
Regulus
Moon Jul 19
Venus
Mercury
18
17
Sun
SW
W
2023 Jul 19 Wednesday, 45m after sunset
9:11 PM CDT = Jul 20, 2:11 Universal Time
view from latitude 40°N, longitude 90°W
sidereal time 240° = 16.02h
Julian Date 2460145.59
2 days after New Moon

JULY

0126.704	Jul	1	SAT	5	Mercury at superior conjunction with the Sun; 1.327 AU from Earth; latitude 5.42°
0126.802	Jul	1	SAT	7	Neptune stationary in right ascension; starts retrograde motion
0126.813	Jul	1	SAT	8	Venus 3.6° W of Mars; 42° and 45° from Sun in evening sky; magnitudes -4.5 and 1.7; quasi-conjunction
0126.875	Jul	1	SAT	9	Moon 1.47° NNE of Antares; 151° and 150° from Sun in evening sky; magnitudes -12.0 and 1.0
0128.986	Jul	3	Mon	11:39	Full Moon
0129.787	Jul	4	Tue	7	Venus at descending node through the ecliptic plane
0130.438	Jul	4	Tue	22:31	Moon at perigee; distance 56.47 Earth-radii
0132.149	Jul	6	Thu	16	Earth at aphelion; 1.0167 AU from the Sun
0132.729	Jul	7	Fri	6	Moon 2.44° SE of Saturn; 127° and 128° from Sun in morning sky; magnitudes -11.4 and 0.8
0133.314	Jul	7	Fri	20	Venus shows greatest illuminated extent, 294 square seconds
0133.492	Jul	7	Fri	24	Mercury at northernmost latitude from the ecliptic plane, 7.0°
0133.521	Jul	8	SAT	1	Mercury 4.9° S of Pollux; 8° and 10° from Sun in evening sky; magnitudes -1.4 and 1.2
0134.167	Jul	8	SAT	16	Moon 1.54° SE of Neptune; 108° and 109° from Sun in morning sky; magnitudes -10.8 and 7.9
0135.247	Jul	9	SUN	18	Venus brightest; magnitude -4.47°
0135.575	Jul	10	Mon	1:48	Last quarter Moon
0135.792	Jul	10	Mon	7	Venus, Mars, and Regulus within circle of diameter 4.65°; about 41° from the Sun in the evening sky; magnitudes -4, 2, 1
0136.250	Jul	10	Mon	18	Mars 0.65° NNE of Regulus; 42° from Sun in evening sky; magnitudes 1.7 and 1.4
0136.559	Jul	11	Tue	1	Moon at ascending node; longitude 30.8°
0137.354	Jul	11	Tue	21	Moon 2.09° NNW of Jupiter; 68° from Sun in morning sky; magnitudes -9.1 and -2.3
0138.208	Jul	12	Wed	17	Moon 2.17° N of Uranus; 58° from Sun in morning sky; magnitudes -8.6 and 5.8
0138.480	Jul	12	Wed	24	Summer solstice for Mars north hemisphere
0138.875	Jul	13	Thu	9	Moon 1.61° SE of Pleiades; 50° and 51° from Sun in morning sky
0139.623	Jul	14	Fri	3	Mars and Neptune at heliocentric opposition; longitudes 175.9° and 355.9°
0140.708	Jul	15	SAT	5	Mercury 0.46° NNE of Beehive Cluster; 15° from Sun in evening sky; magnitudes -0.7 and 3.7
0141.417	Jul	15	SAT	22	Moon 3.5° N of M35 cluster; 21° from Sun in morning sky; magnitudes -5.9 and 5.3
0142.250	Jul	16	SUN	18	Venus 3.5° WSW of Regulus; 34° and 36° from Sun in evening sky; magnitudes -4.5 and 1.4; quasi-conjunction
0142.938	Jul	17	Mon	11	Moon 5.2° S of Castor; 6° and 11° from Sun in morning sky; magnitudes -4.4 and 1.5
0143.188	Jul	17	Mon	17	Moon 1.76° S of Pollux; 5° and 7° from Sun in morning sky; magnitudes -4.3 and 1.2
0143.272	Jul	17	Mon	18:32	New Moon; beginning of lunation 1244

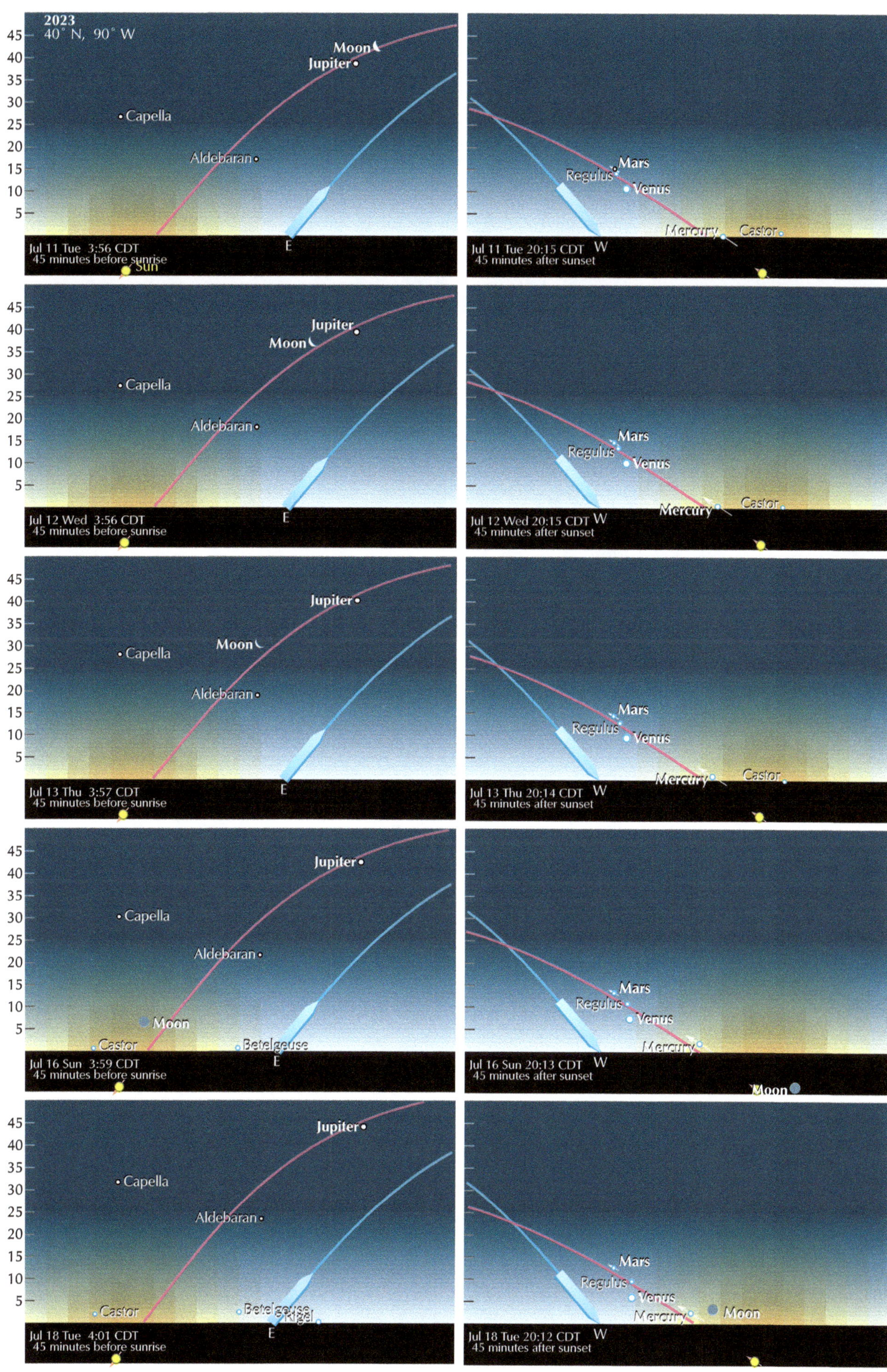
2023
40° N, 90° W
45
40
35
30
25
20
15
10
5
Moon
Jupiter
Capella
Aldebaran
Jul 11 Tue 3:56 CDT
45 minutes before sunrise
Sun
E
Mars
Regulus
Venus
Mercury
Castor
Jul 11 Tue 20:15 CDT
45 minutes after sunset
W
Jul 12 Wed 3:56 CDT
45 minutes before sunrise
Jul 12 Wed 20:15 CDT
45 minutes after sunset
Jul 13 Thu 3:57 CDT
45 minutes before sunrise
Jul 13 Thu 20:14 CDT
45 minutes after sunset
Betelgeuse
Jul 16 Sun 3:59 CDT
45 minutes before sunrise
Jul 16 Sun 20:13 CDT
45 minutes after sunset
Rigel
Jul 18 Tue 4:01 CDT
45 minutes before sunrise
Jul 18 Tue 20:12 CDT
45 minutes after sunset

0144.354	Jul	18	Tue	21	Moon 3.7° NNE of Beehive Cluster; 13° and 12° from Sun in evening sky; magnitudes -5.1 and 3.7
0144.5	Jul	19	Wed		1st day of Muslim year (1445 A.H.)
0145.000	Jul	19	Wed	12	Moon 3.3° NNE of Mercury; 20° and 19° from Sun in evening sky; magnitudes -5.8 and -0.4
0145.806	Jul	20	Thu	7	Moon at apogee; distance 63.70 Earth-radii
0146.125	Jul	20	Thu	15	Moon 7.3° NNE of Venus; 32° and 31° from Sun in evening sky; magnitudes -6.8 and -4.4
0146.271	Jul	20	Thu	19	Moon 3.8° NNE of Regulus; 33° and 32° from Sun in evening sky; magnitudes -6.9 and 1.4
0146.480	Jul	20	Thu	24	Venus stationary in right ascension; starts retrograde motion
0146.813	Jul	21	Fri	8	Moon 2.97° NNE of Mars; 39° and 38° from Sun in evening sky; magnitudes -7.3 and 1.8
0146.813	Jul	21	Fri	8	Sun enters Cancer, at longitude 118.31° on the ecliptic
0147.305	Jul	21	Fri	19	Pluto at opposition in longitude; magnitude 14.4; declination -23.0°
0148.565	Jul	23	SUN	2	Venus stationary in longitude; starts retrograde motion
0148.578	Jul	23	SUN	2	Sun enters the astrological sign Leo, i.e. its longitude is 120°
0150.771	Jul	25	Tue	7	Moon 2.48° NNE of Spica; 82° from Sun in evening sky; magnitudes -9.7 and 1.0
0151.128	Jul	25	Tue	15	Moon at descending node; longitude 209.0°
0151.422	Jul	25	Tue	22:07	First quarter Moon
0151.955	Jul	26	Wed	11	The equation of time is at a minimum of -6.56 minutes
0152.833	Jul	27	Thu	8	Mercury, Venus, and Regulus within circle of diameter 5.34°; about 25° from the Sun in the evening sky; magnitudes 0, -4, 1
0152.979	Jul	27	Thu	12	Mercury 5.1° NNE of Venus; 24° from Sun in evening sky; magnitudes -0.1 and -4.3
0154.292	Jul	28	Fri	19	Moon 1.33° NNE of Antares; 125° from Sun in evening sky; magnitudes -11.3 and 1.0
0154.521	Jul	29	SAT	1	Mercury 0.11° S of Regulus; 25° and 24° from Sun in evening sky; magnitudes 0.0 and 1.4
0155.5	Jul	30	SUN	0	Southern Delta Aquarid meteors; ZHR 25; 3 days before full Moon
0156.894	Jul	31	Mon	9	Mercury at descending node through the ecliptic plane

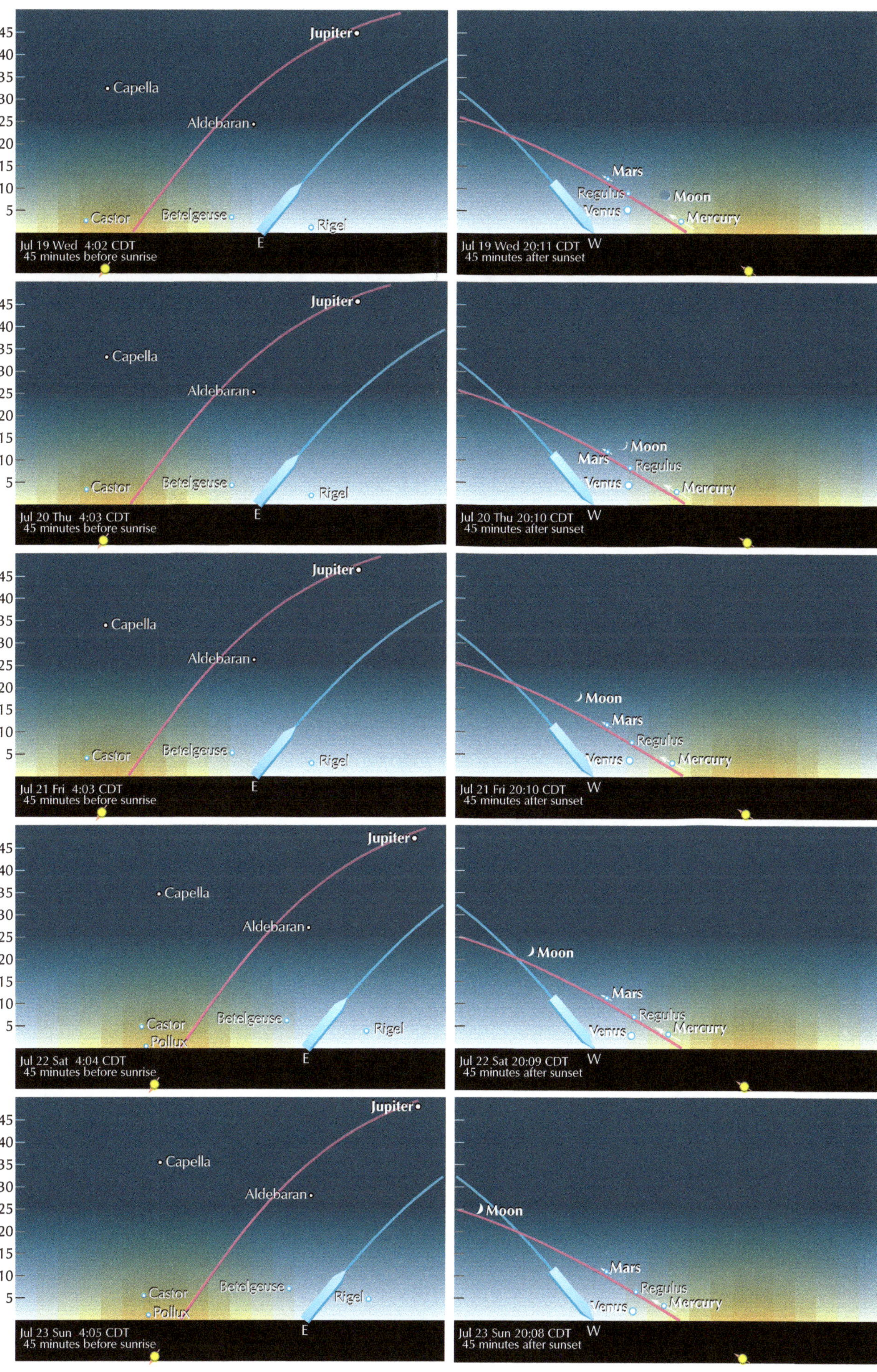
Jupiter
Capella
Aldebaran
Castor
Betelgeuse
Rigel
E
Jul 19 Wed 4:02 CDT
45 minutes before sunrise
Mars
Regulus
Moon
Venus
Mercury
W
Jul 19 Wed 20:11 CDT
45 minutes after sunset
Jul 20 Thu 4:03 CDT
45 minutes before sunrise
Jul 20 Thu 20:10 CDT
45 minutes after sunset
Jul 21 Fri 4:03 CDT
45 minutes before sunrise
Jul 21 Fri 20:10 CDT
45 minutes after sunset
Pollux
Jul 22 Sat 4:04 CDT
45 minutes before sunrise
Jul 22 Sat 20:09 CDT
45 minutes after sunset
Jul 23 Sun 4:05 CDT
45 minutes before sunrise
Jul 23 Sun 20:08 CDT
45 minutes after sunset

AUGUST

SKY DOME

Evening sky
for latitude 40° north

about 10 PM at the 5th,
9 PM at the 20th
of the month

sidereal time
19h

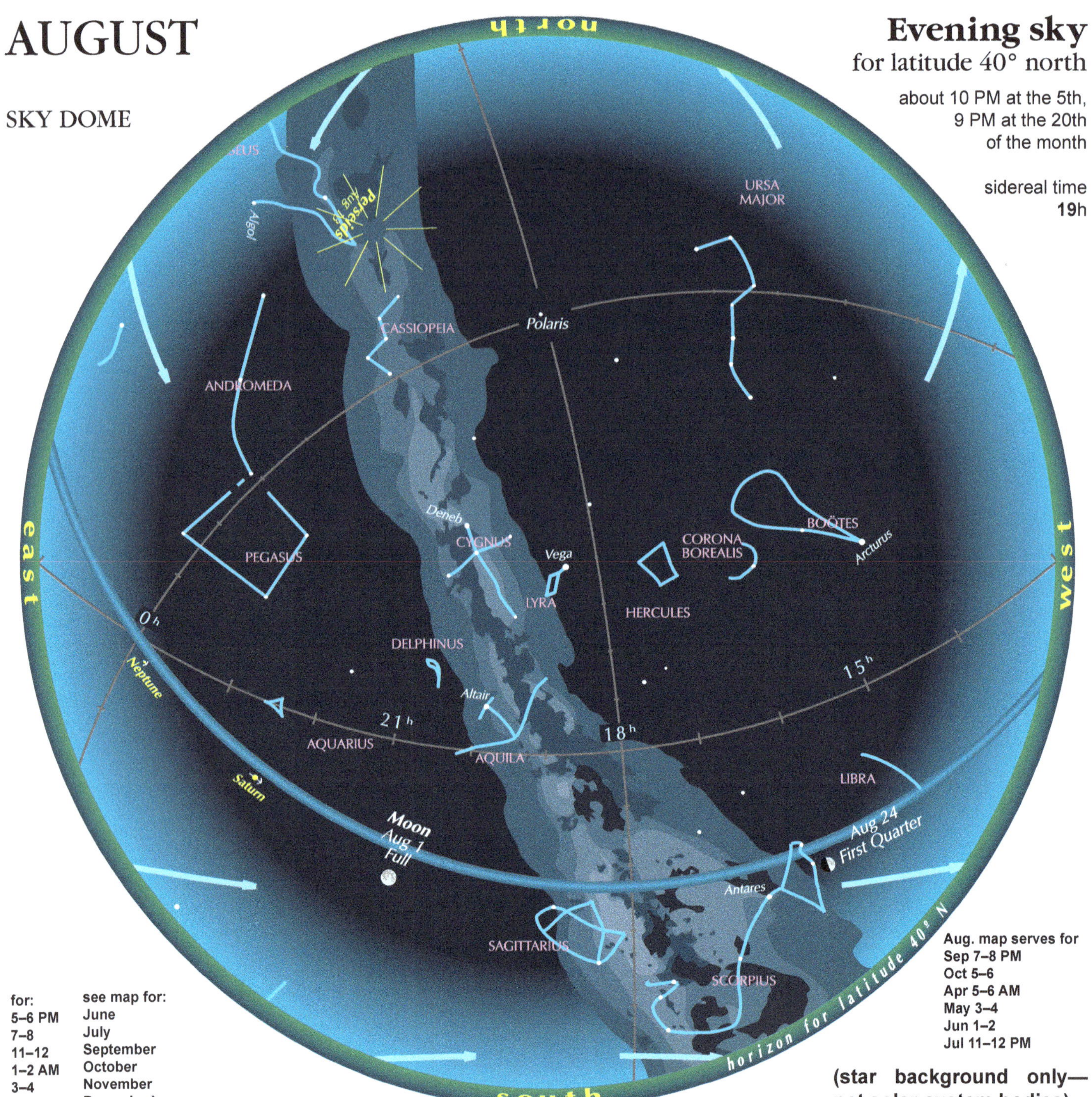

for:
5–6 PM
7–8
11–12
1–2 AM
3–4
5–6

see map for:
June
July
September
October
November
December`

Aug. map serves for
Sep 7–8 PM
Oct 5–6
Apr 5–6 AM
May 3–4
Jun 1–2
Jul 11–12 PM

(star background only—not solar-system bodies)

2023 Aug

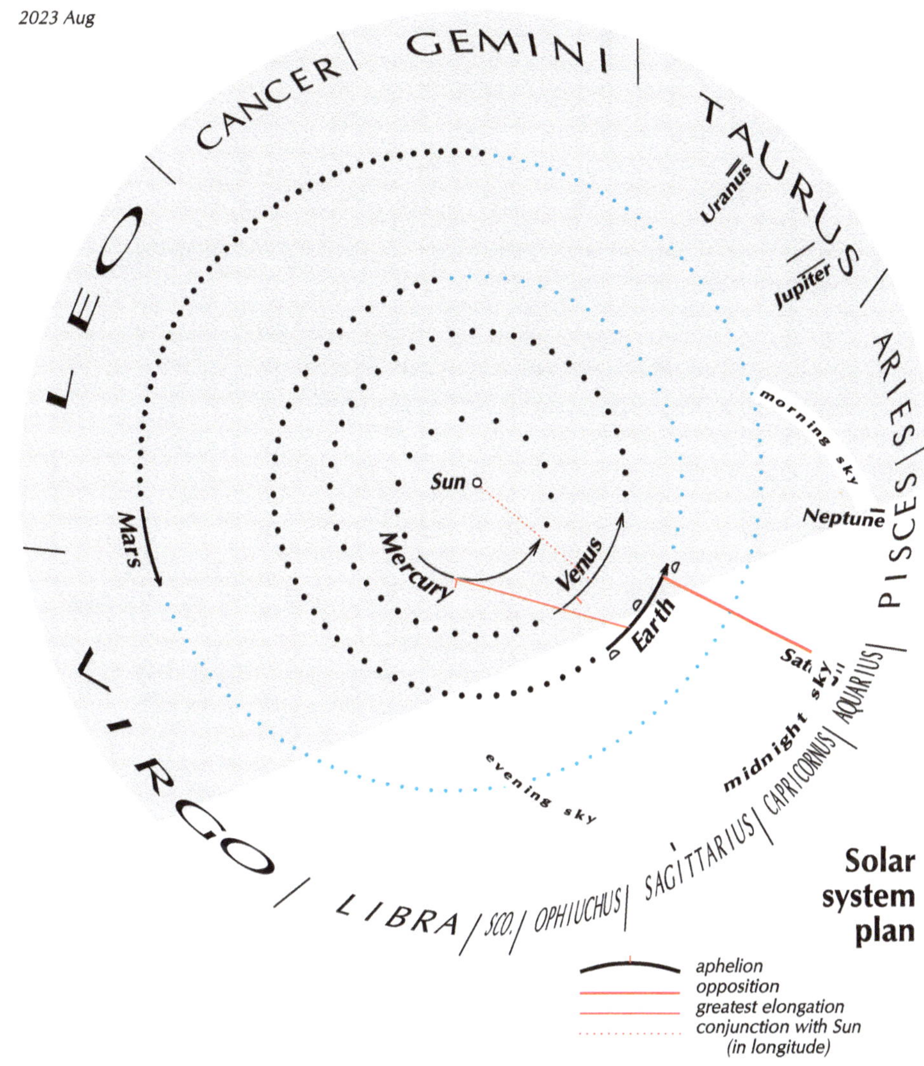
GEMINI
CANCER
LEO
TAURUS
Uranus
Jupiter
ARIES
morning sky
PISCES
Neptune
Sun
Mars
Mercury
Venus
Earth
Sat
midnight sky
AQUARIUS
CAPRICORNUS
SAGITTARIUS
OPHIUCHUS
SCO.
LIBRA
VIRGO
evening sky
Solar system plan
aphelion
opposition
greatest elongation
conjunction with Sun (in longitude)

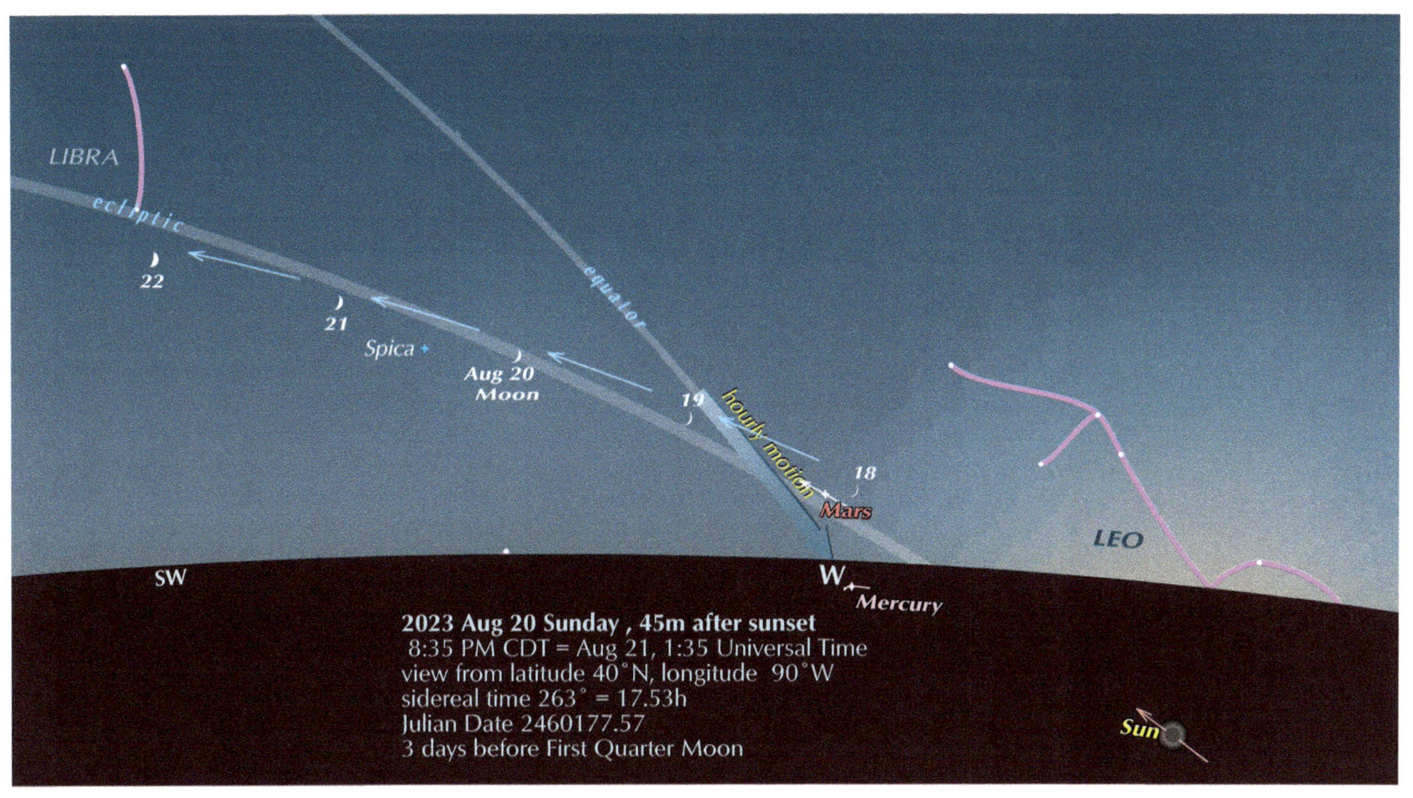
LIBRA
ecliptic
equator
22
21
Spica
Aug 20 Moon
19
hourly motion
18
Mars
LEO
SW
W
Mercury
Sun
2023 Aug 20 Sunday , 45m after sunset
8:35 PM CDT = Aug 21, 1:35 Universal Time
view from latitude 40°N, longitude 90°W
sidereal time 263° = 17.53h
Julian Date 2460177.57
3 days before First Quarter Moon

AUGUST

0158.272	Aug	1	Tue	18:32	Full Moon
0158.749	Aug	2	Wed	5:58	Moon at perigee; distance 56.03 Earth-radii; only 11.4 hours after full Moon
0160.021	Aug	3	Thu	13	Moon 2.27° SE of Saturn; 155° from Sun in morning sky; magnitudes -12.2 and 0.6
0161.479	Aug	4	Fri	24	Moon 1.32° SE of Neptune; 135° from Sun in morning sky; magnitudes -11.6 and 7.8
0163.5	Aug	7	Mon	0	Jupiter at west quadrature, 90° from the Sun
0163.616	Aug	7	Mon	3	Moon at ascending node; longitude 27.9°
0164.502	Aug	8	Tue	0	Venus at aphelion; 0.7282 AU from the Sun
0164.854	Aug	8	Tue	9	Moon 2.69° NNW of Jupiter; 91° from Sun in morning sky; magnitudes -10.1 and -2.4
0164.937	Aug	8	Tue	10:29	Last quarter Moon
0165.5	Aug	9	Wed	0	Moon 2.46° NNW of Uranus; 83° from Sun in morning sky; magnitudes -9.8 and 5.8
0166.104	Aug	9	Wed	15	Moon 1.38° SE of Pleiades; 76° from Sun in morning sky
0166.568	Aug	10	Thu	2	Mercury at easternmost elongation; 27.4° from Sun in evening sky; magnitude 0.4
0167.268	Aug	10	Thu	18	Mercury at aphelion; 0.4667 AU from the Sun
0167.645	Aug	11	Fri	3	Sun enters Leo, at longitude 138.24° on the ecliptic
0168.646	Aug	12	SAT	4	Moon 3.7° N of M35 cluster; 47° from Sun in morning sky; magnitudes -7.9 and 5.3
0169.5	Aug	13	SUN	0	Perseid meteors; ZHR 100; 3 days before new Moon
0169.813	Aug	13	SUN	8	Mercury 4.7° WSW of Mars; 27° and 31° from Sun in evening sky; magnitudes 0.5 and 1.8; quasi-conjunction
0169.964	Aug	13	SUN	11	Venus at inferior conjunction with the Sun; 0.289 AU from Earth; latitude -3.04°
0170.188	Aug	13	SUN	17	Moon 5.1° S of Castor; 30° and 32° from Sun in morning sky; magnitudes -6.7 and 1.5
0170.417	Aug	13	SUN	22	Moon 1.67° S of Pollux; 28° from Sun in morning sky; magnitudes -6.5 and 1.2
0171.604	Aug	15	Tue	3	Moon 3.7° NNE of Beehive Cluster; 15° and 14° from Sun in morning sky; magnitudes -5.3 and 3.7
0172.604	Aug	16	Wed	2	Uranus at west quadrature, 90° from the Sun
0172.901	Aug	16	Wed	9:38	New Moon; beginning of lunation 1245
0172.987	Aug	16	Wed	12	Moon at apogee; distance 63.75 Earth-radii; farthest in year; only 2.1 hours after new Moon
0173.521	Aug	17	Thu	1	Moon 3.8° NNE of Regulus; 8° and 6° from Sun in evening sky; magnitudes -4.6 and 1.4
0175.271	Aug	18	Fri	19	Moon 6.2° NNE of Mercury; 26° and 25° from Sun in evening sky; magnitudes -6.3 and 0.8
0175.583	Aug	19	SAT	2	Moon 1.95° NE of Mars; 29° from Sun in evening sky; magnitudes -6.6 and 1.8
0178.042	Aug	21	Mon	13	Moon 2.27° NE of Spica; 56° from Sun in evening sky; magnitudes -8.5 and 1.0
0178.183	Aug	21	Mon	16	Moon at descending node; longitude 206.4°
0179.697	Aug	23	Wed	5	Mercury stationary in right ascension; starts retrograde motion

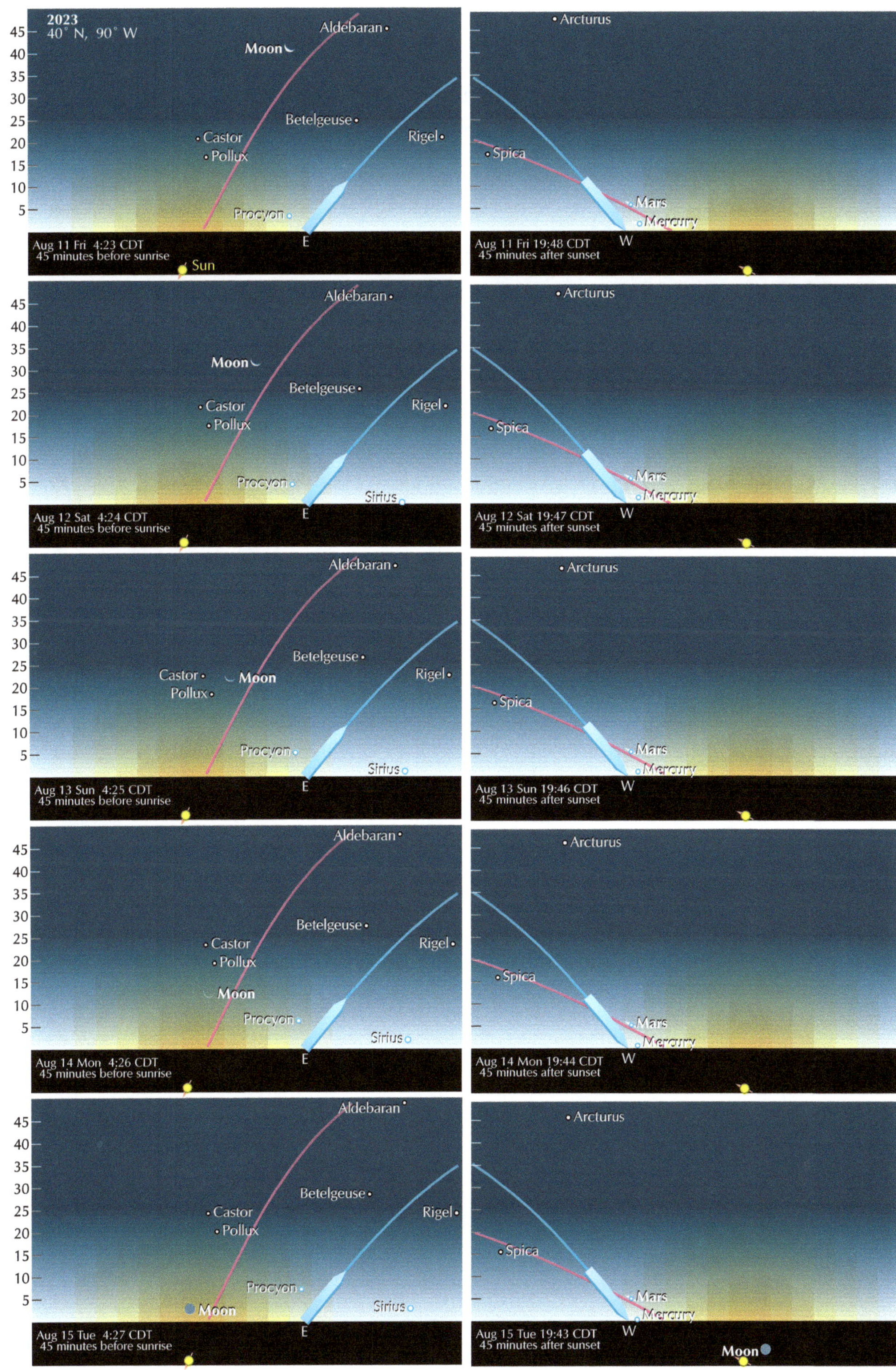
2023
40° N, 90° W
45
40
35
30
25
20
15
10
5
Aldebaran
Moon
Betelgeuse
Castor
Pollux
Rigel
Procyon
E
Aug 11 Fri 4:23 CDT
45 minutes before sunrise
Sun
Arcturus
Spica
Mars
Mercury
W
Aug 11 Fri 19:48 CDT
45 minutes after sunset
Aldebaran
Moon
Betelgeuse
Castor
Pollux
Rigel
Procyon
Sirius
E
Aug 12 Sat 4:24 CDT
45 minutes before sunrise
Arcturus
Spica
Mars
Mercury
W
Aug 12 Sat 19:47 CDT
45 minutes after sunset
Aldebaran
Betelgeuse
Castor
Moon
Pollux
Rigel
Procyon
Sirius
E
Aug 13 Sun 4:25 CDT
45 minutes before sunrise
Arcturus
Spica
Mars
Mercury
W
Aug 13 Sun 19:46 CDT
45 minutes after sunset
Aldebaran
Betelgeuse
Castor
Pollux
Rigel
Moon
Procyon
Sirius
E
Aug 14 Mon 4:26 CDT
45 minutes before sunrise
Arcturus
Spica
Mars
Mercury
W
Aug 14 Mon 19:44 CDT
45 minutes after sunset
Aldebaran
Betelgeuse
Castor
Pollux
Rigel
Procyon
Moon
Sirius
E
Aug 15 Tue 4:27 CDT
45 minutes before sunrise
Arcturus
Spica
Mars
Mercury
W
Aug 15 Tue 19:43 CDT
45 minutes after sunset
Moon

0179.877	Aug	23	Wed	9	Sun enters the astrological sign Virgo, i.e. its longitude is 150°
0180.329	Aug	23	Wed	20	Mercury stationary in longitude; starts retrograde motion
0180.915	Aug	24	Thu	9:57	First quarter Moon
0181.625	Aug	25	Fri	3	Moon 1.07° NNE of Antares; 99° and 98° from Sun in evening sky; magnitudes -10.4 and 1.0; occultation
0182.542	Aug	26	SAT	1	Moon shows maximum libration for the year, 9.01°
0183.845	Aug	27	SUN	8	Saturn at opposition in longitude; magnitude 0.4; declination -11.8°
0185.480	Aug	28	Mon	24	Uranus stationary in longitude; starts retrograde motion
0185.520	Aug	29	Tue	0	Uranus stationary in right ascension; starts retrograde motion
0186.407	Aug	29	Tue	22	Mars crosses equator southward
0186.636	Aug	30	Wed	3	Venus at southernmost latitude from the ecliptic plane, -3.4°
0187.166	Aug	30	Wed	15:60	Moon at perigee; distance 56.00 Earth-radii; only 9.6 hours before full Moon
0187.354	Aug	30	Wed	21	Moon 2.32° SE of Saturn; 175° and 176° from Sun in evening midnight sky; magnitudes -12.7 and 0.4
0187.523	Aug	31	Thu	1	Mercury at southernmost latitude from the ecliptic plane, -7.0°
0187.567	Aug	31	Thu	1:36	Full Moon

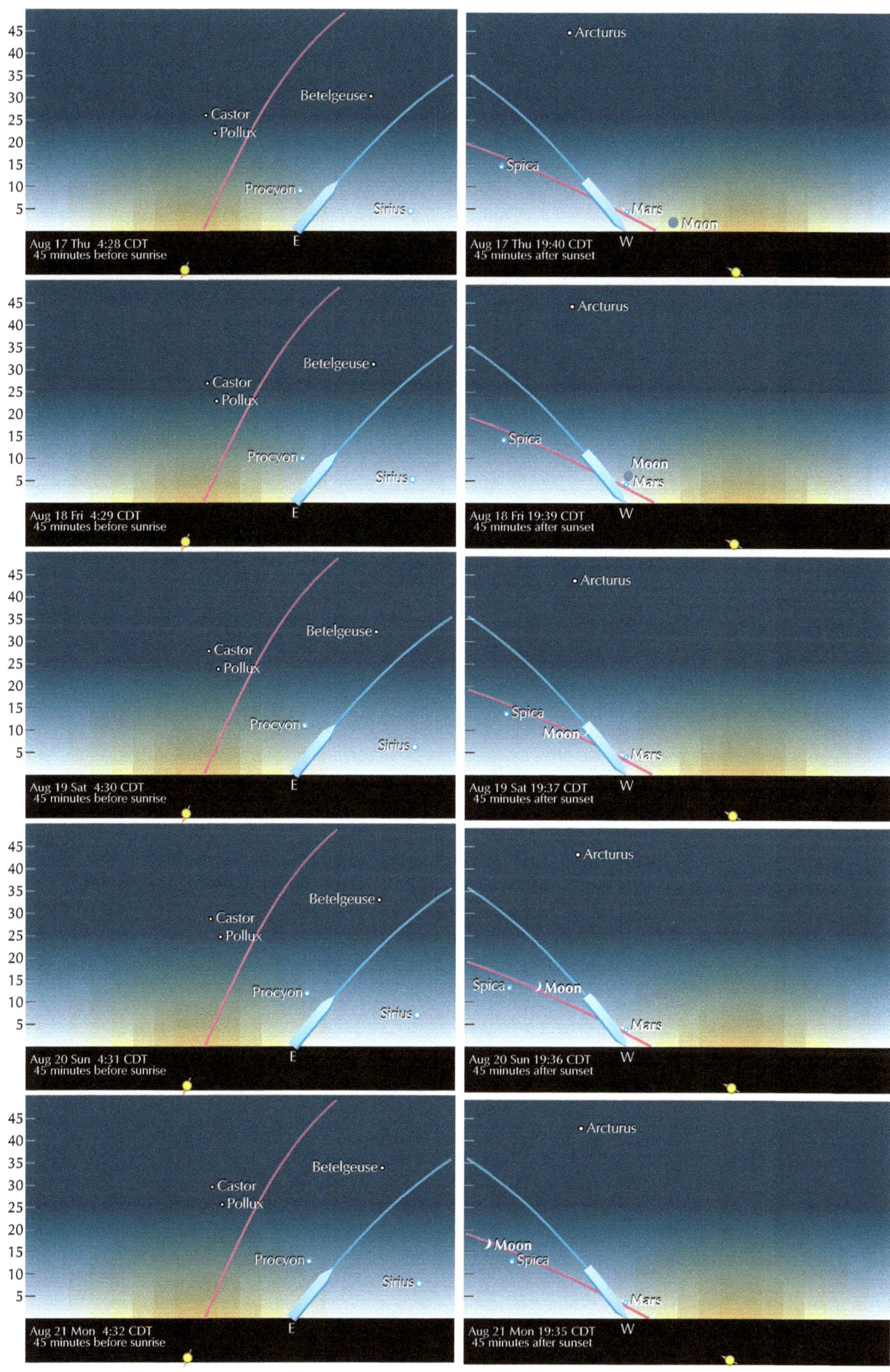
45
40
35
30
25
20
15
10
5
Castor
Pollux
Betelgeuse
Procyon
Sirius
E
Aug 17 Thu 4:28 CDT
45 minutes before sunrise
Arcturus
Spica
Mars
Moon
W
Aug 17 Thu 19:40 CDT
45 minutes after sunset
Aug 18 Fri 4:29 CDT
45 minutes before sunrise
Aug 18 Fri 19:39 CDT
45 minutes after sunset
Aug 19 Sat 4:30 CDT
45 minutes before sunrise
Aug 19 Sat 19:37 CDT
45 minutes after sunset
Aug 20 Sun 4:31 CDT
45 minutes before sunrise
Aug 20 Sun 19:36 CDT
45 minutes after sunset
Aug 21 Mon 4:32 CDT
45 minutes before sunrise
Aug 21 Mon 19:35 CDT
45 minutes after sunset

SEPTEMBER

SKY DOME

Evening sky
for latitude 40° north

about 10 PM at the 5th,
9 PM at the 20th
of the month

sidereal time
21h

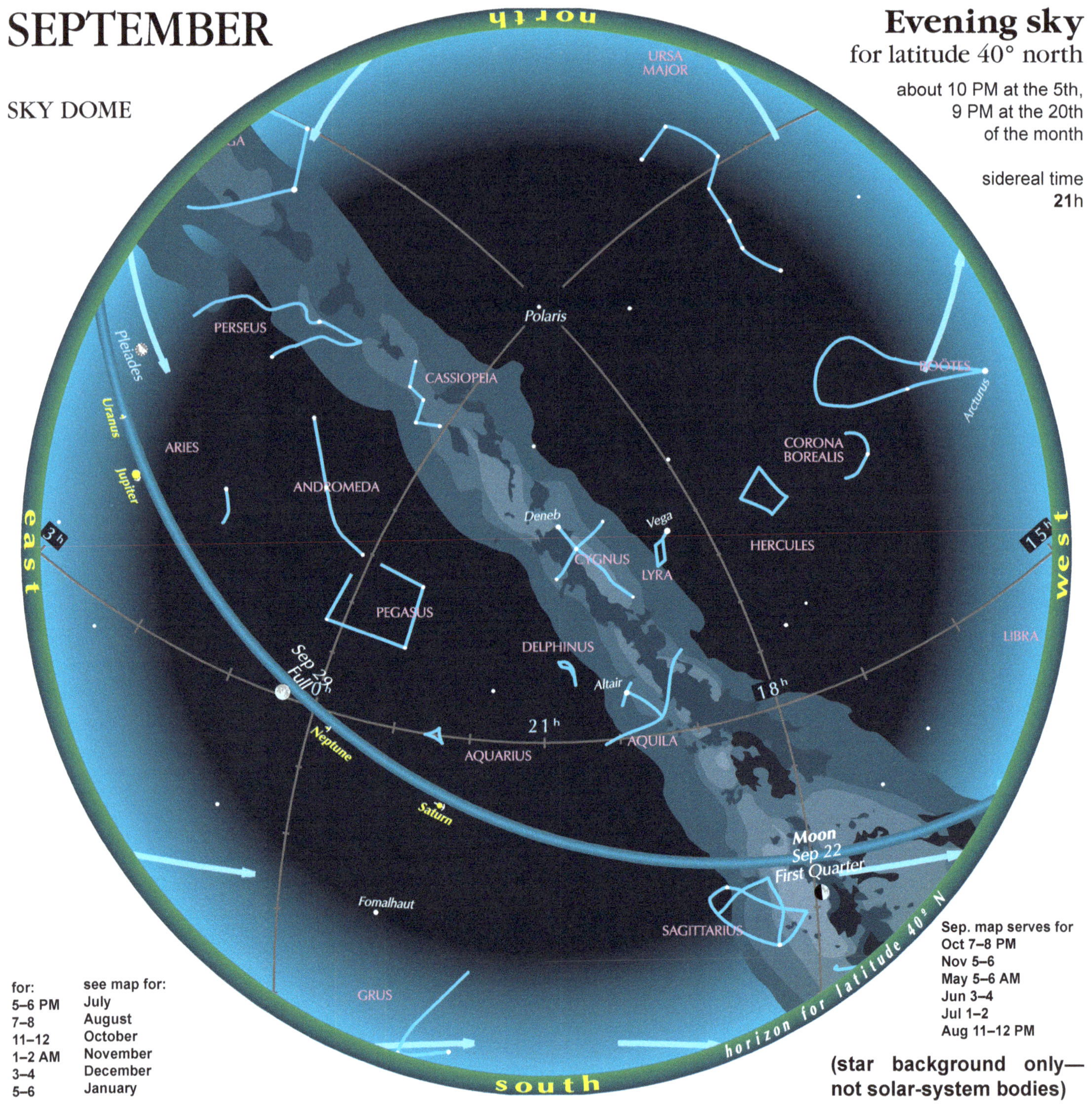

for:	see map for:
5–6 PM	July
7–8	August
11–12	October
1–2 AM	November
3–4	December
5–6	January

Sep. map serves for
Oct 7–8 PM
Nov 5–6
May 5–6 AM
Jun 3–4
Jul 1–2
Aug 11–12 PM

(star background only—not solar-system bodies)

2023 Sep

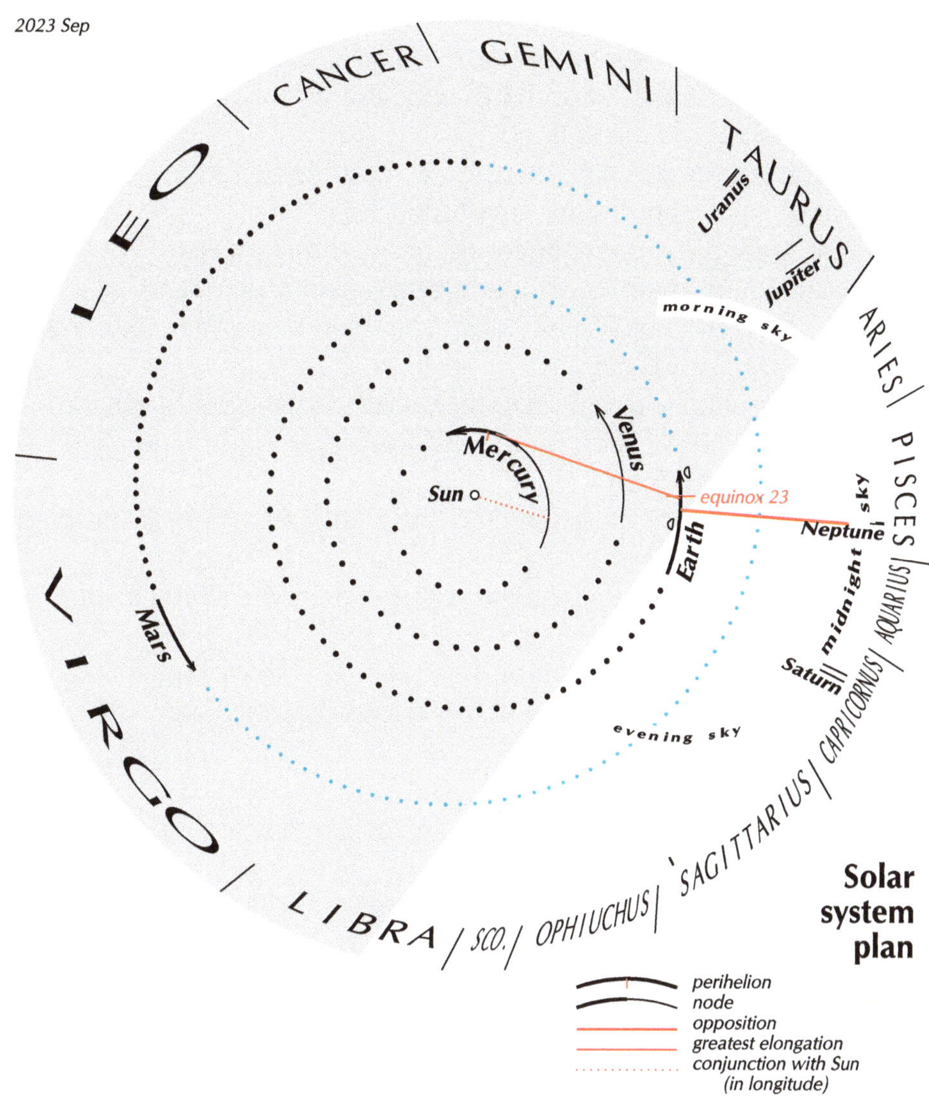

2023 Sep 12 Tuesday , 45m before sunrise
5:53 AM CDT = 10:53 Universal Time
view from latitude 40°N, longitude 90°W
sidereal time 65° = 4.30h
Julian Date 2460199.95
3 days before New Moon

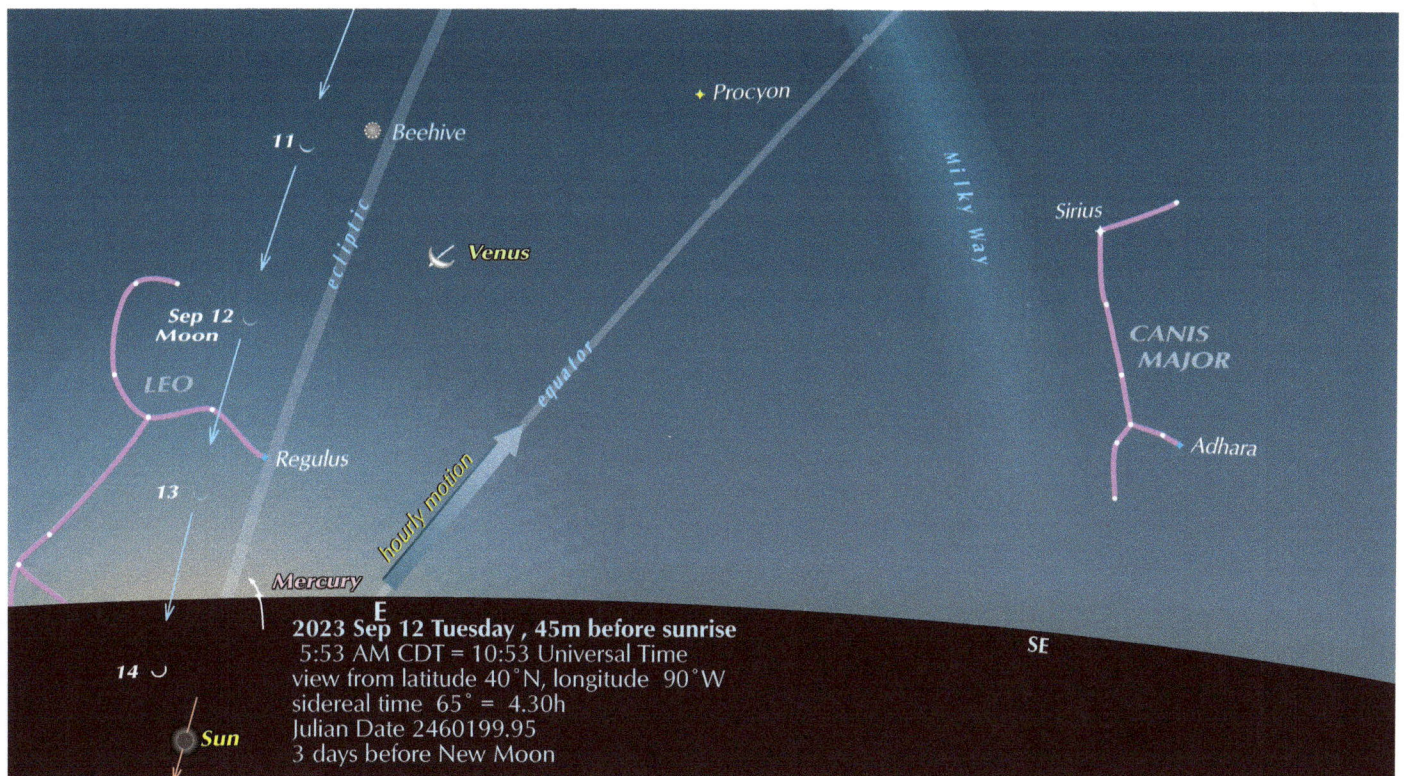

SEPTEMBER

0188.875	Sep	1	Fri	9	Moon 1.28° SE of Neptune; 161° and 162° from Sun in morning sky; magnitudes -12.4 and 7.8; occultation
0189.320	Sep	1	Fri	20	The equation of time is 0
0190.648	Sep	3	SUN	4	Venus stationary in right ascension; resumes direct motion
0190.824	Sep	3	SUN	8	Moon at ascending node; longitude 25.7°
0191.555	Sep	4	Mon	1	Venus stationary in longitude; resumes direct motion
0192.067	Sep	4	Mon	14	Jupiter stationary in longitude; starts retrograde motion
0192.250	Sep	4	Mon	18	Moon 3.1° NNW of Jupiter; 116° from Sun in morning sky; magnitudes -11.0 and -2.6
0192.351	Sep	4	Mon	20	Jupiter stationary in right ascension; starts retrograde motion
0192.813	Sep	5	Tue	8	Moon 2.69° NNW of Uranus; 109° from Sun in morning sky; magnitudes -10.8 and 5.7
0193.396	Sep	5	Tue	22	Moon 1.17° SE of Pleiades; 102° and 103° from Sun in morning sky
0193.961	Sep	6	Wed	11	Mercury at inferior conjunction with the Sun; 0.635 AU from Earth; latitude -6.35°
0194.083	Sep	6	Wed	14	Jupiter 7.5° WSW of Uranus; 118° and 111° from Sun in morning sky; magnitudes -2.6 and 5.7; quasi-conjunction
0194.431	Sep	6	Wed	22:21	Last quarter Moon
0195.917	Sep	8	Fri	10	Moon 3.9° N of M35 cluster; 73° from Sun in morning sky; magnitudes -9.3 and 5.3
0197.417	Sep	9	SAT	22	Moon 5.0° S of Castor; 56° and 57° from Sun in morning sky; magnitudes -8.5 and 1.5
0197.667	Sep	10	SUN	4	Moon 1.53° S of Pollux; 54° from Sun in morning sky; magnitudes -8.3 and 1.2
0198.854	Sep	11	Mon	9	Moon 3.8° NNE of Beehive Cluster; 41° from Sun in morning sky; magnitudes -7.5 and 3.7
0200.149	Sep	12	Tue	16	Moon at apogee; distance 63.70 Earth-radii
0200.771	Sep	13	Wed	7	Moon 3.8° NNE of Regulus; 20° from Sun in morning sky; magnitudes -5.8 and 1.4
0201.479	Sep	13	Wed	24	Moon 5.4° NNE of Mercury; 12° and 13° from Sun in morning sky; magnitudes -5.1 and 1.9
0202.510	Sep	15	Fri	0	Mercury stationary in right ascension; resumes direct motion
0202.569	Sep	15	Fri	1:39	New Moon; beginning of lunation 1246
0203.344	Sep	15	Fri	20	Mercury stationary in longitude; resumes direct motion
0203.5	Sep	16	SAT		Rosh Hashanah, 1st day of Hebrew year 5784 A.M.
0203.708	Sep	16	SAT	5	Mercury 8.0° ESE of Regulus; 15° and 23° from Sun in morning sky; magnitudes 1.1 and 1.4; quasi-conjunction
0204.354	Sep	16	SAT	21	Moon 0.62° NE of Mars; 20° and 19° from Sun in evening sky; magnitudes -5.8 and 1.7; occultation
0204.863	Sep	17	SUN	9	Sun enters Virgo, at longitude 174.21° on the ecliptic
0205.271	Sep	17	SUN	19	Moon 2.14° NE of Spica; 30° from Sun in evening sky; magnitudes -6.7 and 1.0
0205.304	Sep	17	SUN	19	Moon at descending node; longitude 205.0°
0206.007	Sep	18	Mon	12	Venus brightest; magnitude -4.54°
0206.582	Sep	19	Tue	2	Mercury at ascending node through the ecliptic plane
0206.772	Sep	19	Tue	7	Venus shows greatest illuminated extent, 313 square seconds

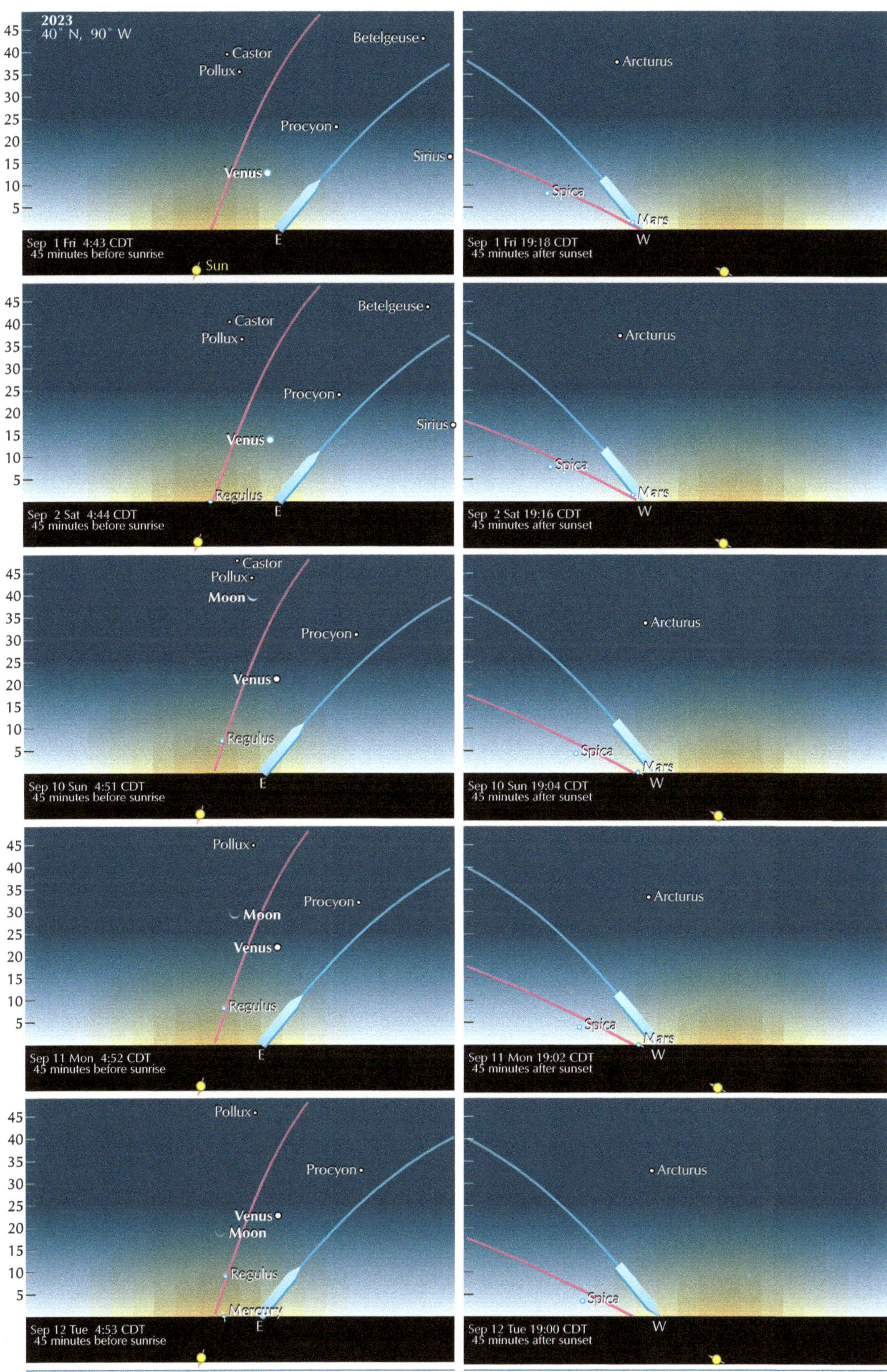
2023
40° N, 90° W
Castor
Pollux
Betelgeuse
Procyon
Sirius
Venus
E
Sep 1 Fri 4:43 CDT
45 minutes before sunrise
Sun
Arcturus
Spica
Mars
W
Sep 1 Fri 19:18 CDT
45 minutes after sunset
Castor
Pollux
Betelgeuse
Procyon
Sirius
Venus
Regulus
E
Sep 2 Sat 4:44 CDT
45 minutes before sunrise
Arcturus
Spica
Mars
W
Sep 2 Sat 19:16 CDT
45 minutes after sunset
Castor
Pollux
Moon
Procyon
Venus
Regulus
E
Sep 10 Sun 4:51 CDT
45 minutes before sunrise
Arcturus
Spica
Mars
W
Sep 10 Sun 19:04 CDT
45 minutes after sunset
Pollux
Procyon
Moon
Venus
Regulus
E
Sep 11 Mon 4:52 CDT
45 minutes before sunrise
Arcturus
Spica
Mars
W
Sep 11 Mon 19:02 CDT
45 minutes after sunset
Pollux
Procyon
Venus
Moon
Regulus
Mercury
E
Sep 12 Tue 4:53 CDT
45 minutes before sunrise
Arcturus
Spica
W
Sep 12 Tue 19:00 CDT
45 minutes after sunset

0206.962	Sep	19	Tue	11	Neptune at opposition in longitude; magnitude 7.8; declination -2.7°
0208.896	Sep	21	Thu	10	Moon 0.95° NE of Antares; 72° from Sun in evening sky; magnitudes -9.4 and 1.0; occultation
0210.047	Sep	22	Fri	13	Mercury at westernmost elongation; 17.9° from Sun in morning sky; magnitude -0.4
0210.314	Sep	22	Fri	19:31	First quarter Moon
0210.785	Sep	23	SAT	6:50	September (northern autumn) equinox
0210.785	Sep	23	SAT	6:50	Sun enters the astrological sign Libra, i.e. its longitude is 180°
0211.253	Sep	23	SAT	18	Mercury at perihelion; 0.3075 AU from the Sun
0214.5	Sep	27	Wed	0:00	Day and night equal, at latitude 40° north
0214.646	Sep	27	Wed	4	Moon 2.42° SE of Saturn; 148° from Sun in evening sky; magnitudes -12.0 and 0.6
0215.543	Sep	28	Thu	1:02	Moon at perigee; distance 56.43 Earth-radii
0216.271	Sep	28	Thu	19	Moon 1.29° SE of Neptune; 171° from Sun in evening midnight sky; magnitudes -12.6 and 7.8; occultation
0216.915	Sep	29	Fri	9:57	Full Moon
0218.202	Sep	30	SAT	17	Moon at ascending node; longitude 24.9°

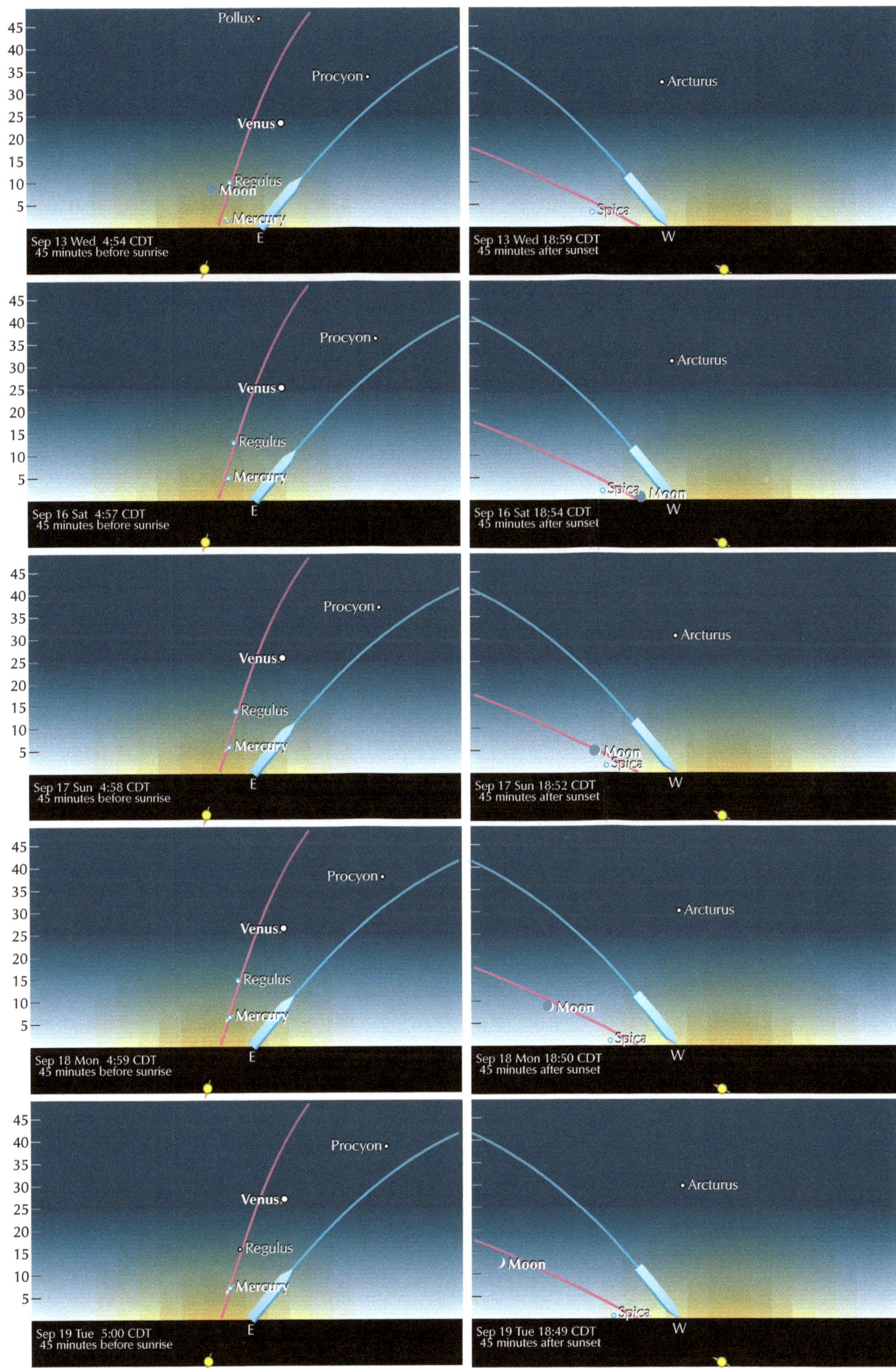

Pollux
Procyon
Venus
Regulus
Moon
Mercury
E
Sep 13 Wed 4:54 CDT
45 minutes before sunrise
Arcturus
Spica
W
Sep 13 Wed 18:59 CDT
45 minutes after sunset
Procyon
Venus
Regulus
Mercury
E
Sep 16 Sat 4:57 CDT
45 minutes before sunrise
Arcturus
Spica
Moon
W
Sep 16 Sat 18:54 CDT
45 minutes after sunset
Procyon
Venus
Regulus
Mercury
E
Sep 17 Sun 4:58 CDT
45 minutes before sunrise
Arcturus
Moon
Spica
W
Sep 17 Sun 18:52 CDT
45 minutes after sunset
Procyon
Venus
Regulus
Mercury
E
Sep 18 Mon 4:59 CDT
45 minutes before sunrise
Arcturus
Moon
Spica
W
Sep 18 Mon 18:50 CDT
45 minutes after sunset
Procyon
Venus
Regulus
Mercury
E
Sep 19 Tue 5:00 CDT
45 minutes before sunrise
Arcturus
Moon
Spica
W
Sep 19 Tue 18:49 CDT
45 minutes after sunset

OCTOBER

Evening sky
for latitude 40° north

SKY DOME

about 10 PM at the 5th,
9 PM at the 20th
of the month

sidereal time
23h

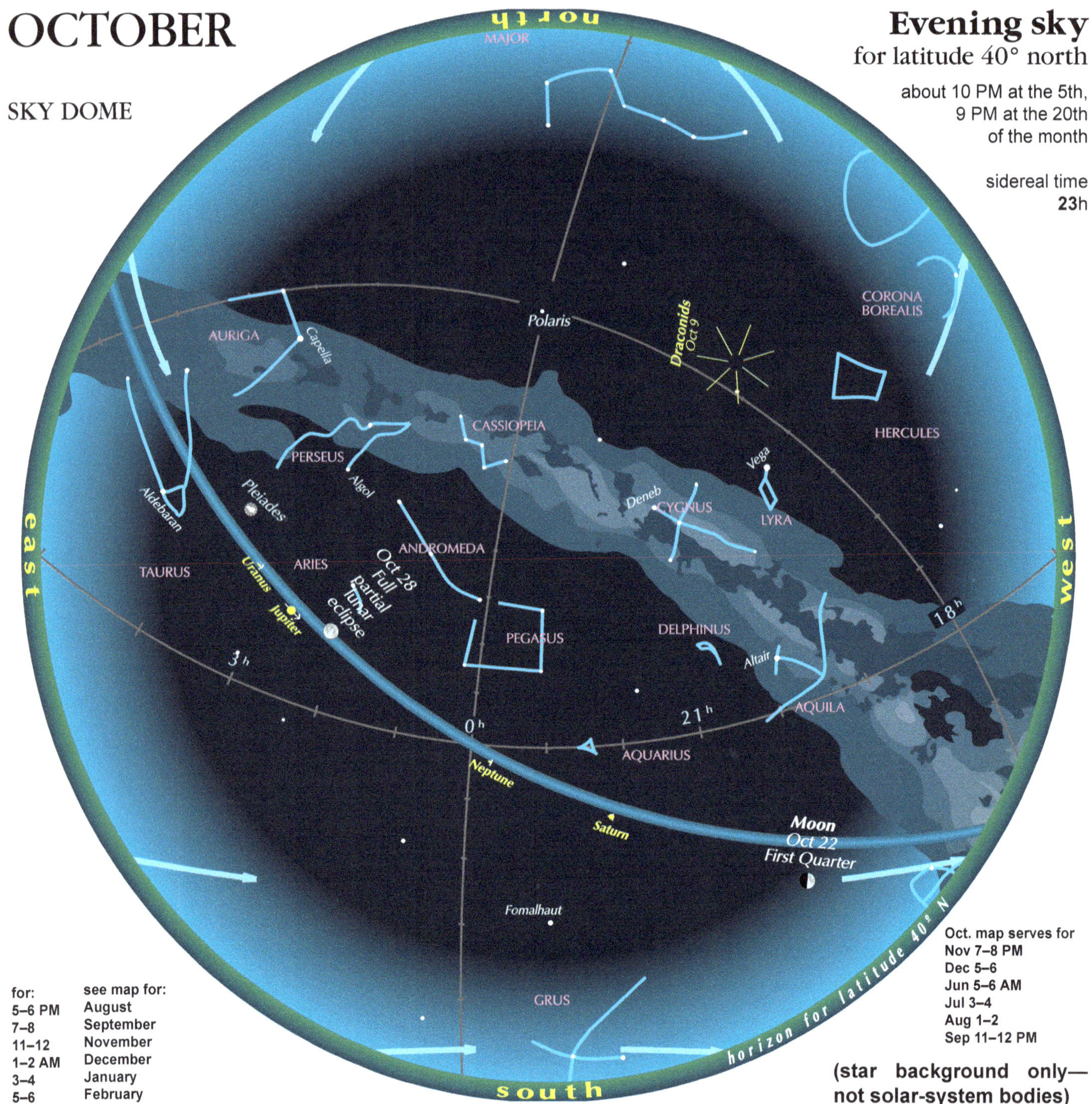

for:
5–6 PM
7–8
11–12
1–2 AM
3–4
5–6

see map for:
August
September
November
December
January
February

Oct. map serves for
Nov 7–8 PM
Dec 5–6
Jun 5–6 AM
Jul 3–4
Aug 1–2
Sep 11–12 PM

(star background only—not solar-system bodies)

2023 Oct

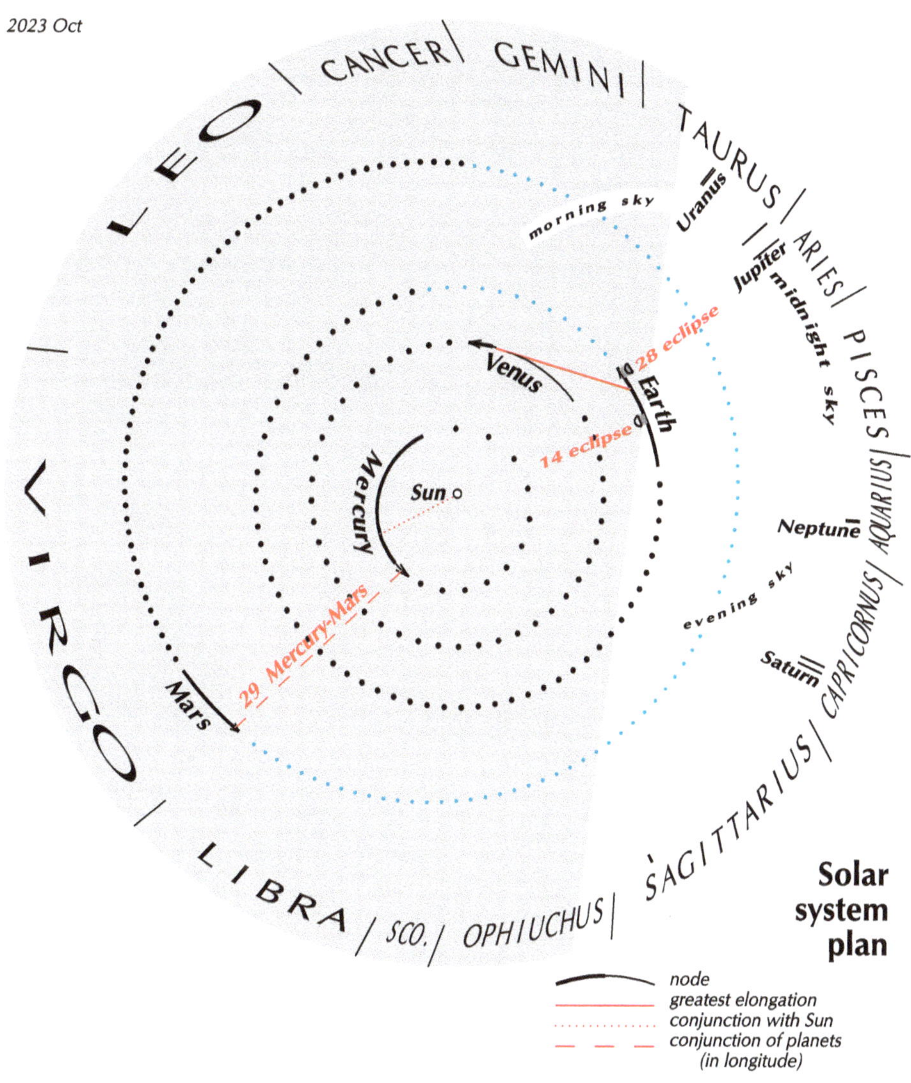
CANCER
GEMINI
TAURUS
LEO
ARIES
PISCES
AQUARIUS
CAPRICORNUS
SAGITTARIUS
OPHIUCHUS
SCO.
LIBRA
VIRGO
morning sky
midnight sky
evening sky
Uranus
Jupiter
Neptune
Saturn
Venus
Earth
Mercury
Mars
Sun
28 eclipse
14 eclipse
29 Mercury-Mars
Solar system plan
node
greatest elongation
conjunction with Sun
conjunction of planets (in longitude)

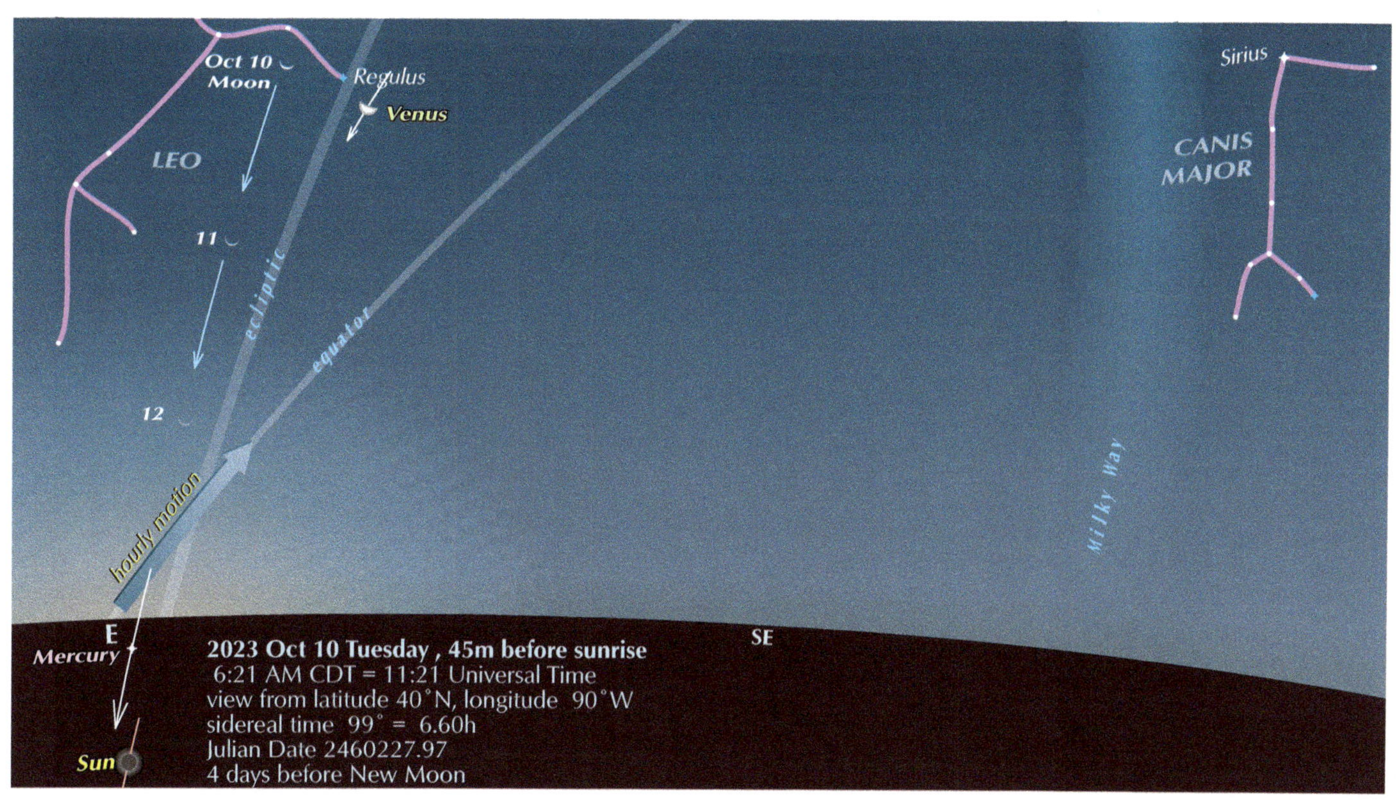
Oct 10 Moon
Regulus
Venus
LEO
11
12
ecliptic
equator
hourly motion
Sirius
CANIS MAJOR
Milky Way
E
SE
Mercury
Sun
2023 Oct 10 Tuesday , 45m before sunrise
6:21 AM CDT = 11:21 Universal Time
view from latitude 40°N, longitude 90°W
sidereal time 99° = 6.60h
Julian Date 2460227.97
4 days before New Moon

OCTOBER

0219.563	Oct	2	Mon	2	Moon 3.2° NNW of Jupiter; 144° from Sun in morning sky; magnitudes -11.8 and -2.8
0220.167	Oct	2	Mon	16	Moon 2.78° NNW of Uranus; 137° from Sun in morning sky; magnitudes -11.6 and 5.7
0220.771	Oct	3	Tue	7	Moon 1.08° SE of Pleiades; 129° from Sun in morning sky
0221.042	Oct	3	Tue	13	Mars 2.39° NNE of Spica; 14° from Sun in evening sky; magnitudes 1.7 and 1.0
0221.461	Oct	3	Tue	23	Mercury at northernmost latitude from the ecliptic plane, 7.0°
0223.208	Oct	5	Thu	17	Moon 4.0° N of M35 cluster; 100° from Sun in morning sky; magnitudes -10.4 and 5.3
0223.375	Oct	5	Thu	21	Moon at northernmost declination in year, 28.30°
0224.075	Oct	6	Fri	13:48	Last quarter Moon
0224.708	Oct	7	SAT	5	Moon 4.8° S of Castor; 83° from Sun in morning sky; magnitudes -9.8 and 1.5
0224.958	Oct	7	SAT	11	Moon 1.43° S of Pollux; 80° from Sun in morning sky; magnitudes -9.6 and 1.2
0226.146	Oct	8	SUN	16	Moon 3.9° NNE of Beehive Cluster; 67° from Sun in morning sky; magnitudes -9.0 and 3.7
0226.5	Oct	9	Mon	0	Draconid meteors; ZHR 5; 2 days after last quarter Moon
0227.146	Oct	9	Mon	16	Venus 2.29° SSW of Regulus; 46° from Sun in morning sky; magnitudes -4.5 and 1.4
0227.663	Oct	10	Tue	4	Moon at apogee; distance 63.57 Earth-radii
0227.903	Oct	10	Tue	10	Pluto stationary in right ascension; resumes direct motion
0228.030	Oct	10	Tue	13	Pluto stationary in longitude; resumes direct motion
0228.063	Oct	10	Tue	14	Moon 3.9° NNE of Regulus; 46° and 47° from Sun in morning sky; magnitudes -7.9 and 1.4
0228.167	Oct	10	Tue	16	Moon 5.9° NNE of Venus; 45° and 46° from Sun in morning sky; magnitudes -7.8 and -4.5
0229.024	Oct	11	Wed	13	Pluto at southernmost declination, -23.26°
0231.896	Oct	14	SAT	10	Moon 0.65° S of Mercury; 4° from Sun in morning sky; magnitudes -4.2 and -1.3
0232.246	Oct	14	SAT	17:54	New Moon; beginning of lunation 1247; annular eclipse of the Sun
0232.521	Oct	15	SUN	1	Moon 2.12° NNE of Spica; 3° from Sun in evening sky; magnitudes -4.2 and 1.0
0232.549	Oct	15	SUN	1	Moon at descending node; longitude 204.9°
0233.006	Oct	15	SUN	12	Mars and Jupiter at heliocentric opposition; longitudes 218.8° and 38.8°
0233.167	Oct	15	SUN	16	Moon 0.94° S of Mars; 11° and 10° from Sun in evening sky; magnitudes -5.0 and 1.6
0235.967	Oct	18	Wed	11	Middle of eclipse season: Sun is at same longitude as Moon's descending node, 204.8°
0236.104	Oct	18	Wed	15	Moon 0.83° NNE of Antares; 45° from Sun in evening sky; magnitudes -7.9 and 1.0; occultation
0236.438	Oct	18	Wed	23	Mercury 2.99° NNE of Spica; 1° and 2° from Sun in morning sky; magnitudes -1.4 and 1.0

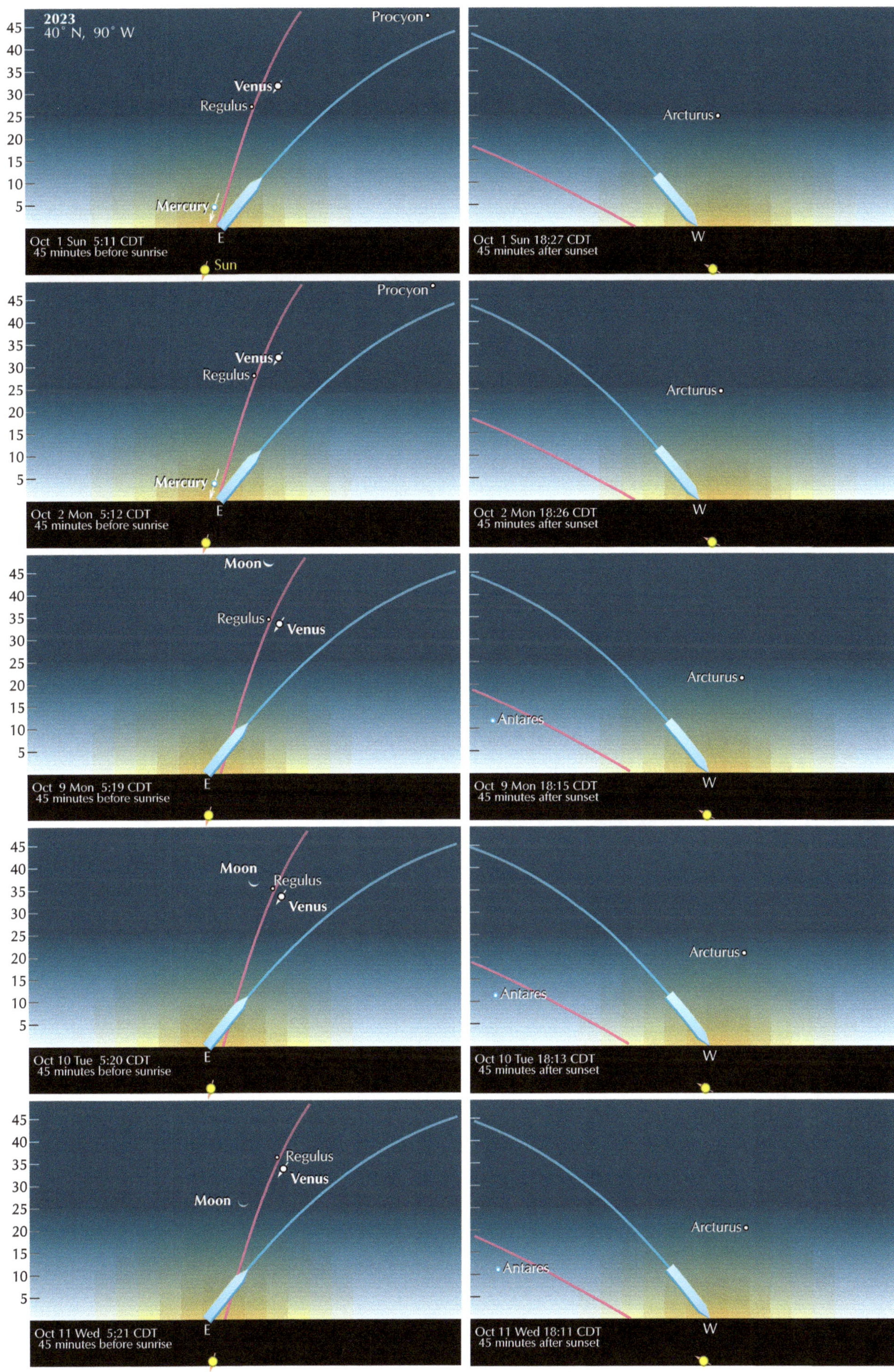
2023
40° N, 90° W
45
40
35
30
25
20
15
10
5
Procyon
Venus
Regulus
Mercury
E
Oct 1 Sun 5:11 CDT
45 minutes before sunrise
Sun
Arcturus
W
Oct 1 Sun 18:27 CDT
45 minutes after sunset
Procyon
Venus
Regulus
Mercury
E
Oct 2 Mon 5:12 CDT
45 minutes before sunrise
Arcturus
W
Oct 2 Mon 18:26 CDT
45 minutes after sunset
Moon
Regulus
Venus
E
Oct 9 Mon 5:19 CDT
45 minutes before sunrise
Arcturus
Antares
W
Oct 9 Mon 18:15 CDT
45 minutes after sunset
Moon
Regulus
Venus
E
Oct 10 Tue 5:20 CDT
45 minutes before sunrise
Arcturus
Antares
W
Oct 10 Tue 18:13 CDT
45 minutes after sunset
Regulus
Venus
Moon
E
Oct 11 Wed 5:21 CDT
45 minutes before sunrise
Arcturus
Antares
W
Oct 11 Wed 18:11 CDT
45 minutes after sunset

0237.724	Oct	20	Fri	5	Mercury at superior conjunction with the Sun; 1.421 AU from Earth; latitude 2.64°
0237.875	Oct	20	Fri	9	Moon at southernmost declination in year, -28.31°
0239.5	Oct	22	SUN	0	Orionid meteors; ZHR 20; near first quarter Moon
0239.645	Oct	22	SUN	3:29	First quarter Moon
0240.396	Oct	22	SUN	22	Venus dichotomy (D-shape)
0241.181	Oct	23	Mon	16	Sun enters the astrological sign Scorpius, i.e. its longitude is 21 0°
0241.456	Oct	23	Mon	23	Venus at westernmost elongation; 46.4° from Sun in morning sky; magnitude -4.4
0241.938	Oct	24	Tue	11	Moon 2.57° SE of Saturn; 120° from Sun in evening sky; magnitudes -11.2 and 0.7
0242.929	Oct	25	Wed	10	Venus at ascending node through the ecliptic plane
0243.625	Oct	26	Thu	3	Moon 1.36° SE of Neptune; 143° from Sun in evening sky; magnitudes -11.9 and 7.8
0243.627	Oct	26	Thu	3:03	Moon at perigee; distance 57.21 Earth-radii
0244.863	Oct	27	Fri	9	Mercury at descending node through the ecliptic plane
0245.636	Oct	28	SAT	3	Moon at ascending node; longitude 24.9°
0246.349	Oct	28	SAT	20:23	Full Moon; partial eclipse of the Moon
0246.5	Oct	29	SUN		Clocks back 1 hour (Europe)
0246.771	Oct	29	SUN	7	Moon 2.89° NNW of Jupiter; 174° from Sun in midnight sky; magnitudes -12.7 and -2.9
0247.083	Oct	29	SUN	14	Mercury 0.33° SSW of Mars; 6° from Sun in evening sky; magnitudes -0.9 and 1.5
0247.521	Oct	30	Mon	1	Moon 2.70° NNW of Uranus; 165° from Sun in morning sky; magnitudes -12.4 and 5.6
0248.188	Oct	30	Mon	17	Moon 1.06° SE of Pleiades; 156° from Sun in morning sky
0248.5	Oct	31	Tue		Halloween
0249.060	Oct	31	Tue	13	Sun enters Libra, at longitude 217.86° on the ecliptic

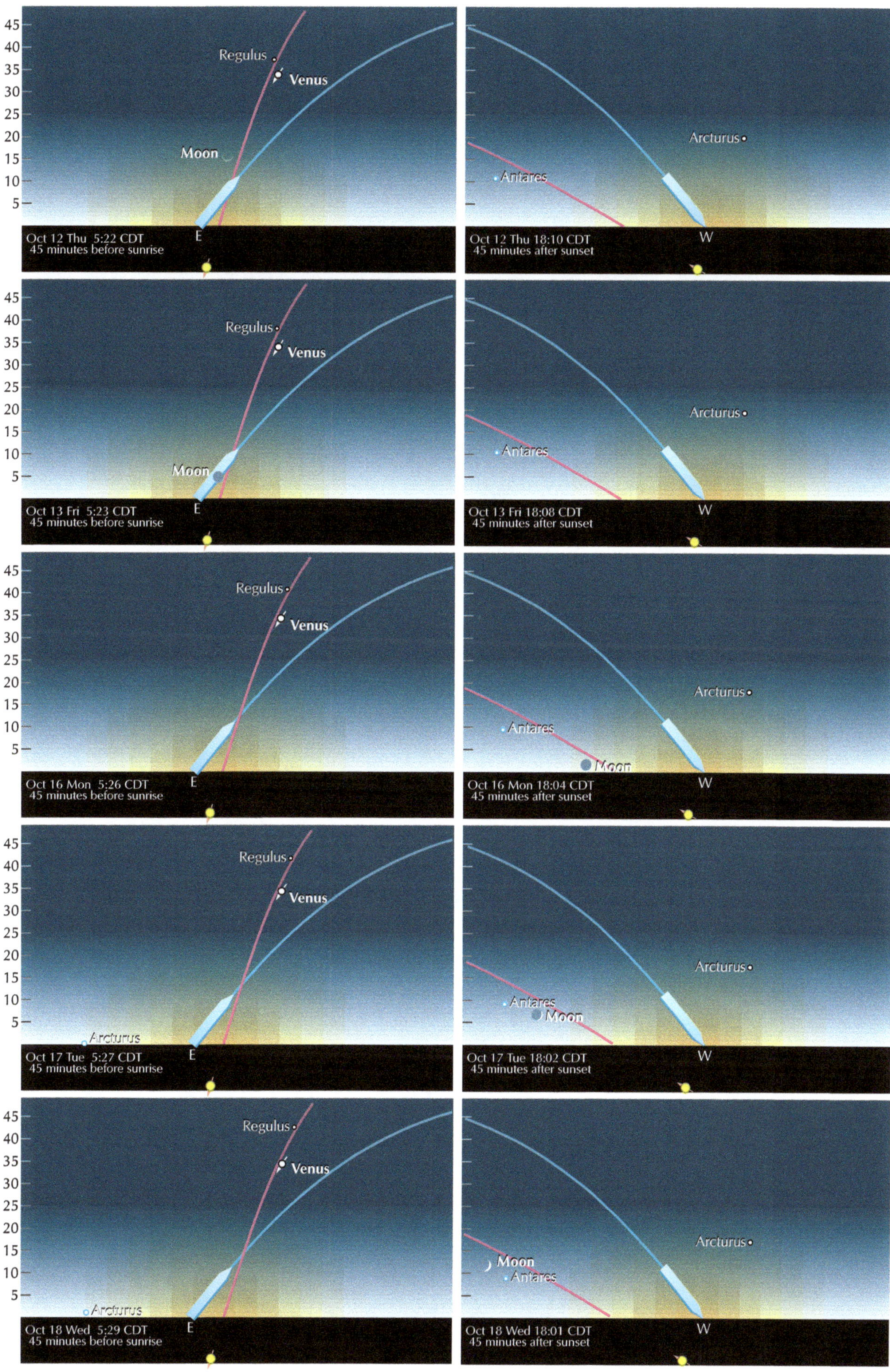
45
40
35
30
25
20
15
10
5
Regulus
Venus
Moon
E
Oct 12 Thu 5:22 CDT
45 minutes before sunrise
Arcturus
Antares
W
Oct 12 Thu 18:10 CDT
45 minutes after sunset
Regulus
Venus
Moon
E
Oct 13 Fri 5:23 CDT
45 minutes before sunrise
Arcturus
Antares
W
Oct 13 Fri 18:08 CDT
45 minutes after sunset
Regulus
Venus
E
Oct 16 Mon 5:26 CDT
45 minutes before sunrise
Arcturus
Antares
Moon
W
Oct 16 Mon 18:04 CDT
45 minutes after sunset
Regulus
Venus
Arcturus
E
Oct 17 Tue 5:27 CDT
45 minutes before sunrise
Arcturus
Antares
Moon
W
Oct 17 Tue 18:02 CDT
45 minutes after sunset
Regulus
Venus
Arcturus
E
Oct 18 Wed 5:29 CDT
45 minutes before sunrise
Arcturus
Moon
Antares
W
Oct 18 Wed 18:01 CDT
45 minutes after sunset

NOVEMBER

SKY DOME

Evening sky

for latitude 40° north

about 10 PM at the 5th,
9 PM at the 20th
of the month

sidereal time
1h

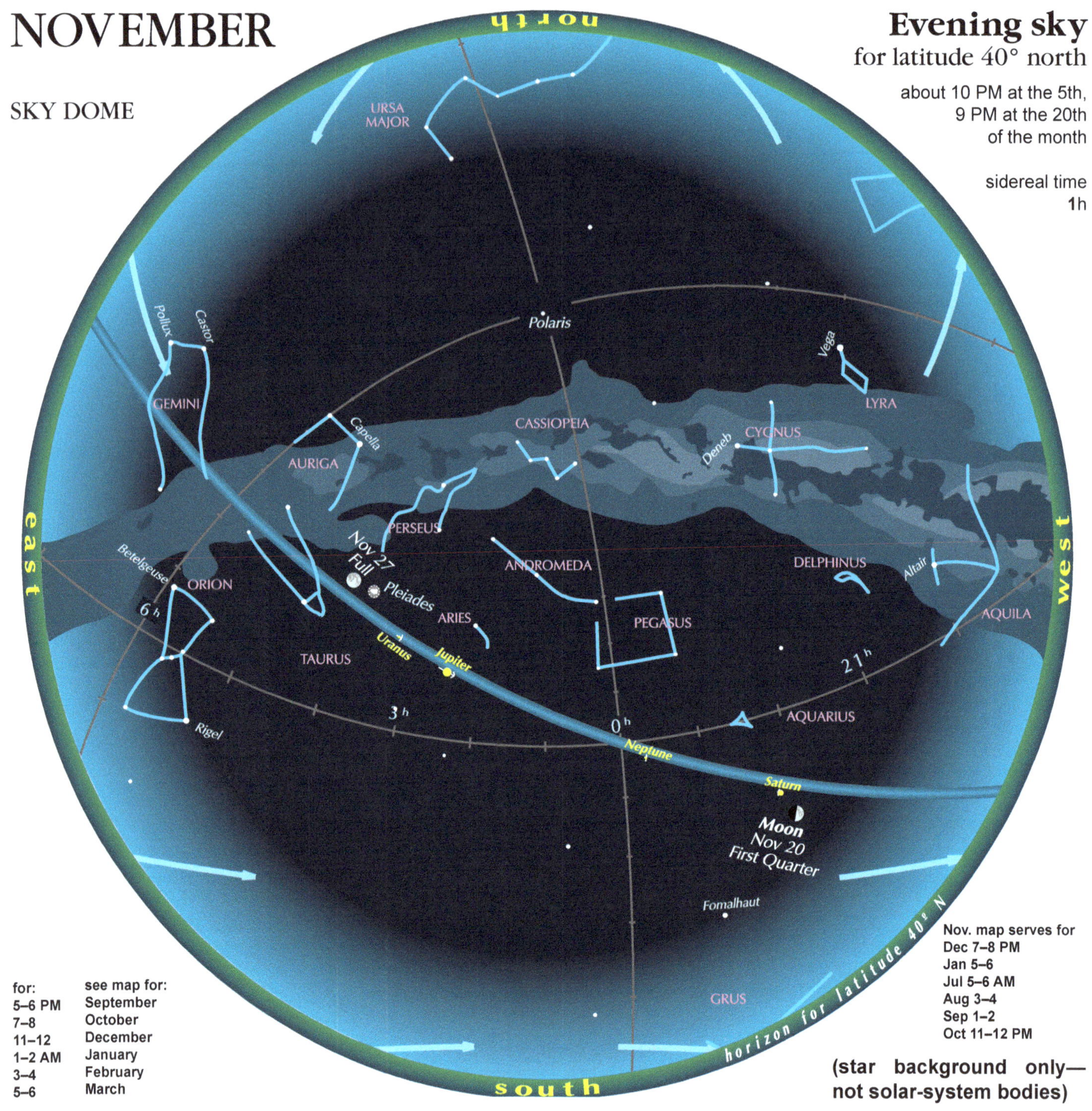

for:	see map for:
5–6 PM	September
7–8	October
11–12	December
1–2 AM	January
3–4	February
5–6	March

Nov. map serves for
Dec 7–8 PM
Jan 5–6
Jul 5–6 AM
Aug 3–4
Sep 1–2
Oct 11–12 PM

(star background only—not solar-system bodies)

2023 Nov

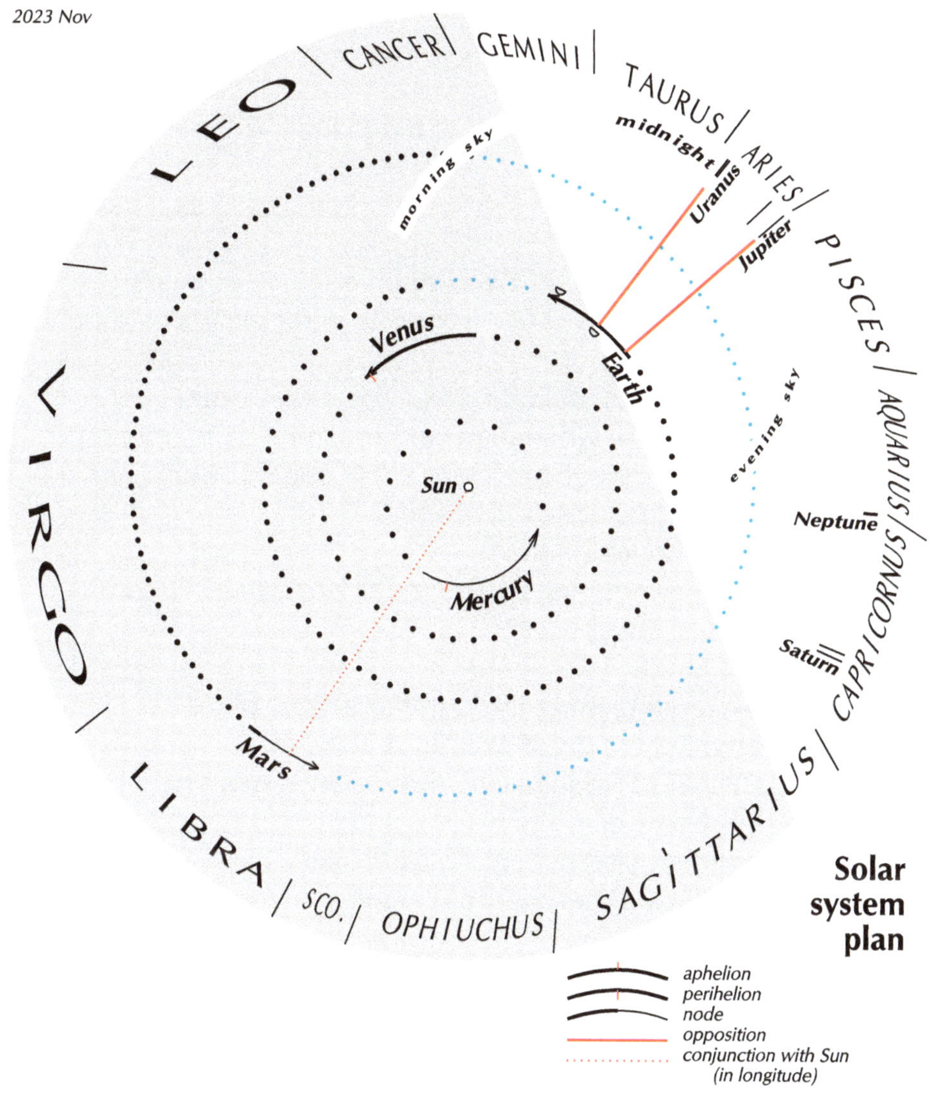

CANCER
GEMINI
TAURUS
ARIES
PISCES
AQUARIUS
CAPRICORNUS
SAGITTARIUS
OPHIUCHUS
SCO.
LIBRA
VIRGO
LEO
midnight
morning sky
evening sky
Uranus
Jupiter
Neptune
Saturn
Earth
Venus
Sun
Mercury
Mars
Solar
system
plan
aphelion
perihelion
node
opposition
conjunction with Sun
(in longitude)

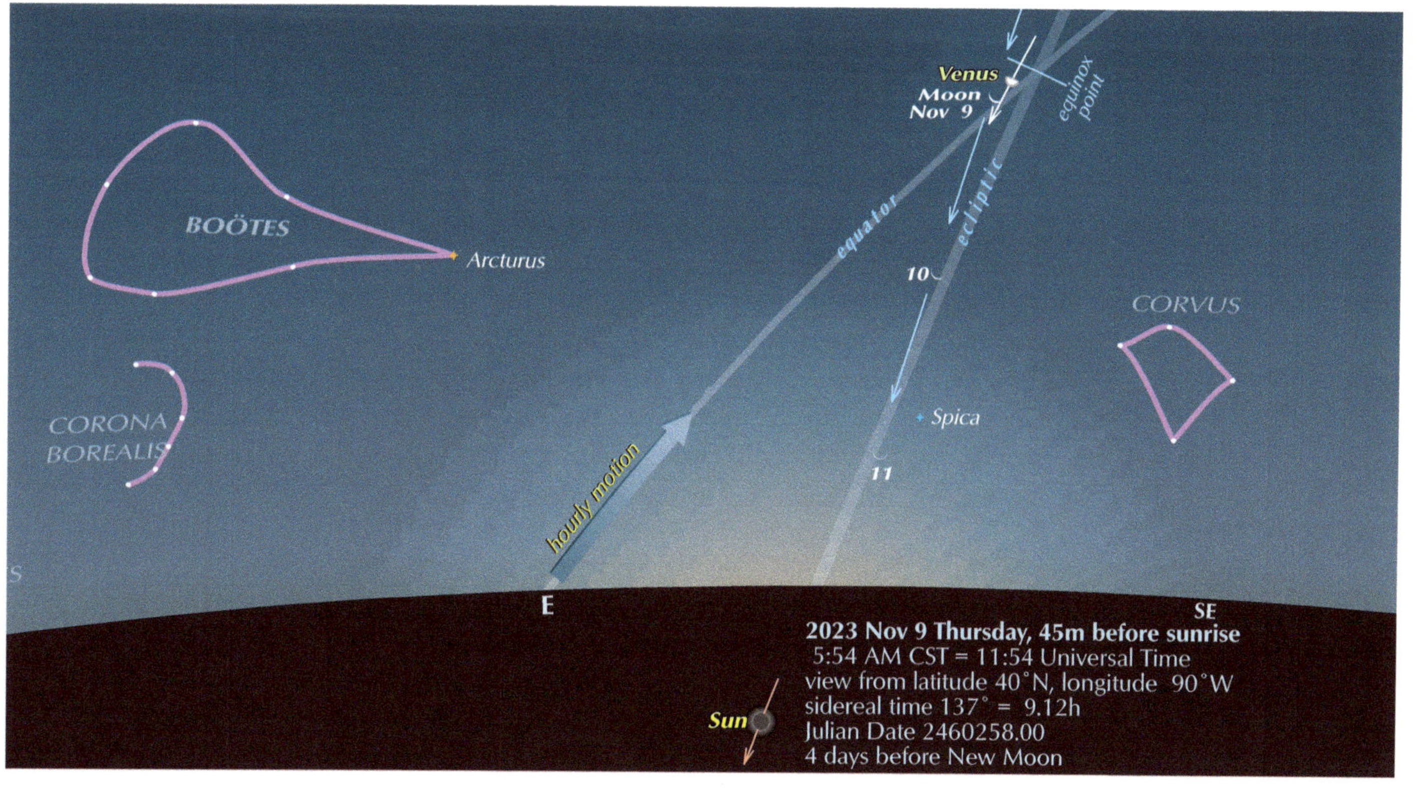

Venus
Moon
Nov 9
equinox
point
equator
ecliptic
10
11
BOÖTES
Arcturus
CORONA
BOREALIS
CORVUS
Spica
hourly motion
E
SE
S
Sun
2023 Nov 9 Thursday, 45m before sunrise
5:54 AM CST = 11:54 Universal Time
view from latitude 40°N, longitude 90°W
sidereal time 137° = 9.12h
Julian Date 2460258.00
4 days before New Moon

NOVEMBER

0250.583	Nov	2	Thu	2	Moon 4.0° N of M35 cluster; 127° from Sun in morning sky; magnitudes -11.3 and 5.3
0251.704	Nov	3	Fri	5	Jupiter at opposition in longitude; magnitude -2.9; declination 13.6°
0252.063	Nov	3	Fri	14	Moon 4.9° S of Castor; 110° from Sun in morning sky; magnitudes -10.7 and 1.5
0252.106	Nov	3	Fri	15	The equation of time is at a maximum of 16.49 minutes
0252.292	Nov	3	Fri	19	Moon 1.44° S of Pollux; 107° from Sun in morning sky; magnitudes -10.7 and 1.2
0252.749	Nov	4	SAT	6	Saturn stationary in longitude; resumes direct motion
0253.162	Nov	4	SAT	16	Saturn stationary in right ascension; resumes direct motion
0253.479	Nov	4	SAT	24	Moon 3.9° NNE of Beehive Cluster; 94° and 95° from Sun in morning sky; magnitudes -10.2 and 3.7
0253.5	Nov	5	SUN		Clocks back 1 hour (America)
0253.5	Nov	5	SUN		Southern Taurid meteors; ZHR 5; near last quarter Moon
0253.859	Nov	5	SUN	8:38	Last quarter Moon
0255.145	Nov	6	Mon	15	Mars at descending node through the ecliptic plane
0255.237	Nov	6	Mon	18	Mercury at aphelion; 0.4667 AU from the Sun
0255.375	Nov	6	Mon	21	Moon 3.8° NNE of Regulus; 74° from Sun in morning sky; magnitudes -9.3 and 1.4
0255.409	Nov	6	Mon	22	Moon at apogee; distance 63.43 Earth-radii
0257.654	Nov	9	Thu	4	Mars and Uranus at heliocentric opposition; longitudes 231.0° and 51.0°
0257.958	Nov	9	Thu	11	Moon 0.90° NE of Venus; 45° and 46° from Sun in morning sky; magnitudes -7.8 and -4.3; occultation
0259.5	Nov	11	SAT		Veterans' or Remembrance Day
0259.854	Nov	11	SAT	9	Moon 2.14° NE of Spica; 24° and 25° from Sun in morning sky; magnitudes -6.3 and 1.0
0259.867	Nov	11	SAT	9	Moon at descending node; longitude 204.8°
0260.5	Nov	5	SUN		Northern Taurid meteors; ZHR 5; 1 day before new Moon
0261.893	Nov	13	Mon	9:26	New Moon; beginning of lunation 1248
0262.021	Nov	13	Mon	13	Moon 2.36° SSW of Mars; 3° and 1° from Sun in evening sky; magnitudes -4.2 and 1.5
0262.214	Nov	13	Mon	17	Uranus at opposition in longitude; magnitude 5.6; declination 17.7°
0263.104	Nov	14	Tue	15	Moon 1.65° S of Mercury; 15° from Sun in evening sky; magnitudes -5.5 and -0.4
0263.358	Nov	14	Tue	21	Moon, Mercury, and Antares within circle of diameter 4.17°; about 17° from the Sun in the evening sky; magnitudes -6, 0, 1
0263.375	Nov	14	Tue	21	Moon 0.88° NNE of Antares; 18° from Sun in evening sky; magnitudes -5.8 and 1.0; occultation
0265.625	Nov	17	Fri	3	Mercury 2.53° NNE of Antares; 16° from Sun in evening sky; magnitudes -0.4 and 1.0
0266.5	Nov	18	SAT	0	Leonid meteors; ZHR 10; 2 days before first quarter Moon
0266.755	Nov	18	SAT	6	Mars at conjunction with the Sun; 2.526 AU from Earth; latitude -0.19°

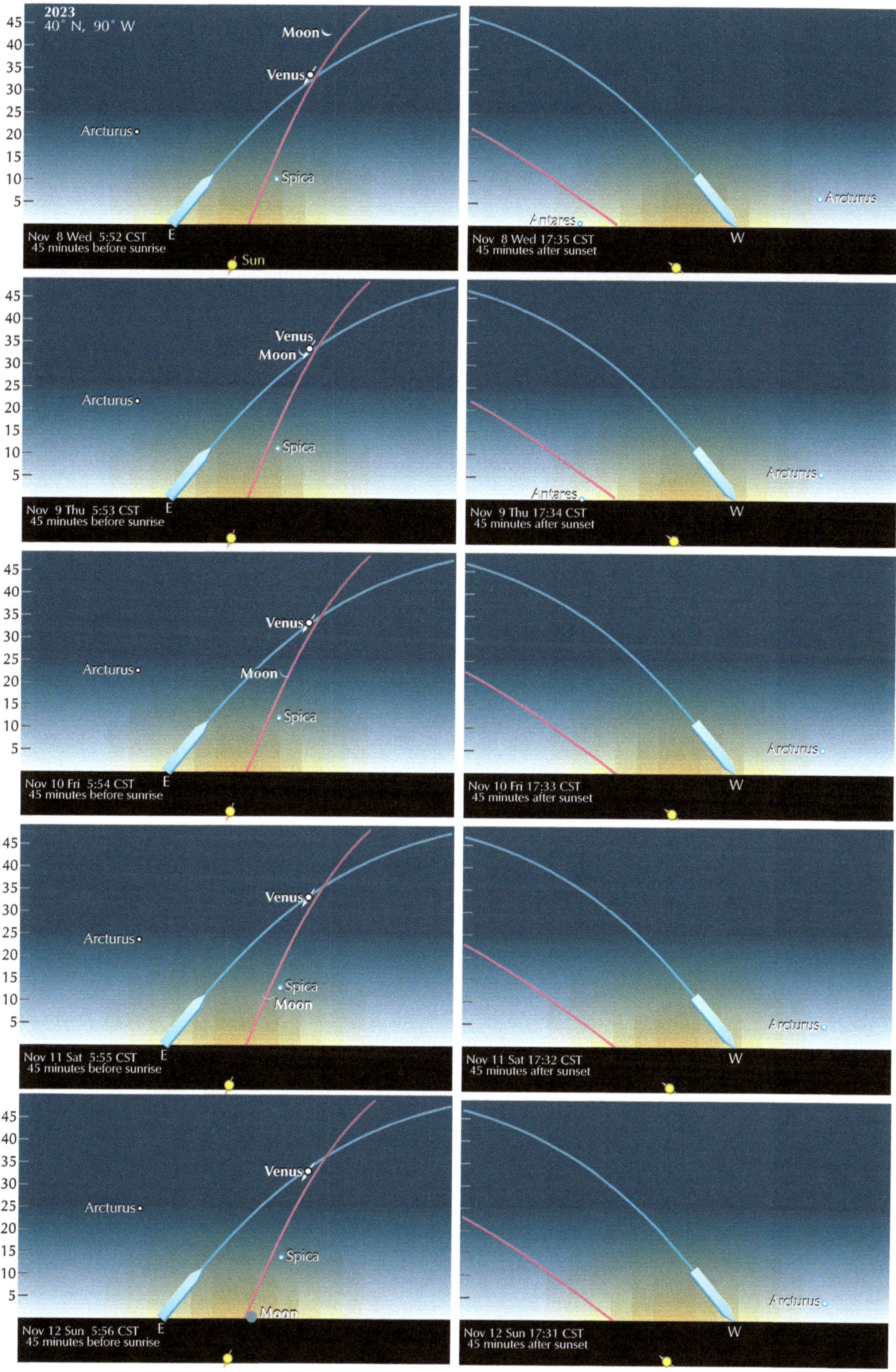
2023
40° N, 90° W
Moon
Venus
Arcturus
Spica
Sun
Antares
Nov 8 Wed 5:52 CST
45 minutes before sunrise
E
Nov 8 Wed 17:35 CST
45 minutes after sunset
W
Nov 9 Thu 5:53 CST
45 minutes before sunrise
Nov 9 Thu 17:34 CST
45 minutes after sunset
Nov 10 Fri 5:54 CST
45 minutes before sunrise
Nov 10 Fri 17:33 CST
45 minutes after sunset
Nov 11 Sat 5:55 CST
45 minutes before sunrise
Nov 11 Sat 17:32 CST
45 minutes after sunset
Nov 12 Sun 5:56 CST
45 minutes before sunrise
Nov 12 Sun 17:31 CST
45 minutes after sunset

0268.951	Nov	20	Mon	10:50	First quarter Moon
0269.188	Nov	20	Mon	17	Moon 2.51° SE of Saturn; 93° from Sun in evening sky; magnitudes -10.3 and 0.9
0270.386	Nov	21	Tue	21:16	Moon at perigee; distance 57.98 Earth-radii
0270.896	Nov	22	Wed	10	Moon 1.34° SE of Neptune; 116° and 115° from Sun in evening sky; magnitudes -11.1 and 7.9
0271.086	Nov	22	Wed	14	Sun enters the astrological sign Sagittarius, i.e. its longitude is 240°
0271.905	Nov	23	Thu	10	Saturn at east quadrature, 90° from the Sun
0272.125	Nov	23	Thu	15	Moon shows minimum libration for the year, 2.77°
0272.271	Nov	23	Thu	18	Sun enters Scorpius, at longitude 241.20° on the ecliptic
0272.961	Nov	24	Fri	11	Moon at ascending node; longitude 24.5°
0273.917	Nov	25	SAT	10	Moon 2.56° NNW of Jupiter; 155° from Sun in evening sky; magnitudes -12.2 and -2.8
0274.833	Nov	26	SUN	8	Moon 2.59° NNW of Uranus; 167° from Sun in evening sky; magnitudes -12.4 and 5.6
0275.493	Nov	26	SUN	24	Mercury at southernmost latitude from the ecliptic plane, -7.0°
0275.583	Nov	27	Mon	2	Moon 1.08° SE of Pleiades; 175° and 174° from Sun in evening midnight sky
0275.886	Nov	27	Mon	9:15	Full Moon
0277.021	Nov	28	Tue	13	Venus at perihelion; 0.7184 AU from the Sun
0277.979	Nov	29	Wed	12	Moon 3.9° N of M35 cluster; 154° and 155° from Sun in morning sky; magnitudes -12.0 and 5.3
0278.229	Nov	29	Wed	18	Venus 4.2° NNE of Spica; 43° from Sun in morning sky; magnitudes -4.2 and 1.0
0278.353	Nov	29	Wed	20	Mercury at southernmost declination, -25.87°
0279.090	Nov	30	Thu	14	Sun enters Ophiuchus, at longitude 248.09° on the ecliptic
0279.438	Nov	30	Thu	23	Moon 5.0° S of Castor; 137° from Sun in morning sky; magnitudes -11.6 and 1.5

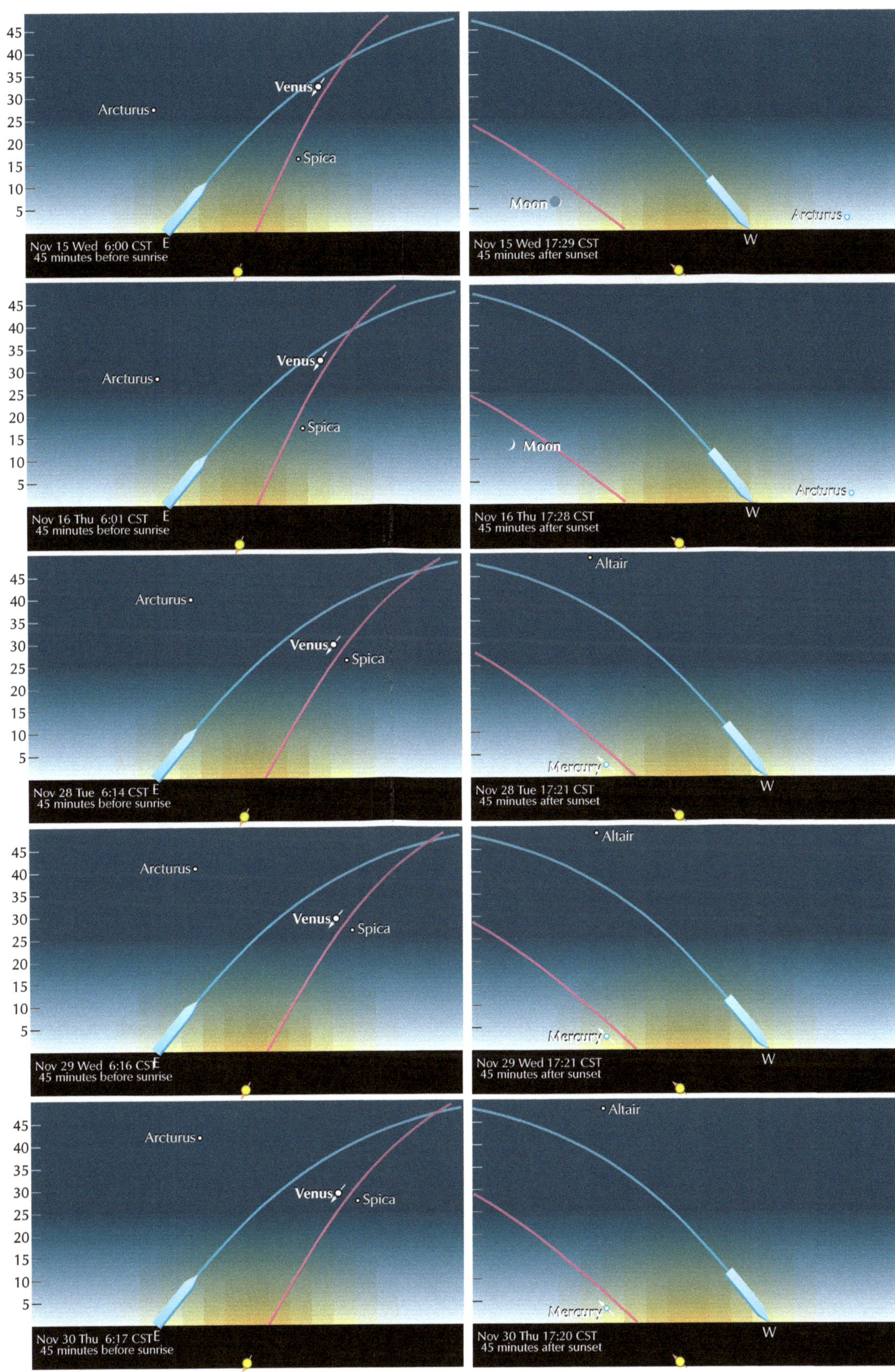
Nov 15 Wed 6:00 CST
45 minutes before sunrise
E
Venus
Arcturus
Spica
Nov 15 Wed 17:29 CST
45 minutes after sunset
W
Moon
Arcturus
Nov 16 Thu 6:01 CST
45 minutes before sunrise
E
Venus
Arcturus
Spica
Nov 16 Thu 17:28 CST
45 minutes after sunset
W
Moon
Arcturus
Nov 28 Tue 6:14 CST
45 minutes before sunrise
E
Arcturus
Venus
Spica
Nov 28 Tue 17:21 CST
45 minutes after sunset
W
Altair
Mercury
Nov 29 Wed 6:16 CST
45 minutes before sunrise
E
Arcturus
Venus
Spica
Nov 29 Wed 17:21 CST
45 minutes after sunset
W
Altair
Mercury
Nov 30 Thu 6:17 CST
45 minutes before sunrise
E
Arcturus
Venus
Spica
Nov 30 Thu 17:20 CST
45 minutes after sunset
W
Altair
Mercury

DECEMBER

Evening sky

for latitude 40° north

about 10 PM at the 5th,
9 PM at the 20th
of the month

sidereal time
3h

SKY DOME

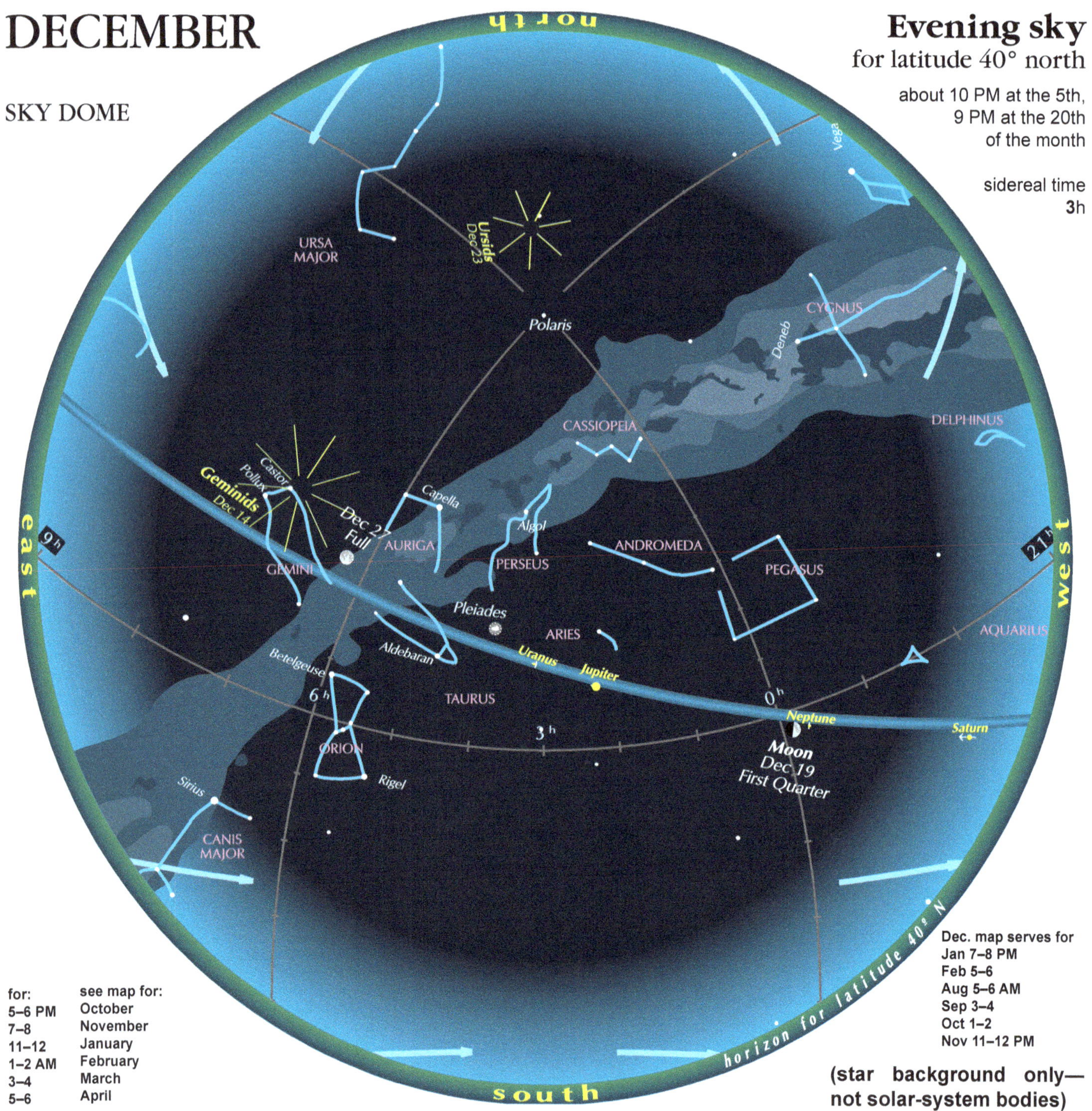

for:	see map for:
5–6 PM	October
7–8	November
11–12	January
1–2 AM	February
3–4	March
5–6	April

Dec. map serves for
Jan 7–8 PM
Feb 5–6
Aug 5–6 AM
Sep 3–4
Oct 1–2
Nov 11–12 PM

(star background only—not solar-system bodies)

2023 Dec

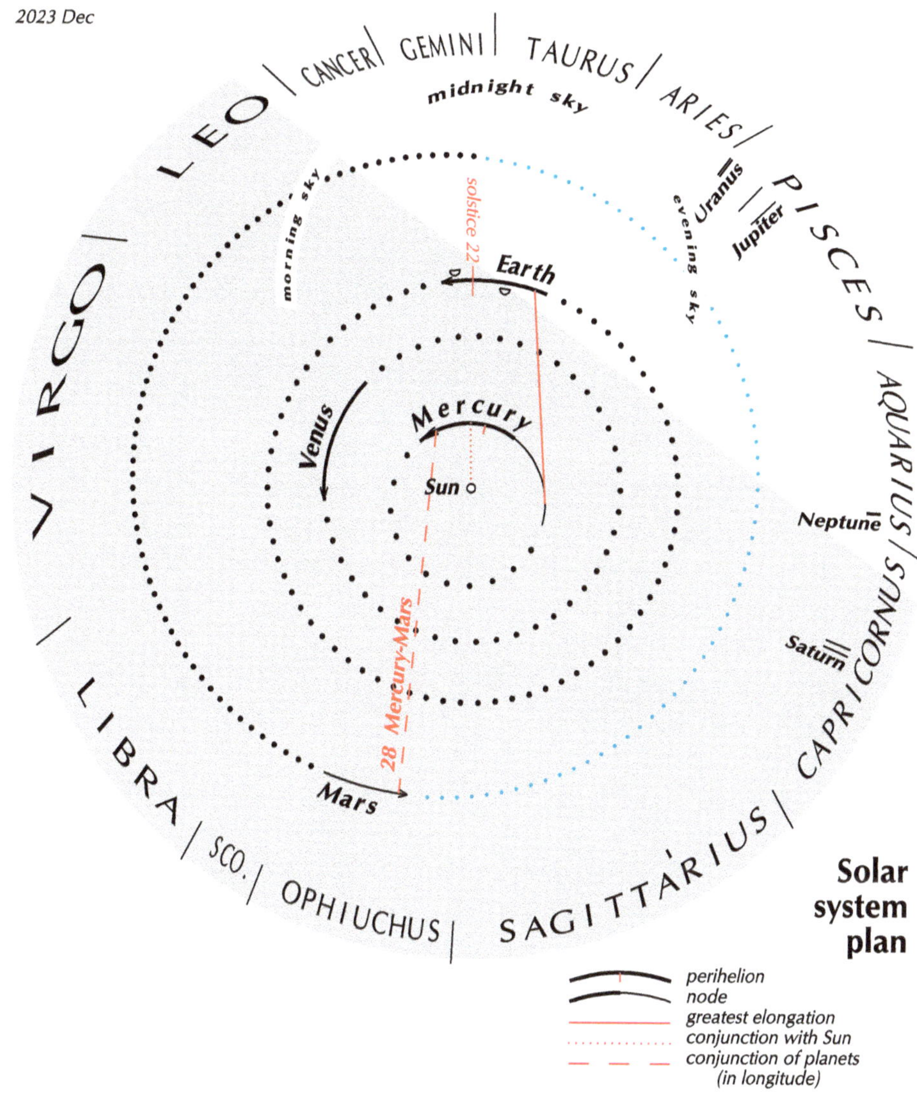
CANCER
GEMINI
TAURUS
ARIES
PISCES
AQUARIUS
CAPRICORNUS
SAGITTARIUS
OPHIUCHUS
SCO.
LIBRA
VIRGO
LEO
midnight sky
morning sky
evening sky
Uranus
Jupiter
solstice 22
Earth
Venus
Mercury
Sun
Neptune
Saturn
28 Mercury-Mars
Mars
Solar system plan
perihelion
node
greatest elongation
conjunction with Sun
conjunction of planets
(in longitude)

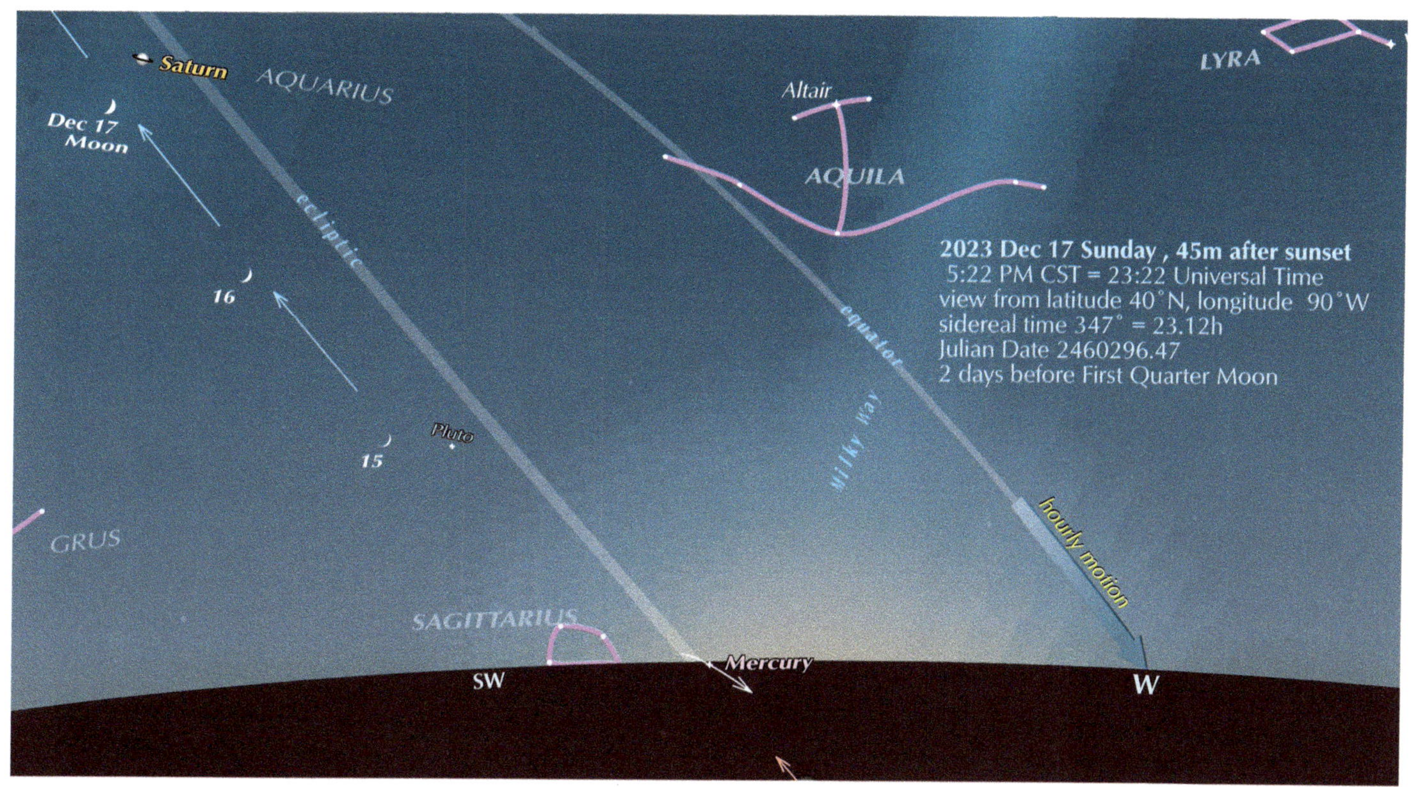
Saturn
AQUARIUS
Dec 17 Moon
16
15
Pluto
ecliptic
Altair
AQUILA
LYRA
equator
Milky Way
hourly motion
GRUS
SAGITTARIUS
Mercury
SW
W
2023 Dec 17 Sunday , 45m after sunset
5:22 PM CST = 23:22 Universal Time
view from latitude 40°N, longitude 90°W
sidereal time 347° = 23.12h
Julian Date 2460296.47
2 days before First Quarter Moon

DECEMBER

0279.667	Dec	1	Fri	4	Moon 1.58° S of Pollux; 135° from Sun in morning sky; magnitudes -11.5 and 1.2
0280.813	Dec	2	SAT	8	Moon 3.7° NNE of Beehive Cluster; 122° from Sun in morning sky; magnitudes -11.1 and 3.7
0282.708	Dec	4	Mon	5	Moon 3.7° NNE of Regulus; 101° and 102° from Sun in morning sky; magnitudes -10.4 and 1.4
0283.096	Dec	4	Mon	14	Mercury at easternmost elongation; 21.3° from Sun in evening sky; magnitude -0.4
0283.275	Dec	4	Mon	19	Moon at apogee; distance 63.40 Earth-radii
0283.744	Dec	5	Tue	5:51	Last quarter Moon
0284.858	Dec	6	Wed	9	Neptune stationary in longitude; resumes direct motion
0285.297	Dec	6	Wed	19	Neptune stationary in right ascension; resumes direct motion
0287.063	Dec	8	Fri	14	Mars 4.3° N of Antares; 6° and 8° from Sun in morning sky; magnitudes 1.4 and 1.0
0287.142	Dec	8	Fri	15	Moon at descending node; longitude 203.7°
0287.191	Dec	8	Fri	16:35	Earliest sunset, at latitude 40° north
0287.208	Dec	8	Fri	17	Moon 2.01° NNE of Spica; 52° from Sun in morning sky; magnitudes -8.3 and 1.0
0288.104	Dec	9	SAT	15	Moon 3.3° SSW of Venus; 42° from Sun in morning sky; magnitudes -7.7 and -4.1
0290.658	Dec	12	Tue	4	Moon, Mars, and Antares within circle of diameter 4.99°; about 10° from the Sun in the morning sky; magnitudes -5, 1, 1
0290.729	Dec	12	Tue	6	Moon 0.89° NNE of Antares; 10° and 11° from Sun in morning sky; magnitudes -5.1 and 1.0
0290.917	Dec	12	Tue	10	Moon 3.5° S of Mars; 8° and 7° from Sun in morning sky; magnitudes -4.8 and 1.4
0291.480	Dec	12	Tue	23:31	New Moon; beginning of lunation 1249
0291.700	Dec	13	Wed	5	Mercury stationary in right ascension; starts retrograde motion
0291.793	Dec	13	Wed	7	Mercury stationary in longitude; starts retrograde motion
0292.5	Dec	14	Thu	0	Geminid meteors; ZHR 150; 1 day after new Moon
0292.750	Dec	14	Thu	6	Moon 4.4° S of Mercury; 17° and 16° from Sun in evening sky; magnitudes -5.8 and 0.6
0294.551	Dec	16	SAT	1	Mercury at ascending node through the ecliptic plane
0295.281	Dec	16	SAT	18:44	Moon at perigee; distance 57.68 Earth-radii
0296.521	Dec	18	Mon	1	Moon 2.30° SE of Saturn; 67° and 66° from Sun in evening sky; magnitudes -9.3 and 1.0
0297.366	Dec	18	Mon	21	Sun enters Sagittarius, at longitude 266.66° on the ecliptic
0298.125	Dec	19	Tue	15	Moon 1.19° SE of Neptune; 88° from Sun in evening sky; magnitudes -10.2 and 7.9; occultation
0298.277	Dec	19	Tue	18:39	First quarter Moon
0298.433	Dec	19	Tue	22	Venus at northernmost latitude from the ecliptic plane, 3.4°
0299.222	Dec	20	Wed	17	Mercury at perihelion; 0.3075 AU from the Sun
0300.016	Dec	21	Thu	12	Asteroid 4 Vesta at opposition in longitude; magnitude 6.4
0300.080	Dec	21	Thu	14	Moon at ascending node; longitude 22.6°
0300.645	Dec	22	Fri	3:28	December (northern winter) solstice

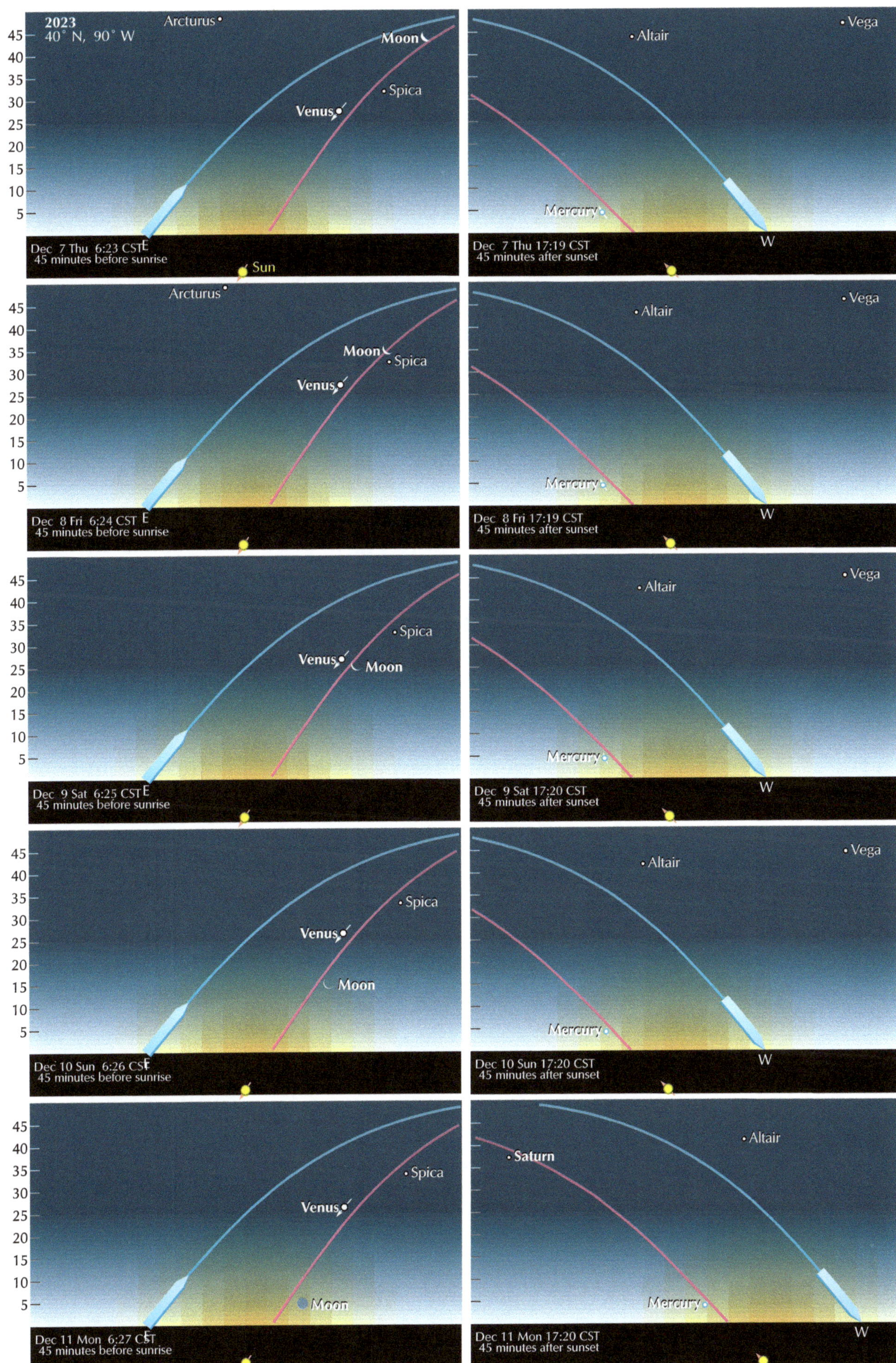
2023
40° N, 90° W
Arcturus
Moon
Spica
Venus
Dec 7 Thu 6:23 CST E
45 minutes before sunrise
Sun
Altair
Vega
Mercury
Dec 7 Thu 17:19 CST
45 minutes after sunset
W
Dec 8 Fri 6:24 CST E
45 minutes before sunrise
Dec 8 Fri 17:19 CST
45 minutes after sunset
Dec 9 Sat 6:25 CST E
45 minutes before sunrise
Dec 9 Sat 17:20 CST
45 minutes after sunset
Dec 10 Sun 6:26 CST E
45 minutes before sunrise
Dec 10 Sun 17:20 CST
45 minutes after sunset
Saturn
Dec 11 Mon 6:27 CST E
45 minutes before sunrise
Dec 11 Mon 17:20 CST
45 minutes after sunset

0300.645	Dec	22	Fri	3:28	Sun enters the astrological sign Capricornus, i.e. its longitude is 270°
0301.042	Dec	22	Fri	13	Moon 2.39° NNW of Jupiter; 125° from Sun in evening sky; magnitudes -11.3 and -2.7
0301.283	Dec	22	Fri	19	Mercury at inferior conjunction with the Sun; 0.676 AU from Earth; latitude 4.70°
0301.5	Dec	23	SAT	0	Ursid meteors; ZHR 10; 3 days after first quarter Moon
0302.063	Dec	23	SAT	14	Moon 2.61° NNW of Uranus; 138° from Sun in evening sky; magnitudes -11.7 and 5.7
0302.896	Dec	24	SUN	10	Moon 1.00° SE of Pleiades; 148° from Sun in evening sky
0303.5	Dec	25	Mon		Christmas
0304.140	Dec	25	Mon	15	The equation of time is 0
0305.333	Dec	26	Tue	20	Moon 3.8° N of M35 cluster; 175° and 177° from Sun in evening midnight sky; magnitudes -12.6 and 5.3
0305.523	Dec	27	Wed	0:33	Full Moon
0306.417	Dec	27	Wed	22	Mercury 3.6° N of Mars; 12° from Sun in morning sky; magnitudes 2.0 and 1.4
0306.771	Dec	28	Thu	7	Moon 5.1° S of Castor; 165° and 163° from Sun in morning sky; magnitudes -12.3 and 1.5
0307.021	Dec	28	Thu	13	Moon 1.71° S of Pollux; 162° from Sun in morning sky; magnitudes -12.2 and 1.2
0308.167	Dec	29	Fri	16	Moon 3.5° NNE of Beehive Cluster; 149° and 150° from Sun in morning sky; magnitudes -11.9 and 3.7
0309.430	Dec	30	SAT	22	Mercury at northernmost latitude from the ecliptic plane, 7.0°
0309.589	Dec	31	SUN	2	Jupiter stationary in longitude; resumes direct motion
					JANUARY
0310.5	Jan	1	Mon		Gregorian calendar Jan 1 = Julian calendar 2023 Dec 19

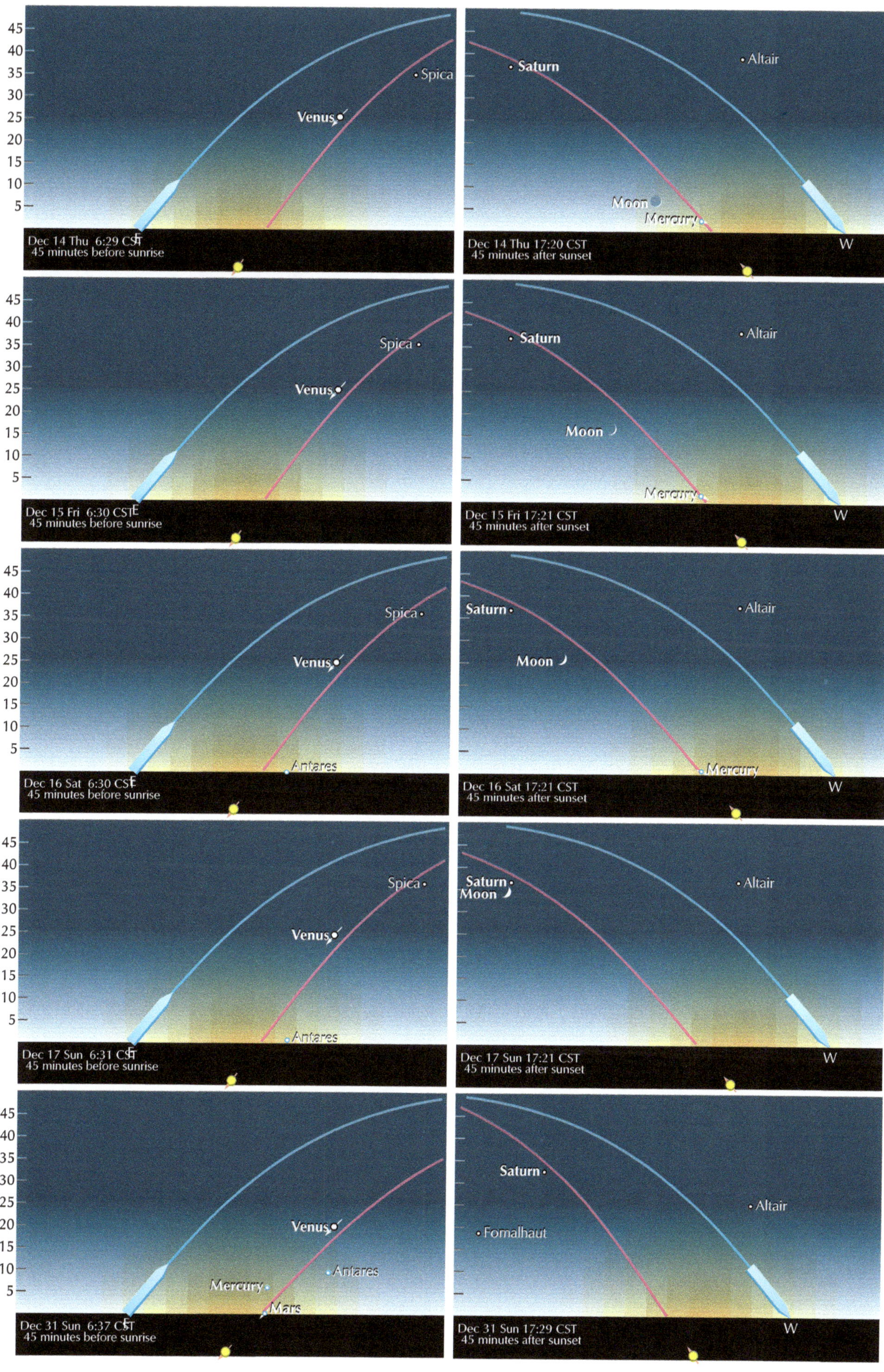

45
40
35
30
25
20
15
10
5
Spica
Venus
Dec 14 Thu 6:29 CST
45 minutes before sunrise
Saturn
Altair
Moon
Mercury
W
Dec 14 Thu 17:20 CST
45 minutes after sunset
Spica
Venus
Dec 15 Fri 6:30 CST
45 minutes before sunrise
Saturn
Altair
Moon
Mercury
W
Dec 15 Fri 17:21 CST
45 minutes after sunset
Spica
Venus
Antares
Dec 16 Sat 6:30 CST
45 minutes before sunrise
Saturn
Altair
Moon
Mercury
W
Dec 16 Sat 17:21 CST
45 minutes after sunset
Spica
Venus
Antares
Dec 17 Sun 6:31 CST
45 minutes before sunrise
Saturn
Moon
Altair
W
Dec 17 Sun 17:21 CST
45 minutes after sunset
Venus
Mercury
Antares
Mars
Dec 31 Sun 6:37 CST
45 minutes before sunrise
Saturn
Altair
Fomalhaut
W
Dec 31 Sun 17:29 CST
45 minutes after sunset

SUN, EARTH, AND SEASONS

The Sun's daily arc across your sky shifts gradually northward from December until June, then back south from June to December. The cause is that the spinning Earth maintains an attitude tilted by 23.4° to the ecliptic plane in which it travels around the Sun. (This is called the obliquity of Earth's rotational axis.)

At the March equinox, the Sun, appearing to travel along the ecliptic, reaches the point (in Pisces) where it crosses the celestial equator into the northern celestial hemisphere. It passes overhead at noon for all places along Earth's equator. Night and day are everywhere of equal length—hence "equi-nox." The hemispheres receive equal sunlight. Earth's two poles are equidistant from the Sun, the north pole leaning backward from the direction of travel. This is the spring or vernal equinox for our northern hemisphere; but it is the fall or autumn equinox for the southern.

At the June (or northern summer) solstice, Earth's north pole is tilted toward the Sun at the maximal 23.4° angle. The Sun as seen from Earth reaches the point on the ecliptic that is farthest north (23.4°) of the celestial equator; its northward progress "stalls" (sol-stitium). It passes overhead for places along the Tropic of Cancer. For the northern hemisphere, days are longest, nights shortest.

At the September equinox (autumn or fall equinox for the northern hemisphere), the Sun reaches the other point (in Virgo) where it crosses the equator, into the southern celestial hemisphere. It again passes overhead along Earth's equator. The two poles are again equidistant from the Sun, the north pole now leaning forward. Again the hemispheres receive equal sunlight, and day and night are equal.

At the December (or northern winter) solstice, the Sun appears farthest south of the celestial equator. It passes overhead along the Tropic of Capricorn. Earth's north pole leans maximally away from the Sun. For northern lands, days are shortest.

For our north pole, the Sun at the equinoxes has no altitude, appearing to run all around the horizon; it is in the sky permanently from March to September; and not at all from September to March. For lands poleward of the Arctic and Antarctic circles, these 24-hour days and 24-hour nights persist for up to 6 months. For lands between the tropics, the Sun is sometimes north and sometimes south of overhead. For north-temperate latitudes (such as 40° north), the Sun makes a slanting arch which always passes south of overhead. From March to September this arch is longer and higher, so that the Sun is in the sky more than half the time; it is higher than average at each time of day; its light arrives through less atmosphere and at a steeper angle, so is more concentrated per area of ground.

These four cardinal events fall this year on March 20, June 21, Sep. 23, Dec. 21. The March equinox was till 2007 on March 21 or 20; is now only on the 20th; in 2044 it will begin falling sometimes on the 19th. The June solstice was till 1975 on June 22 or 21; then only on June 21; in 2012 it began falling sometimes on June 20. The September equinox till 1931 fell sometimes on Sep. 24; since 1968 only on Sep. 23 or 22. The December solstice till 1697 fell on Dec. 20 or 21; from 1702 it has been on Dec. 21 or 22; will next slip back to Dec. 20 in 2080.

Daylight begins and ends when the top of the Sun is visible. And the Sun's apparent height is raised by refraction when it is near the horizon. So the actual dates when day and night hours are most nearly equal are before the spring equinox and after the autumn equinox, by about 3 days at latitude 40° north; and the total of daytime in the year is longer than nighttime.

The Sun's arc through the daytime sky, at the equinoxes and solstices, for places at latitude 40°. The Sun's disk is shown at hourly intervals, and is exaggerated 4 times in size. The projection, based simply on altitude and azimuth, makes the horizon appear as a straight line; if instead we used a polar projection based on the middle of the sky, the horizon would be a circle and the Sun's arcs would be less curved.

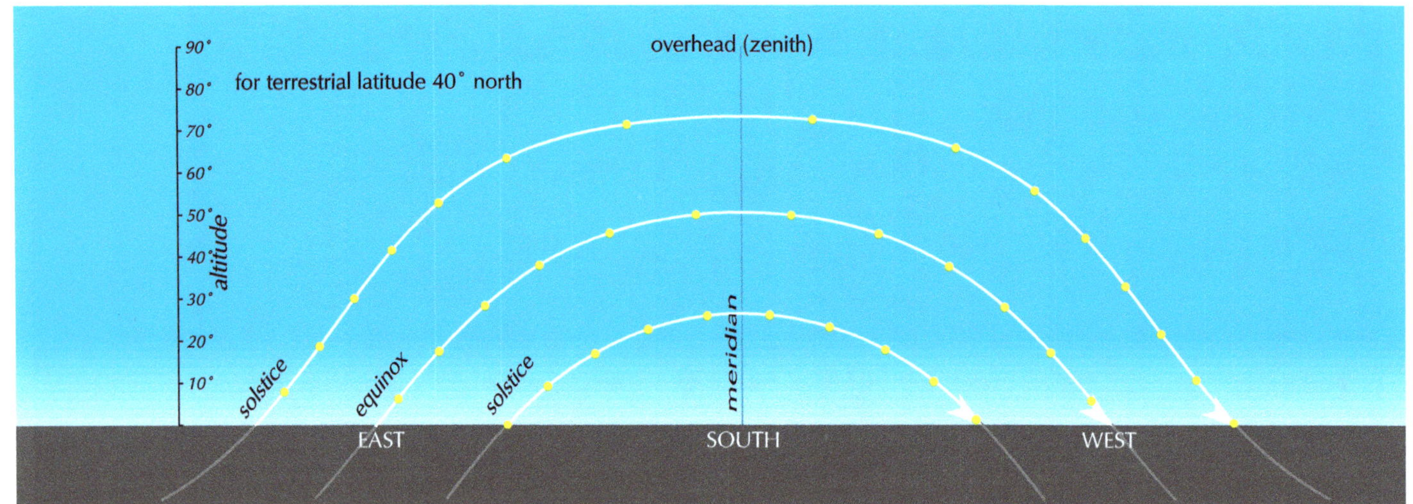

Earth at the equinoxes and solstices. The Sun's size is exaggerated 20 times, the Earth's 1,500 times.

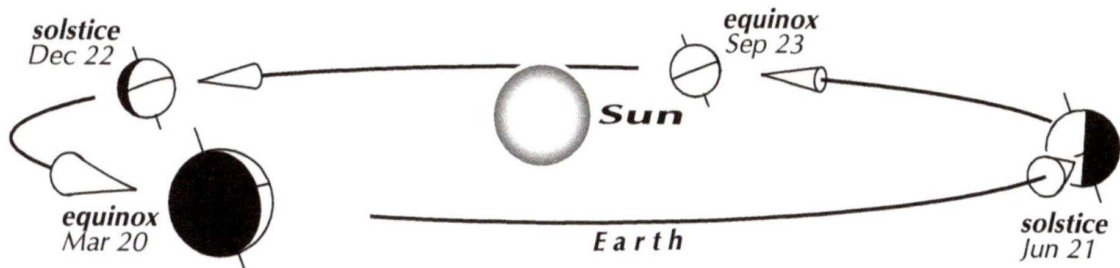

Earth at the equinoxes and solstices, seen from the direction of the Sun (actually, from 35° above the Sun's viewpoint, so as to be able to show some of Earth's night side) and from a distance of 60 Earth-radii. The thick arrow is a "rail" along which the planet is riding in its orbit at its speed of about 2,574,000 km per day; each projecting part of the arrow is a distance the Earth advances in one minute (around 1,800 km). An arrow above the equator shows how fast Earth rotates (15° per hour) around its axis (shown by sticks at the poles). A trident represents the vertical beam of sunlight, striking where the Sun is at the zenith at noon.

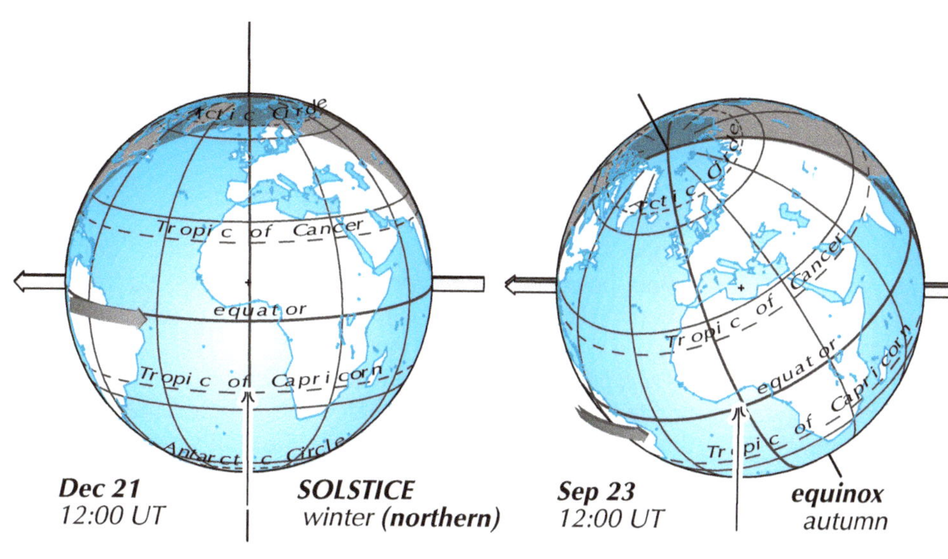

We show the Earth at 12h UT on the day of the event: the hour when the 0° longitude line (Greenwich) faces the Sun. This is 6 PM local time at longitude 90° west, in the Central time-zone of the USA: well after sunrise in summer, well before it in winter. The terminator (day-night boundary) at both solstices touches the Arctic and Antarctic circles—giving 24-hour sunshine in summer, 24-hour night in winter. The slope of the terminator shows that, in summer, daylight lasts longer for places at higher latitudes (such as Canada) than for those nearer the equator (such as the USA); but sunlight arrives at lower angles.

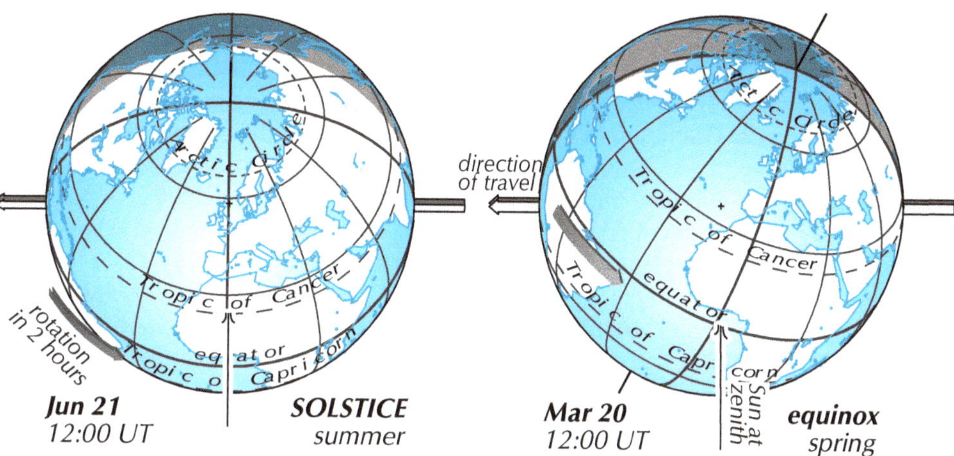

Sun-Earth distance, perihelion and aphelion

In its slightly elliptical orbit, Earth reaches these innermost and outermost points at Jan. 4 16h UT and July 6 20h UT.

The average distance between the centers of Sun and Earth (the astronomical unit, AU) is 149,597,871 kilometers (92,955,807 miles). The eccentricity of the orbit is 0.017. So at perihelion and aphelion the Sun-Earth distance is less or more by only 0.017 AU).

This is about 2,540,000 km, or 270 times the width of the Earth—small compared with the average distance. Light takes about 8 minutes to travel from Sun to Earth, and about 8 seconds for the extra distance at aphelion.

Mainly because of the swinging of Earth and Moon around their barycenter or common center of mass (which is 1,738 km under Earth's surface), the minimum Sun-Earth distance varies: up to about 0.00005 AU or 7,500 km greater (near new Moon) or smaller (near full Moon). This year it is 0.98329 AU (147,098,000 km).

And the date of perihelion varies, from about Jan. 1 22h at last quarter Moon (when the Moon is ahead of us) as in 1989, to Jan. 5 8h near first quarter as in 2020.

The aphelion distance varies to about 0.0001 AU more or less, and the date from July 3 6 UT to July 6 24 UT. This year it is 1.01671 AU (152,097,650 km).

All this relatively slight change in distance from the Sun has little effect on warmth and is not the cause of our seasons. Indeed, we happen to be nearest to the Sun in the middle of our north-hemisphere winter. By contrast, Mars varies greatly in distance from the Sun and this has great effect on its seasons.

Sun's progress through the constellations

The Sun appears—or would, if we could see past it to the stars—to progress from Sagittarius into Capricornus on Jan. 20; into Aquarius on Feb. 16; and so on through the 12 constellations of the zodiac. That is, it crosses the constellation boundaries as used in astronomy. These lines, giving definition to the traditional areas of the celestial sphere's 88 constellations, were partly worked out by Benjamin Gould in 1875, and in 1930 were completed by Eugène Delporte and adopted by the International Astronomical Union.

The constellations have irregular shapes and sizes. Scorpius lies mostly south of the Sun's path, more of which passes through Ophiuchus; so Ophiuchus is in effect a 13th zodiacal constellation. The lengths of the ecliptic lying inside the constellations are: Capricornus 28.18°, Aquarius 23.68°, Pisces 37.52°, Aries 24.38°, Taurus 36.96°, Gemini 27.83°, Cancer 19.92°, Leo 35.98°, Virgo 43.64°, Libra 23.34°, Scorpius 6.90°, Ophiuchus 18.57°, Sagittarius 33.1°. (Total, 360°.) Virgo is easily the longest. The Sun spends proportionate time in each constellation.

Because of precession—the shifting westward, at about 14° per thousand years, of the point where the Sun appears at the March equinox—the longitudes of these crossing-points from the equinox point slowly increase. So the dates when the Sun reaches them become, each year, a few hours later.

By a different system, still used in astrology, the zodiac is divided into 12 equal 30°-wide "signs," fixed in relation to the moving March equinox. When the Sun's longitude is 0°, 90°, 180°, and 270°—that is, at the March equinox, June solstice, September equinox, and December solstice—it is said to enter the signs Aries, Cancer, Libra, Capricorn. This is why the March equinox point is traditionally called the First Point of Aries. The tropics of Cancer and Capricorn, lines of latitude around the Earth at 23.4° north and south, were so named because the Sun was overhead on them as it reached the June and December solstices.

But precession has by now made this system out of date with the Sun's actual position by about two thousand years; that is, by the width of about one whole constellation. When astrology says the Sun is entering the sign Aquarius, it is actually entering the next constellation back westward, Capricornus; and so on, roughly.

"Roughly," as the diagram shows, because of the unequal

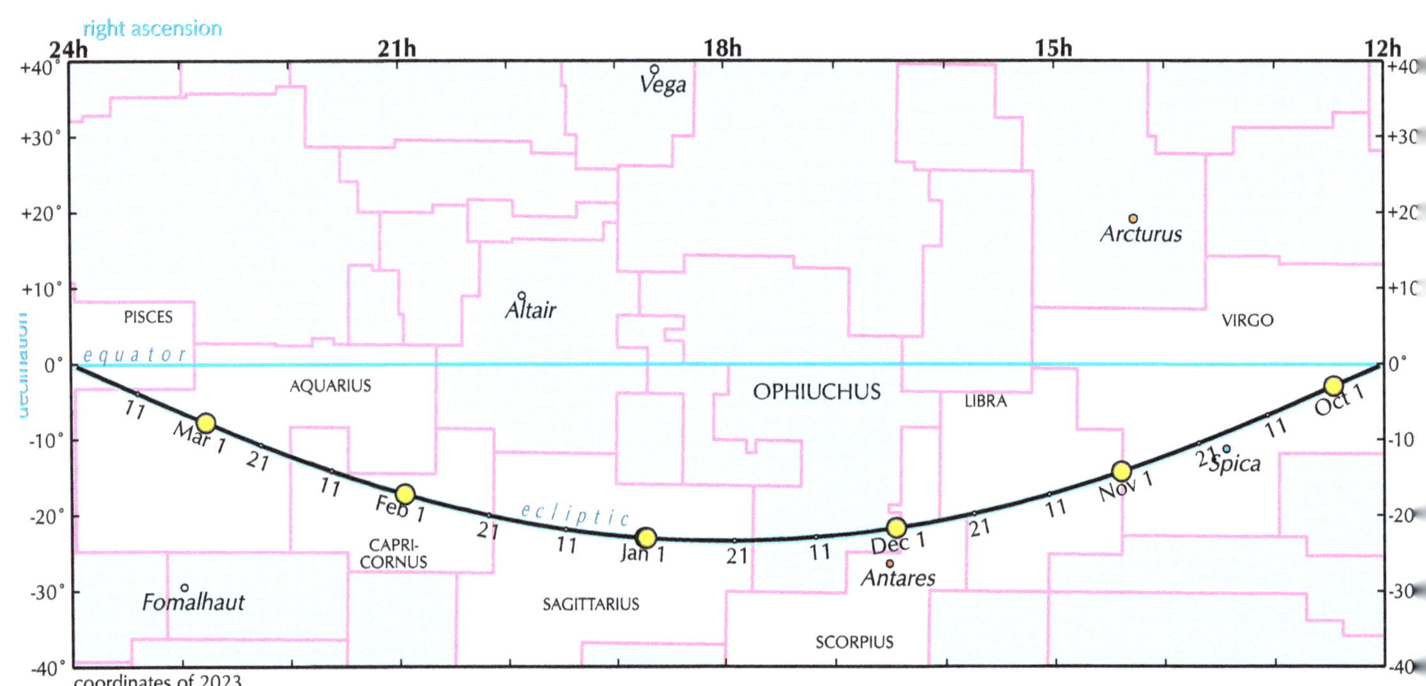

lengths of the astronomical constellations. Most interesting is what has recently happened near the June solstice. The Sun, reaching longitude 90° and its most northerly point, is said to enter the sign Cancer. The real position of this point moved back through Gemini to the Taurus border. In 1990 (Jan. 1, according to the calculation of Jean Meeus) the longitude of the constellation boundary crossed 90°. So up to 1989, the Sun at the instant of the solstice was in Gemini, one constellation back from what astrology says, as with all the other correspondences; but from 1990 on, the Sun at this instant is still briefly in Taurus, the second constellation back. With the centuries, the rest of the signs will similarly become two constellations out of date; then three; and so on over the 25,800-year cycle of precession.

Dates when the Sun enters the 30-degree-wide signs, and the astronomical constellations.

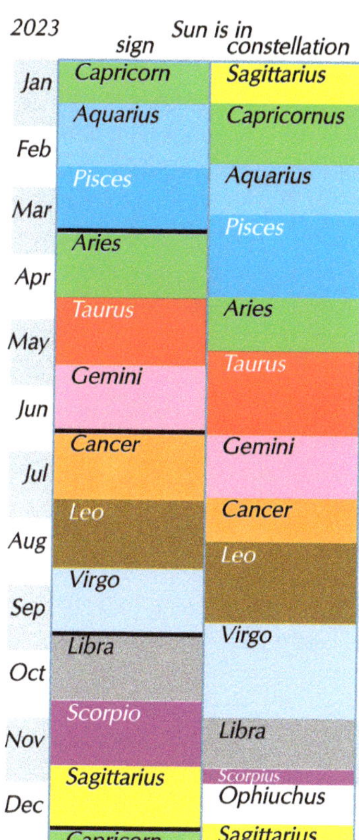

Equation of time

This is the difference between apparent solar time (the time when the Sun arrives at, for instance, the noon meridian) and mean time: the time it would arrive there if it went around the sky at constant speed—which it does not, because of the ellipticity and inclination of Earth's orbit. For more about this, see www.universalworkshop.com/the-equation-of-time/

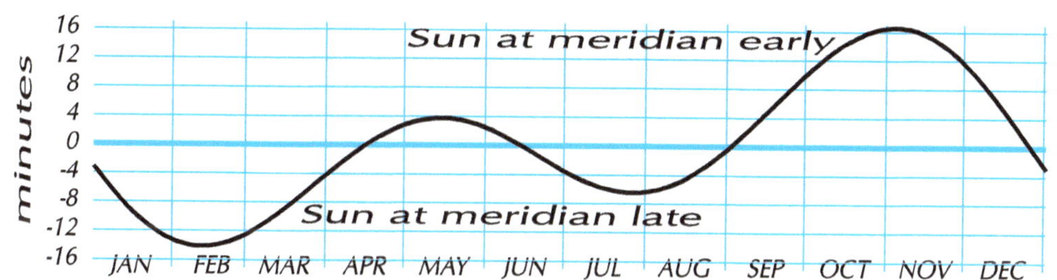

Clock shifting

Governments require us to change our clocks by one hour, from standard time to "daylight saving time," between a date in spring and a date in autumn. The mnemonic is: "Spring forward, Fall back." In the USA, the dates are the 2nd Sunday in March and the first Sunday in November. In Europe: last Sunday in March, last Sunday in October. About 300 countries and territories have their own rules. This year:

USA: 2023 March 12, November 5   Europe: 2023 March 26, October 29

For the history of this artificial time, its relation to natural time and to latitudes, and my opinion of it, see www.universalworkshop.com/clock-shifting-times/

The Sun's tilt

Its north pole is most inclined (7.25°) away from Earth on March 6, and most toward us on Sep. 8.

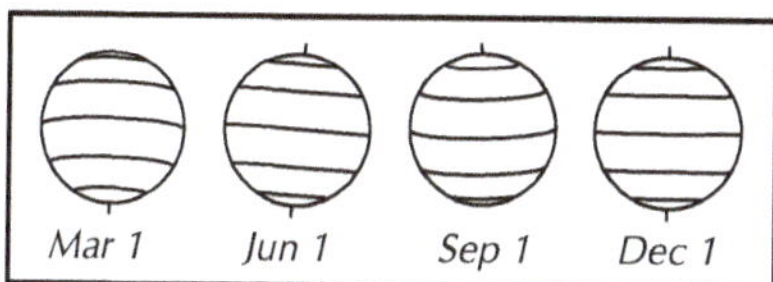

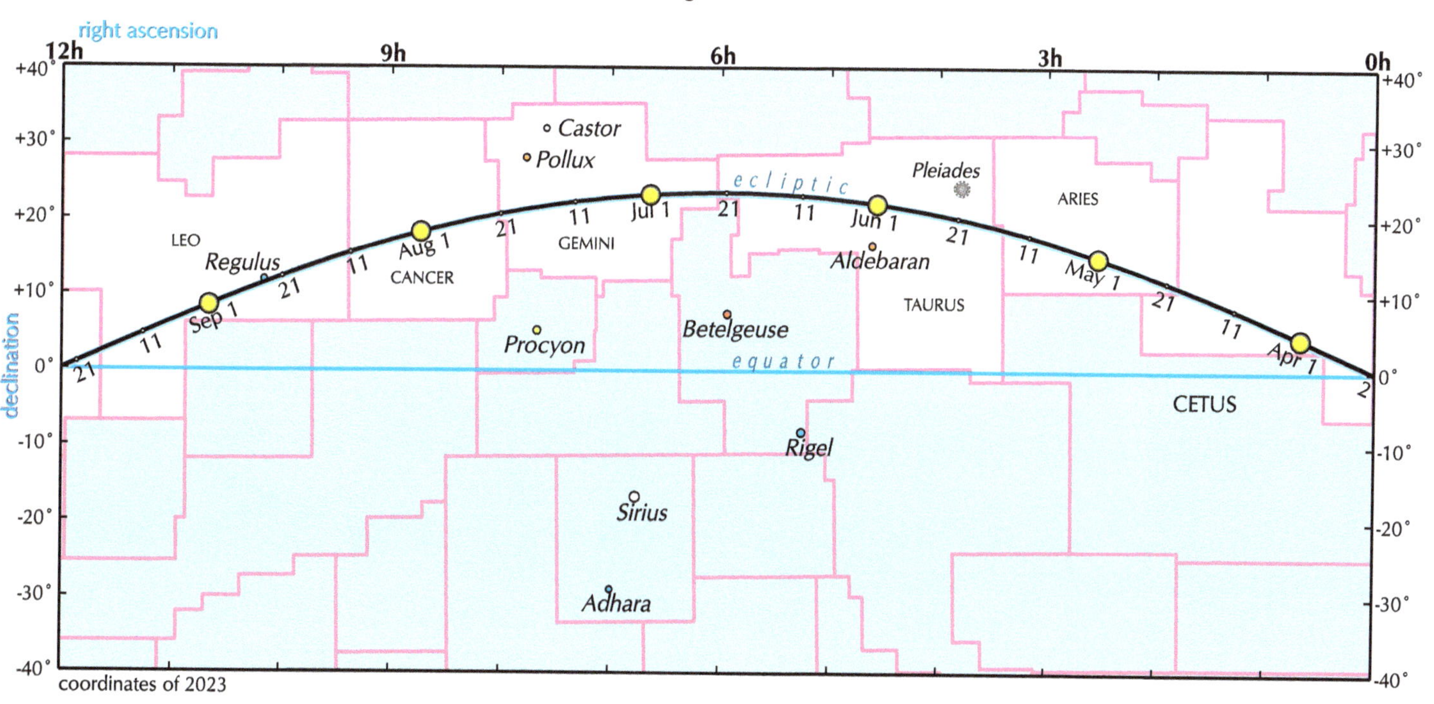

MOON

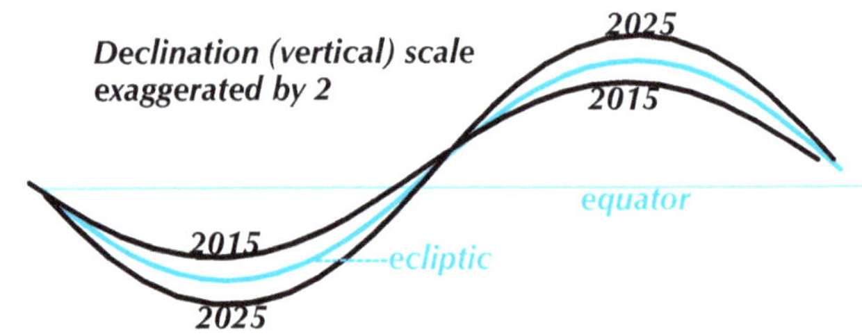

The Moon moves across the starry background each hour a little more than its own width of half a degree. Each day, it moves on average 13.2° (not much less than an hour of right ascension, which is 15°). Each month, it moves on average 401°, or 1$^1/_9$ times around the sky—thus, from its January 1 position around and past this position to the February 1 position.

So in the year it travels 13.38 times around the sky (passing most stars 13 and some 14 times). But from our moving viewpoint as we orbit the Sun, the Moon seems to travel around us only 12.3 times. That is, it passes 12 or 13 times through each of its phases in relation to the Sun, such as new Moon or full Moon.

It does not circle around Earth's equator (as many small satellites circle around larger planets) but behaves more like a companion planet, traveling roughly in the same plane as the Earth. Its orbit is inclined about 5° to the Earth's (varying between 4.995° and 5.295° with a period of about half a year). So it appears to follow the ecliptic, departing up to about 5° north or south of it.

But this orbital plane continually twists (precesses), so that the two nodes, where the orbit crosses the ecliptic, migrate back (westward) at about 19° a year (migrating all around in 18.61 years). This year the ascending node moves from Aries back into Pisces; and, in the opposite region, the descending node regresses from Libra into Virgo.

This determines how far north and south the Moon ranges in the sky.

There are "flat" years, such as 2015 and 2034, when the ascending node is in Virgo where the ecliptic descends through the celestial equator, so that the Moon's path curves to less than 19° north and south (roughly the 23.5° inclination of the ecliptic *minus* the 5° inclination of the Moon).

About 9 years later comes a "hilly" year such as 2006 or 2025, when the ascending node is in Pisces where the ecliptic also ascends, so that the Moon reaches its farthest possible north and south, 28.72° (the obliquity of the ecliptic *plus* that of the Moon). (It attained those extremes near the equinoxes: e.g. on 2006 March 22 (south) and Sep. 15 (north).)

Halfway between these are "ecliptic-like" years, when the ascending node is near the top or bottom of the ecliptic, so that the Moon's path is like a copy of the ecliptic: rising and falling like it about 23° north and south, but displaced west (2011) or east (2019).

2023, like 2022, is hilly, though not yet at the extreme of hilliness to be reached in 2025. Because the ascending node is in coming into Pisces, the Moon reaches northern and southern extreme declinations of 28.3° on Oct. 5 and Oct. 20.

Chart of the Moon's path. Each year it travels 13.4 times around the sky, from west to east (right to left), through the 12 constellations of the zodiac, also Ophiuchus (it can touch Cetus, Orion, Auriga, Hydra, Sextans, Crater, Corvus, Scutum, Pegasus). The two nodes—points where the path crosses the ecliptic—shift gradually westward. We show only the paths for January (thick line) and December (thin). The Moon itself is shown at the instants when it is new (black) and full (white), at 5 times its true size. In each synodic month or lunation (29.5 days) the Moon goes from a new-Moon position all around the sky and on to the next new-Moon position; in each calendar month of 30 or 31 days its journey overlaps by a bit more.

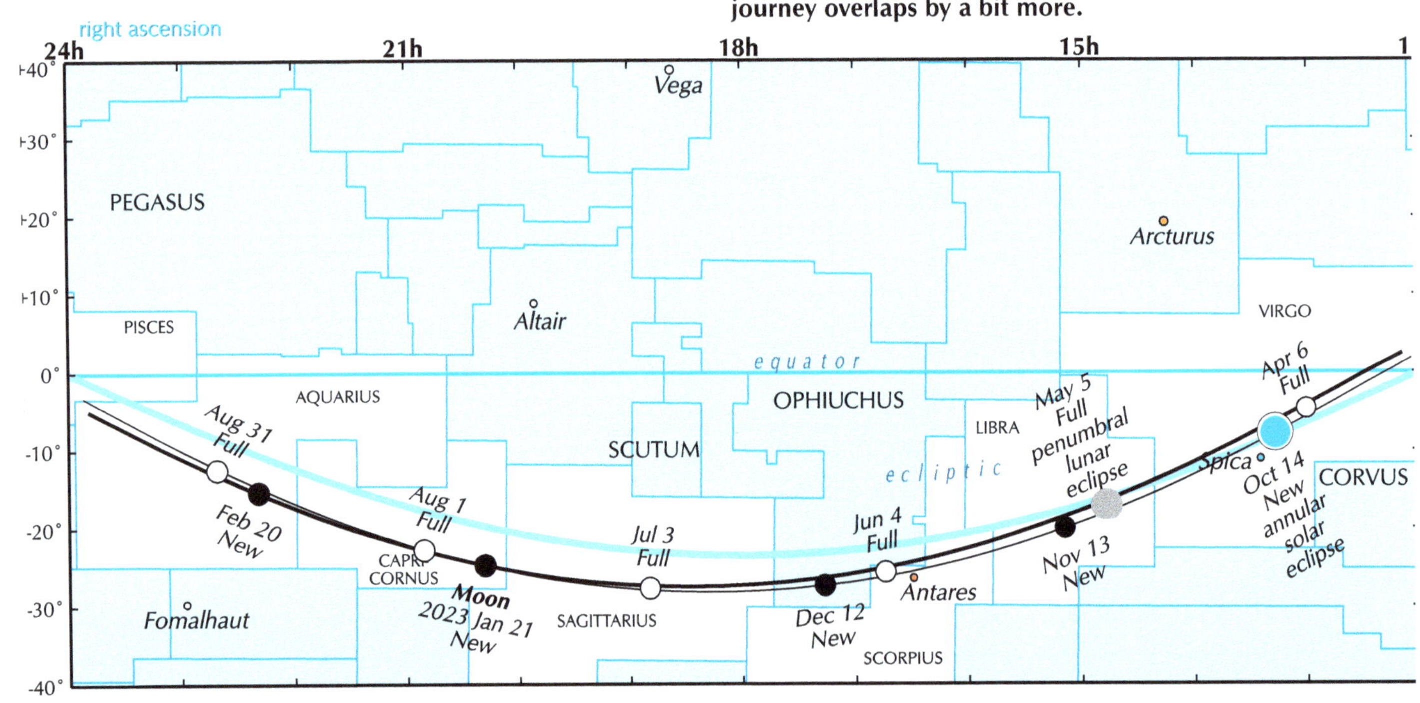

The traveling Moon cuts out a swath, which is widest near the nodes where the orbit is being dragged along, narrow halfway between the nodes. But in 2022 the Moon's path misses all of them.

The swath is really wider than in the chart, for two reasons: the width of the Moon itself, about half a degree; and parallax. That is, as seen from places at the north end of the Earth, the Moon appears nearly a degree farther south (or, from the south of the Earth, nearly a degree farther north). Stars within this swath can get occulted (hidden) by the Moon, as seen from at least some part of Earth. Four of the stars in the occultable band—Aldebaran, Regulus, Spica, Antares—are of first magnitude. In 2023 the Moon path passes north of them all until, from August to November, it hits Antares.

The orbit is not only tilted but eccentric (elliptical), as the plan shows. Whereas the nodes move backward, the near and far positions, perigee and apogee, migrate generally forward—about twice as fast, but irregularly: forward for 5-7 perigees, then backward for 1-3. They progress all the way around in 8.85 years, thus on average 40° a year. This year perigee oscillates between 302° and 356°: the Moon is nearest, largest, and moving fastest when traveling in the Capricornus region.

Thus the egg-shape of the orbit rotates jerkily counterclockwise (as seen from the north). Looked at perpendicularly as well as edge on, the Moon during the year sweeps out a swath. It is narrow near apogee, because the Moon's apogee distance scarcely changes; wide near perigee, because the perigee distance varies markedly; it is also wide in the quadrants between, where the precessing orbit slopes in and out. Approaching perigee, the Moon is on the inner edge of the swath in January, the outer edge in December; after perigee, vice versa.

Putting the two swaths together, we realize that the Moon carves out a torus ("doughnut") of space whose width varies in two dimensions.

As the two pairs of positions—the two nodes and the two apsides (perigee and apogee)—revolve in opposite directions, they meet and pass quite frequently. Through most of last year they were fairly close together; this year they are apart, about at right angles to each other.

Eclipses have to take place near the nodes. This year, the April and October solar eclipses find the Moon at middling and more-than-middling distance, so its cone of total shadow scarcely reaches us.

All this describes the Moon's orbit as it appears from the Earth: as if the Earth were fixed. In this frame of reference, the Moon's motion looks like a near-circle many times repeated. In the wider frame in which Earth and Moon together are journeying around the Sun, the same motion looks like a single near-circle, so much vaster that the deflections in the Moon's path are ironed out and the whole path is never even convex toward the Sun! (See the MOON'S ORBIT section in the *Astronomical Companion*.)

There is, as the distance-graph shows, a remarkable relation between the extremes of distance and the syzygies (new and full Moons).

The average period from new to new is the synodic month of 29.53 days. The average period of the distance-wave (perigee to perigee) is the anomalistic month—two days shorter, at 27.55. The result is like a "beat" between two trains of sound-waves. There is a time of year when new Moon is coinciding with perigee, and full Moon with apogee. Last year it came in early January and again late December. This year it has shifted to late January. Then, 6½ months later, comes a time when the reverse happens: *full* Moon occurs near perigee (last year in June, this year in August). Each year, each of these times shifts later by a bit more than a month.

At the syzygies, the three bodies are in a straight line, Sun-Moon-Earth or Sun-Earth-Moon. The Moon's instantaneous orbit is squeezed toward this line: it becomes a more eccentric ellipse, with nearer-in perigee, farther-out apogee. The effect is less when the Moon is farther out, Earth's pull

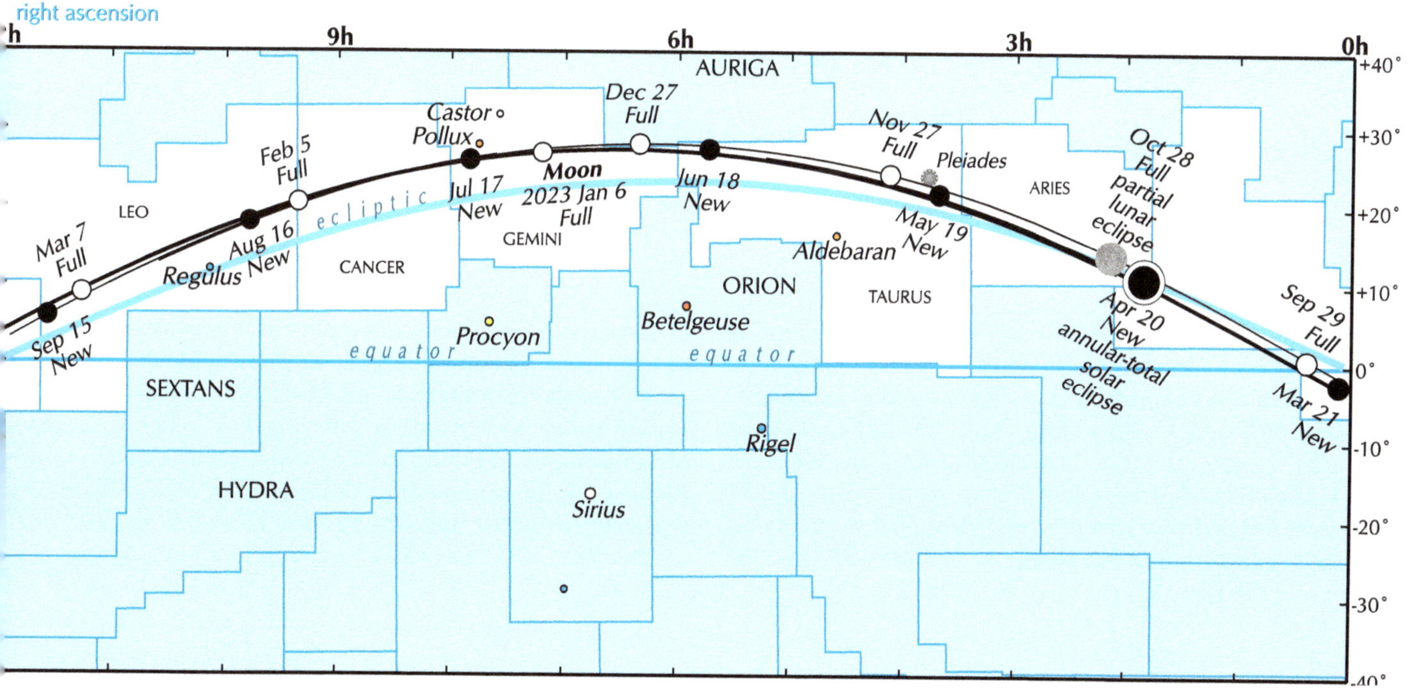

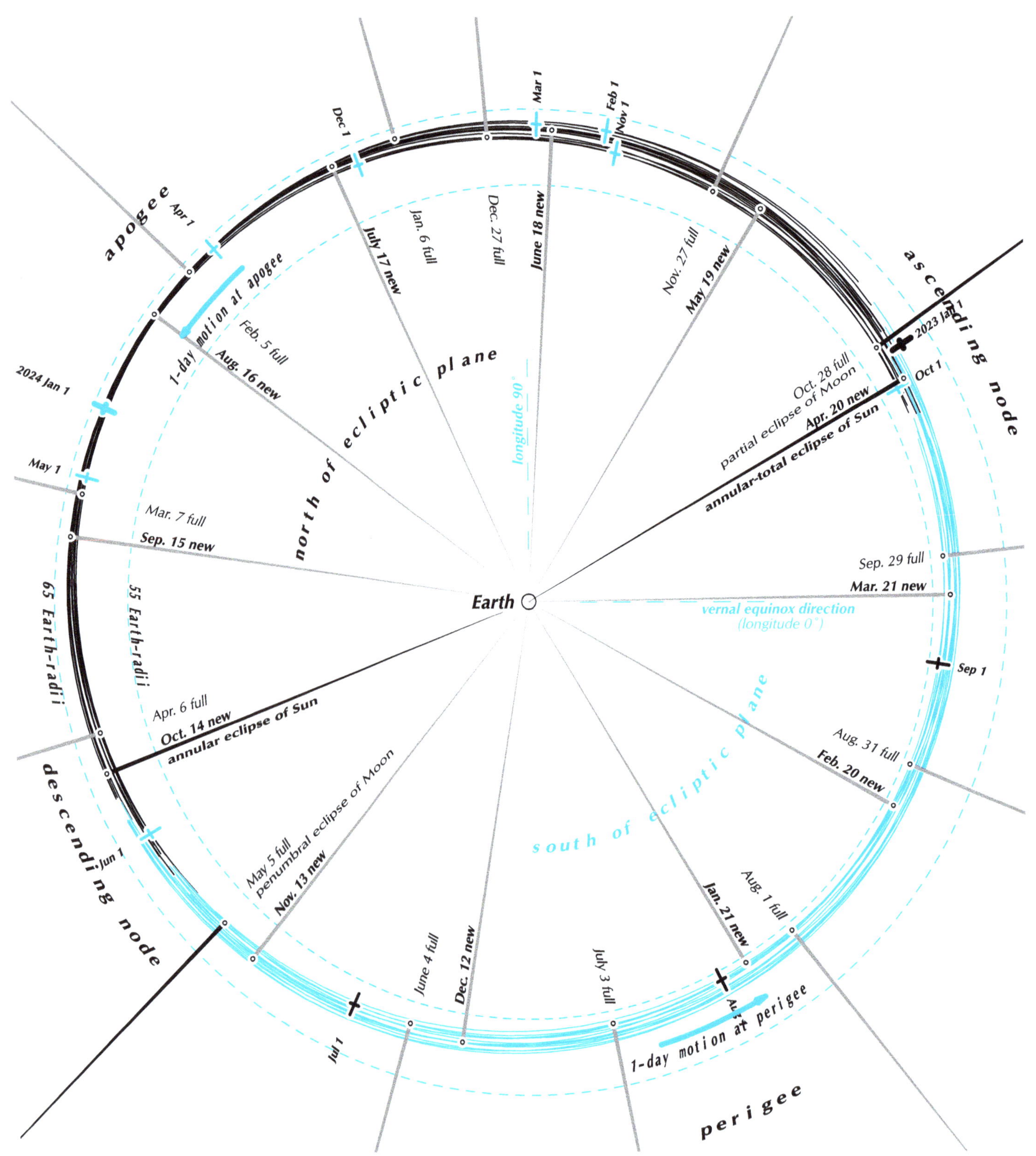

Plan of the Moon's movements projected on the plane of its orbit, as seen from the north; with the Earth held still (so that the Sun is imagined to be going around also, once in the year, 389 times farther away than the Moon's average distance). Where the Moon is north of the ecliptic plane, its path is shown black; where south, blue. At 0h UT on the 1st of each month, a bar across the orbit shows the Moon's position (and a short bar along shows its motion over the hour before and the hour after).

At instants of new and full Moon, the Moon is drawn, true to scale, plus its umbra, or cone of total shadow. This points radially away from Earth at full Moon, and toward its center at new Moon. Notice the pattern made by the tips of the new-Moon umbrae, reaching to or through Earth only when the Moon is on the perigee (near-in) side of its orbit. This shows that eclipses of the Sun can become total only on that side; on the other side, they can be no more than annular.

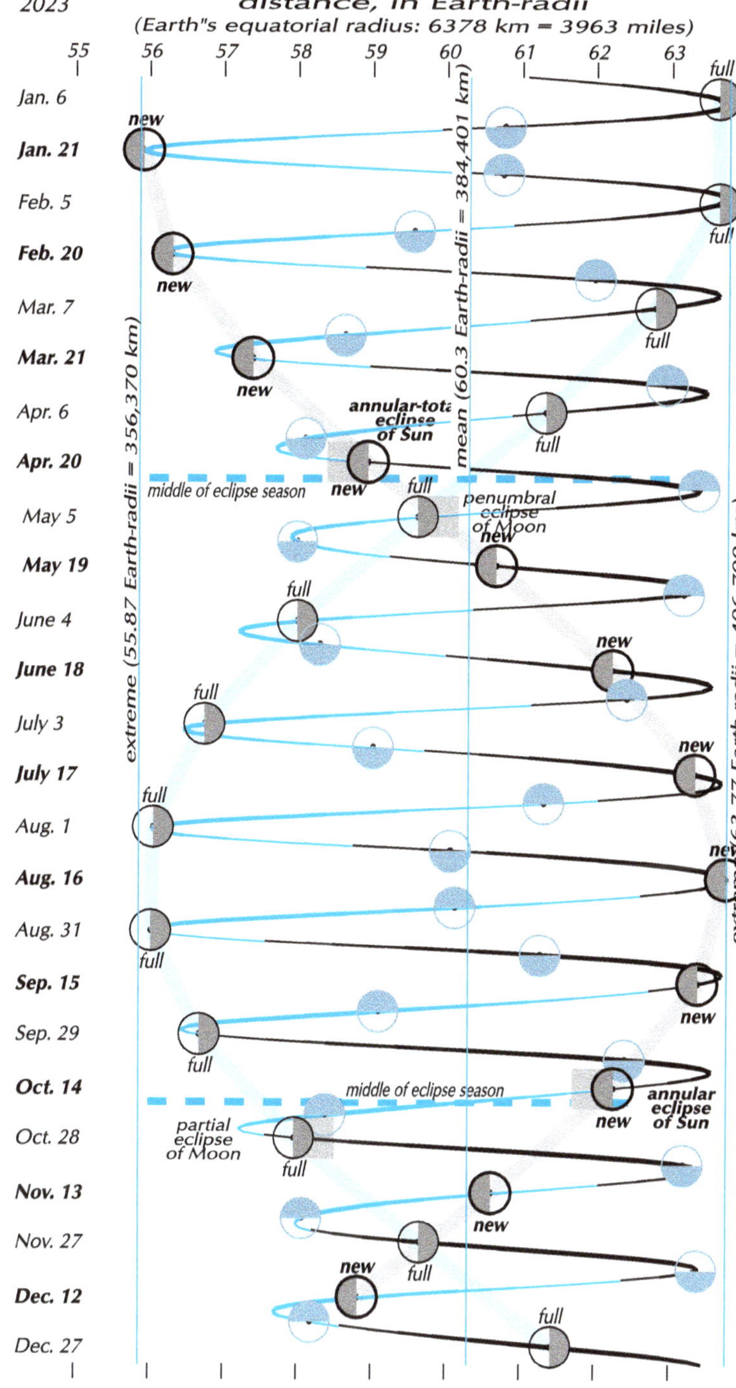

Graph of the Moon's varying distance. The Moon is drawn, to scale, at the moments of its cardinal phases—new (dark side toward us), first quarter (sunlit side to west), full (sunlit side toward us), last quarter (sunlit side to east). The curve from Moon to Moon is black when the Moon is north of the ecliptic, blue when south; and thicker when farther north or south.

For the mean distance, Allen's *Astrophysical Quantities* (1999) gives 384,401±1 km (60.27 Earth-radii ER). Jean Meeus in *Mathematical Astronomy Morsels* (1997) gives 10 values, ranging from 381,546 to 385,001 km (59.82 to 60.36 ER), depending on what we mean by "mean distance"!

on it being less; the apogee distance increases, but not by much. At perigee, however, the Moon, if it happens to be directly toward or away from the Sun, comes nearer in toward us than its average distance by about twice the diameter of the Earth.

This enhances any total solar eclipse that occurs near to perigee. This year, however, none do.

Tides, too, are strongly affected by the relation between perigee and syzygy. High tide comes twice a day, once under the Moon (actually, behind it, because friction with the seabed and coasts delays the water by irregular amounts) and once when the Moon is on the Earth's opposite side.

The Sun, too, has a tidal pull on the Earth, rather less than half that of the Moon. So at or just after the new and full Moons of each month, when Sun, Moon and Earth are in line, the tide is amplified into a "spring tide" (from the "jump" sense of *spring*).

There is a tidal swelling on *both* sides of the Earth. So, whether spring tide is at a new or a full Moon, there is one flood tide under the noonday Sun, the other at midnight. But one kind of midnight spring tide is moonlit, the other swells in darkness.

When the Moon is near—at perigee—its tidal pull is greater. Moreover, perigees, as we have seen, are much nearer perigees when they coincide with new or full Moon. Thus when these times of coincidence arrive, we expect the tides of greatest amplitude: *highest high tides* and lowest low tides of the year. These perigean spring tides may cause coastal flooding, especially if coinciding also with storms at sea; possibly they trigger earthquakes and volcanoes.

Dates of closest coincidence, with difference in hours, and Moon distance.

perigee	syzygy		hours	km
Jan 21 20:55	Jan 21 20:54	new	-0.0	356,596
Feb 19 8:57	Feb 20 7: 8	new	22.2	358,274
Aug 2 5:58	Aug 1 18:32	full	-11.4	357,338
Aug 30 15:60	Aug 31 1:36	full	9.6	357,203

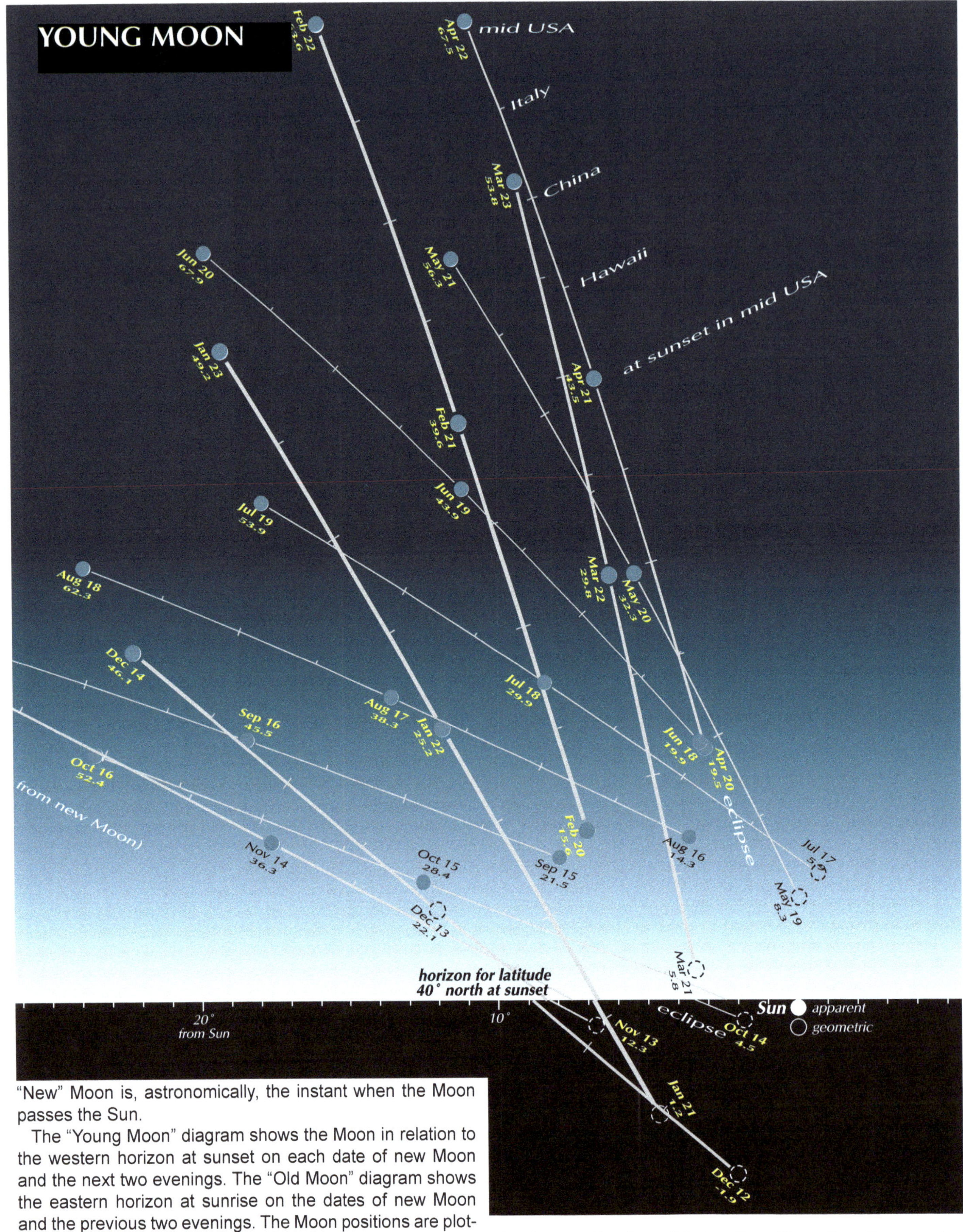

"New" Moon is, astronomically, the instant when the Moon passes the Sun.

The "Young Moon" diagram shows the Moon in relation to the western horizon at sunset on each date of new Moon and the next two evenings. The "Old Moon" diagram shows the eastern horizon at sunrise on the dates of new Moon and the previous two evenings. The Moon positions are plotted by altitude from the horizon and azimuth (distance parallel to the horizon) from the Sun.

The view is from an eastern American location, latitude 40° north, longitude 75° west. The Moon's position is corrected for the parallax from this location. (Seen from the equator, the Moon would appear about 0.6° farther north.) And the position is corrected for atmospheric refraction, which, near the horizon, raises it as much as half a degree.

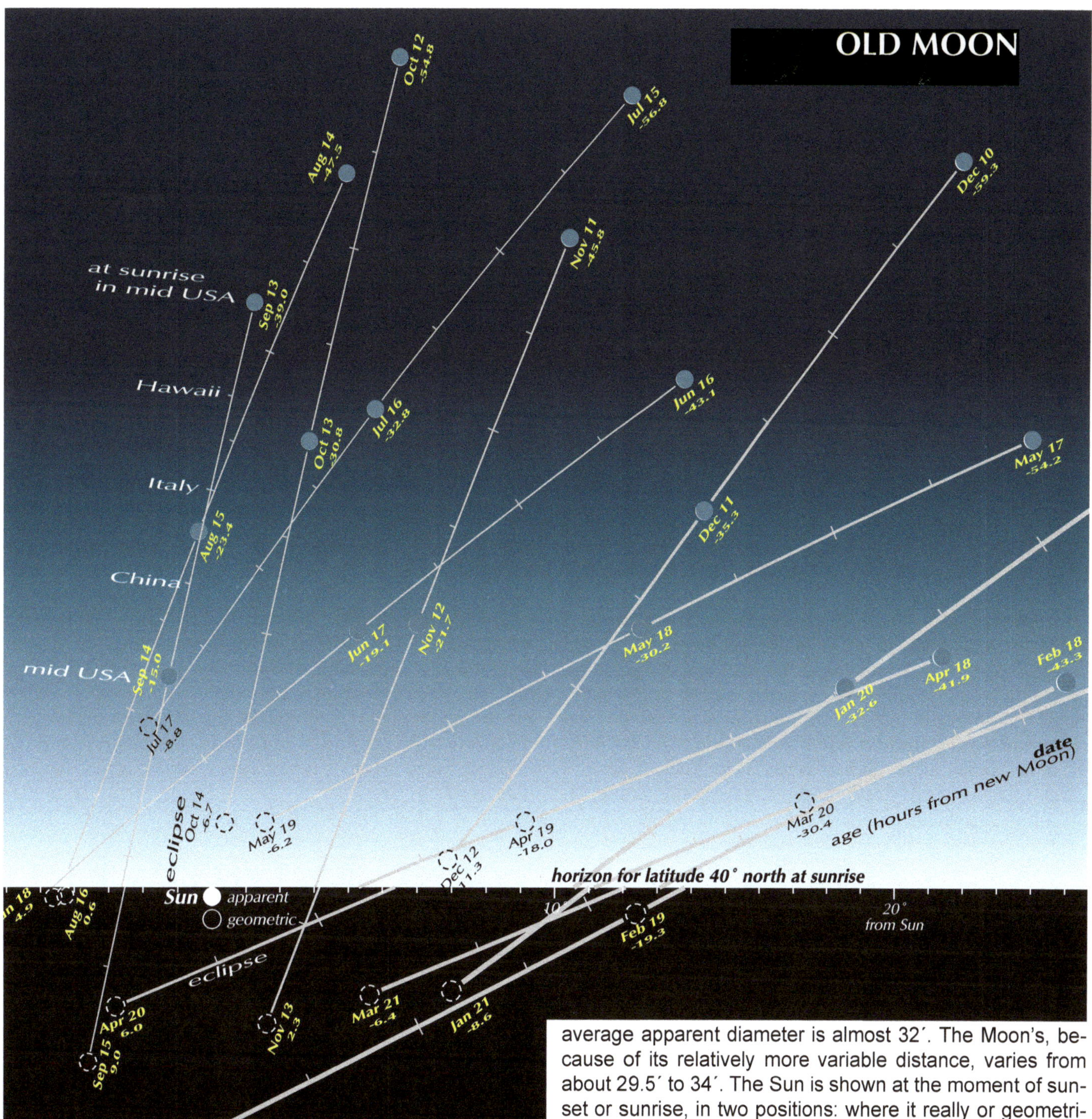

An hour later, the sunset horizon will be considerably higher (or the sunrise horizon lower).

The Moon's illuminated crescent is drawn with the correct thickness. The un-sunlit side is shown in gray, as if visible by earthshine—which it often is in a clear sky when neither too far down in the glare, nor too far in the other direction where the Earth does not reflect enough light to it. Only a dashed circle is drawn when the Moon is less than 7° from the Sun or 3° above the horizon, assuming these minimum limits for detectability.

The lines of the Moon's track onward from day to day are drawn thicker when the Moon is nearer. Sun and Moon are both drawn to the general scale of 6 mm to 1°. The Sun's average apparent diameter is almost 32´. The Moon's, because of its relatively more variable distance, varies from about 29.5´ to 34´. The Sun is shown at the moment of sunset or sunrise, in two positions: where it really or geometrically is and where it appears to be. Because of atmospheric refraction, which is strongest at the horizon, the Sun's rays come around the surface of the Earth curvingly to us, making the Sun appear about half a degree higher. And day prevails if the merest speck of the Sun shows at the horizon; so sunset and sunrise are defined as the instants when the *top* of the Sun, not its center, appears to be on the horizon.

For more northerly latitudes on Earth, imagine the horizon tilted down on the left. For the north pole, it would be vertical. At the equator, Sun and Moon rise and set about vertically. In the southern hemisphere, their tracks slope the opposite way.

Trying to see the Moon as near as possible to its new moment is a fine sport, and is useful. Calendars such as the Jewish and Muslim have lunar months, beginning with the evening when the Moon can first be seen.

ECLIPSES

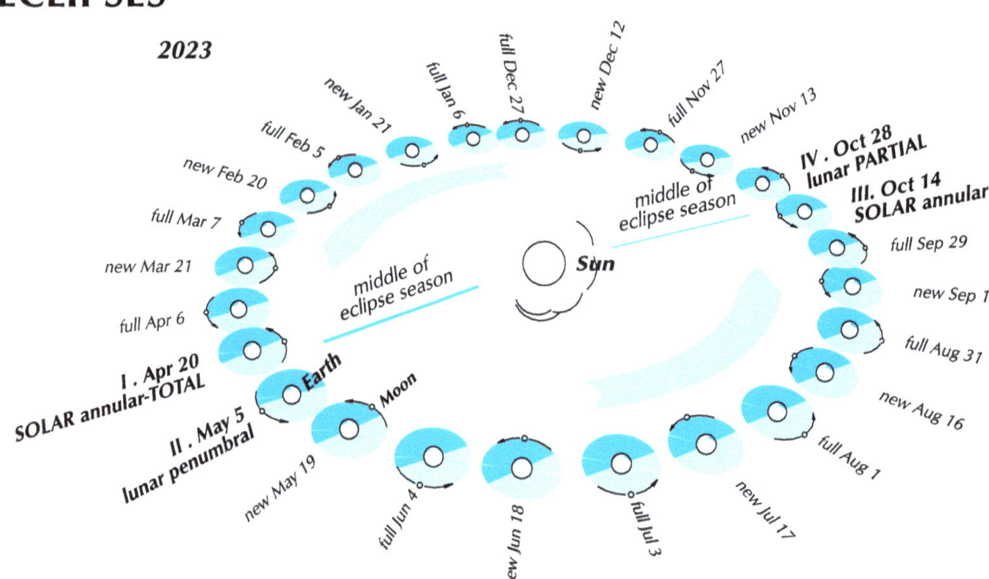

Earth at each date of "syzygy": Moon sunward from it (new Moon) or outward (full). The plane of the Moon's orbit is shown, darker blue for the half north of the ecliptic. This plane gradually rotates backward. There is an eclipse if the Moon is new or full when near ascending or descending node through the ecliptic plane. Small arrows show the Moon's course over 7 days.

The Sun's size is exaggerated by 15, Earth's and Moon's by 600; the Earth-Moon distance by 40; the inclination of the Moon's orbit is exaggerated from 5° to 10°.

In 2023 there are two eclipse seasons, each with a solar and a lunar eclipse, making a total of four. This is the most common number.

An eclipse season is a span of days during which new and full Moon occur close to the ascending or descending node of the Moon's orbit. Sometimes there can be extra eclipses, for two reasons: part of a third eclipse season overlaps the beginning or end of the year; or an eclipse is near enough to the center of a season that it is flanked by two marginal eclipses. So there can be 5 eclipses, as in 2016, 2018, and 2019; or 6, as in 2009 and 2020; 7, as in 1982 and 2038. (See the long "bead curtain" chart in our *Under-Standing of Eclipses*.)

2023 is a slight year for eclipses. The April solar eclipse yields only brief totality, on remote coasts in Australia and Indonesia; the October one makes a broad track across the western U.S. but is completely annular (ring-shaped, not total). The May lunar eclipse, seen from the continents around the Indian Ocean, is only partial; the October one is theoretically visible from most of Asia and Africa and much of America but is of the penumbral kind that is scarcely detectable at all.

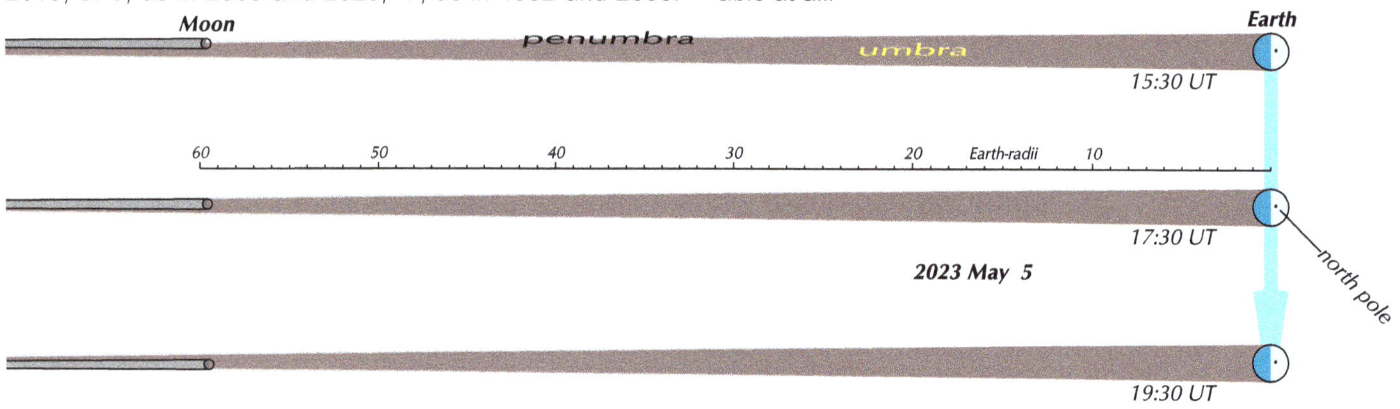

Plan views of the lunar eclipses. As the full Moon overtakes Earth on the outside, it encounters Earth's penumbra and then umbra.

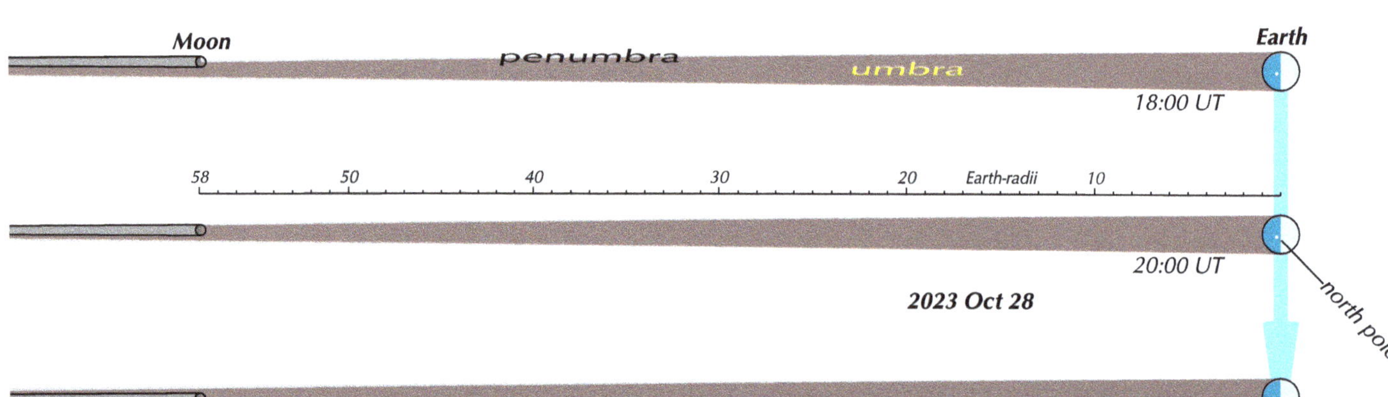

Earth

penumbra umbra

Moon

2:30 UT

Plan views of the eclipse, from north of the ecliptic plane. The scale (25 mm for 1 Earth-radius) is true for Earth and Moon and the distance between them. (The Sun is about 400 times farther than the Moon.) But the distance between successive positions should be greater. In the similar plan views for the other eclipses, the vertical scale is compressed even more.

Each hour the Earth travels forward 8½ times its own width in its orbit around the Sun. The Moon travels forward with it, but, around new Moon, slightly slower, because in an Earth-centered frame it is moving backward. Its penumbra is about half as wide as the Earth and goes on widening.

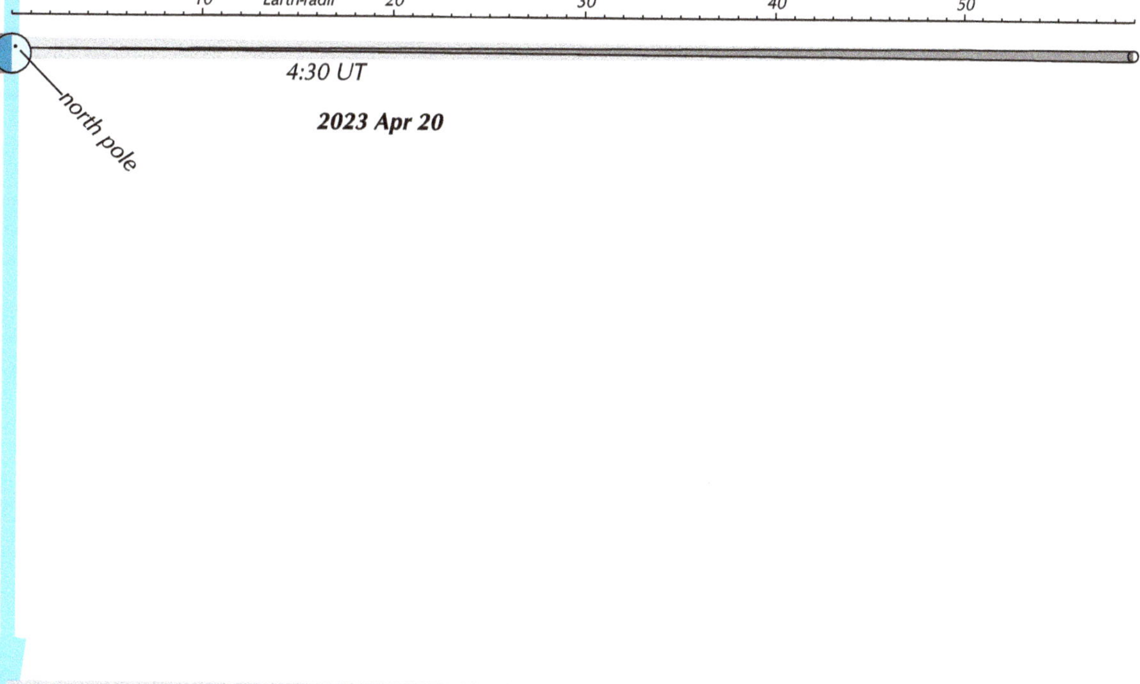

6:30 UT

The 2023 Apr. 20 eclipses is annular-total because the Moon's umbra reaches Earth's surface but not Earth's center. The Oct. 14 eclipse is annular because the umbra does not reach Earth's surface.

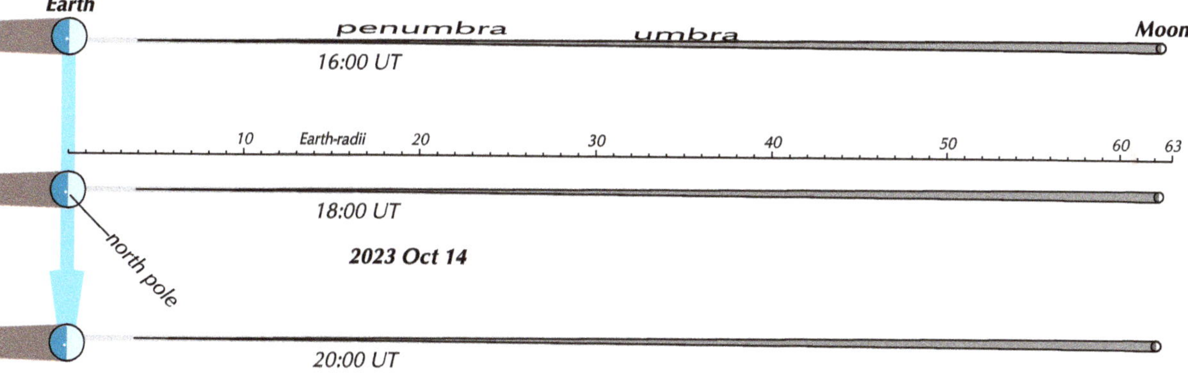

The April-May eclipse season

I—Annular-total eclipse of the Sun, April 20

The Moon comes around to its new position, toward the Sun, about 8 hours before ascending northward through the ecliptic plane, so its shadow sweeps across the southern and equatorial parts of the Earth.

(*Gamma*, the quantity that ranks eclipses, is -0.393, meaning that the axis of the shadow passes south of Earth's center by that fraction of Earth's radius.)

But the Moon is at a distance of 58.9 Earth-radii, only slightly short of its average distance. So the cone of the umbra, or total shadow, reaches Earth's surface but not quite as far as Earth's central plane. The result is that the tip of the umbra passes above the surface during the first and last hour (those looking up from ships would see the Moon surrounded by a hair-thin annulus or ring of Sun). For the rest of the umbra's course, it reaches the surface; the Moon can be seen fitting over the Sun, very tightly especially at beginning and end, along a track that is narrow, especially at beginning and end.

At 1:34 Universal Time, the brow of the penumbra strikes the Indian Ocean, near the French-owned Île Saint-Paul and Île Amsterdam, far south of India and far west of Australia. Then the penumbra spreads onto the ocean till, just before 2:00 UT, it reaches the west coast of Australia, where watchers (using careful eye protection) start to see a nick in the Sun's eastern rim. This partial eclipse spreads over Australia and Indonesia. Along the region's edges, as in Vietnam and Taiwan on the north, Tasmania on the south, very little of the Sun is covered.

At 2:37, the tip of the umbra touches down, west of where the penumbra did because the Earth has meanwhile rotated eastward. The resulting annular eclipse touches land only along a bit of the coast of Western Australia (at 3:30, which is 11:30 by local standard time), and Timor and the Indonesian-ruled part of Papua. The magnitude of the eclipse is 1.013, which means that, at the most, the Moon appears only that much larger than the Sun it is covering.

Since ancient times it has been known that any eclipse is followed by a similar one after an interval called the saros: 18.03 years, or, more exactly, 18 years plus 10, 11, or 12 days (depending on leap days) and a third of a day. Successive eclipses of a series evolve slowly in geometry, but are displaced a third of the way around the Earth.

This is eclipse 52 of the 80 in solar saros series 129, which happen at the Moon's ascending node and move successively southward across Earth. The series began on 1103 Oct. 3 and ends on 2546 Mar. 3, with short partial eclipses in the Arctic and Antarctic. There were 20 partial eclipses; from 1464 they became annular; the present is the last of three annular-total; then there will be 9 total eclipses from 2041 to 2185; and the last 19 will again be partial.

Timetables of the eclipse seasons of 2023

All times are in Universal Time (UT)

Apr 16 02:31--Moon at perigee

I: annular-total eclipse of the Sun, April 20

Apr 20 01:34--first contact of Moon's penumbra with Earth, at local sunrise
Apr 20 02:37--central eclipse begins: first contact of axis of Moon's shadow with Earth, at local sunrise
Apr 20 03:56--mid eclipse; Moon's center north of Sun's as measured perpendicularly to Earth's equator
Apr 20 04:15--new Moon (conjunction of Moon with Sun in ecliptic longitude): Moon's center is north of Sun's as measured perpendicularly to the ecliptic
Apr 20 05:57--central eclipse ends: last contact of axis of Moon's shadow with Earth, at local sunset
Apr 20 06:59--partial eclipse ends: last contact of penumbra with Earth, at local sunset
Apr 20 11:32--Moon's center reaches ascending node through the ecliptic

Apr 24 15--middle of eclipse season: Sun at same longitude as Moon's ascending node
Apr 28 6:52--Moon at apogee

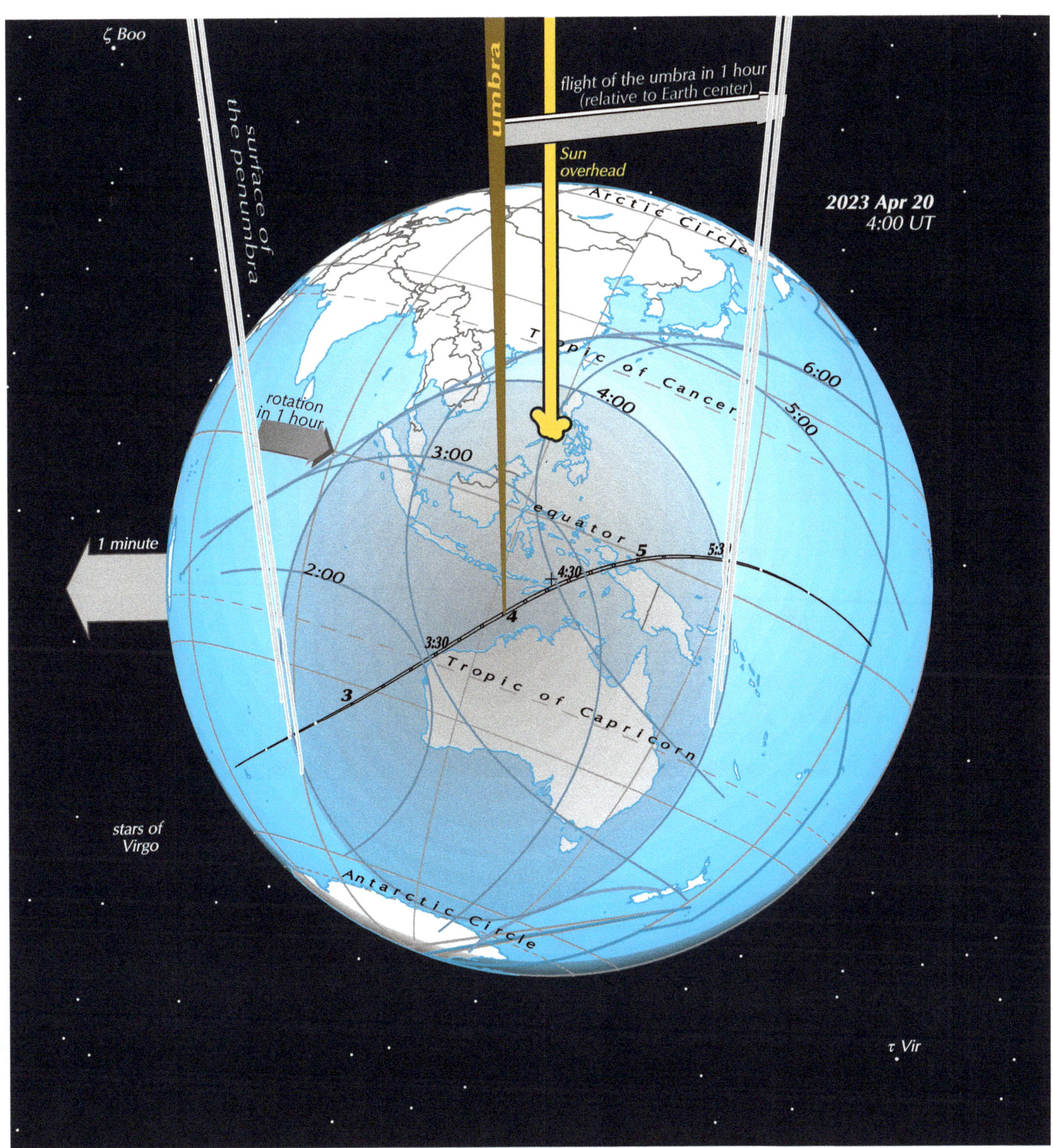

Earth as the Moon's shadow crosses it. The viewpoint is 12 Earth-radii from the Earth's center. So the Moon is about 5 times farther back, along the shadow which it casts; the Sun is about 400 times farther. The cone of the penumbra or partial shadow spreads as it goes away from the Moon. The cone of the umbra or total shadow tapers; on April 20 it ends 0.145 Earth-radii short of the "fundamental plane," the plane through Earth's center perpendicular to the shadow. On Oct. 14 it ends 3.9 Earth-radii short. Dashed lines show its continuation, the antumbra, from within which a ring of Sun would be seen to surround the Moon.

The large gray patch, faint at its outer edge, is the footprint of the penumbra at the time of the picture. It is about twice as wide as the Moon and a little more than half as wide as the Earth. Curves mark its outline at other times. Together these curves define the limit of partial eclipse.

The arrow showing Earth's flight, and the "Sun overhead" arrow, are in the ecliptic plane. The umbra and the "Sun overhead" arrow travel in opposite directions; where they pass at the same meridian of longitude, mid eclipse happens at local noon. Compare "flight of the Earth in 1 minute" and "flight of the umbra in 1 hour"; you see that really, in relation to the Sun, the Moon is moving forward with the Earth, but slightly slower.

II—Penumbral eclipse of the Moon, May 5

The middle moment of the eclipse season (when, if an eclipse happened, it would be exactly central) is on April 24, only 4 days after the solar eclipse; so when the Moon comes around on the outside 11 days later it almost misses Earth's shadow.

It descends across the ecliptic on May 4 at 22 UT, and, starting as much as 17 hours later, travels for 4 hours through the penumbra, the outer shadow. From within this at least part of the Sun is visible; it's like daylight, and the shadow is pale. Only in the central few minutes (just before 11 PM by Indian standard time) are watchers likely to discern a slight grayness on the northern region of the Moon. The penumbral magnitude of the eclipse is 0.989, meaning that at mid eclipse the penumbra reaches across this fraction of the Moon's diameter; only a sliver of Moon is outside it.

This subtle spectacle is centered over the Indian Ocean, and few out of the millions from Iran to New Zealand are likely to notice any change in the full Moon.

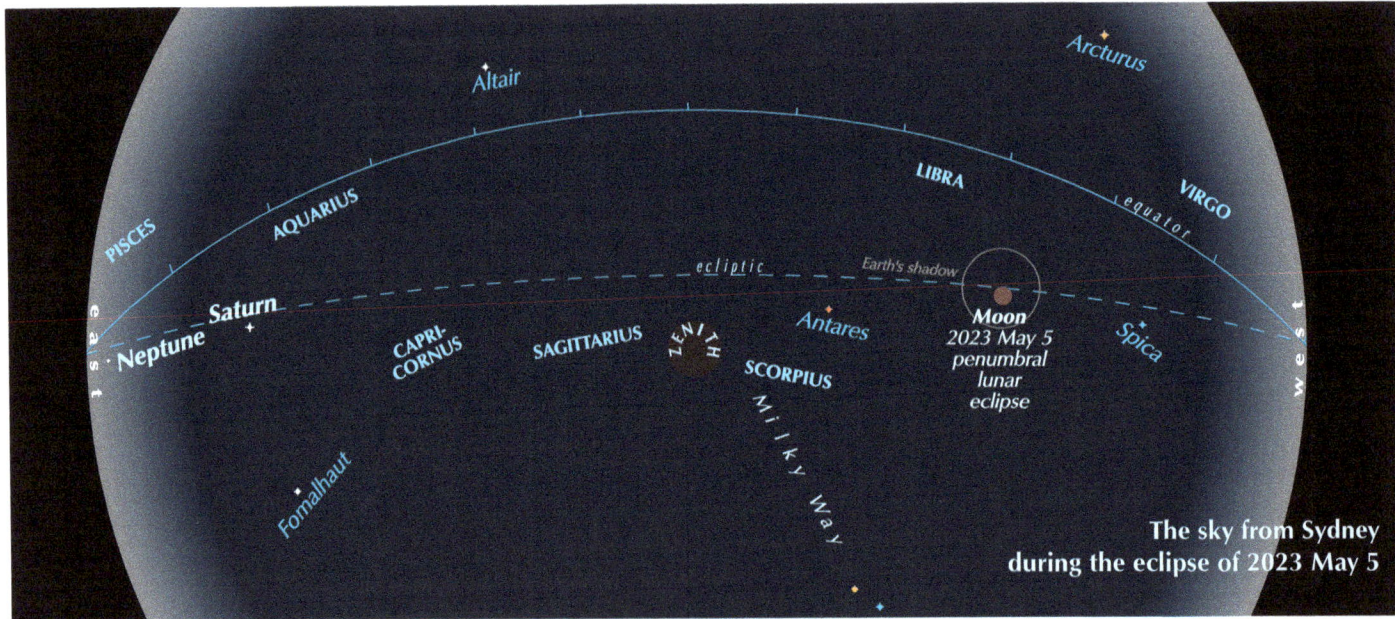

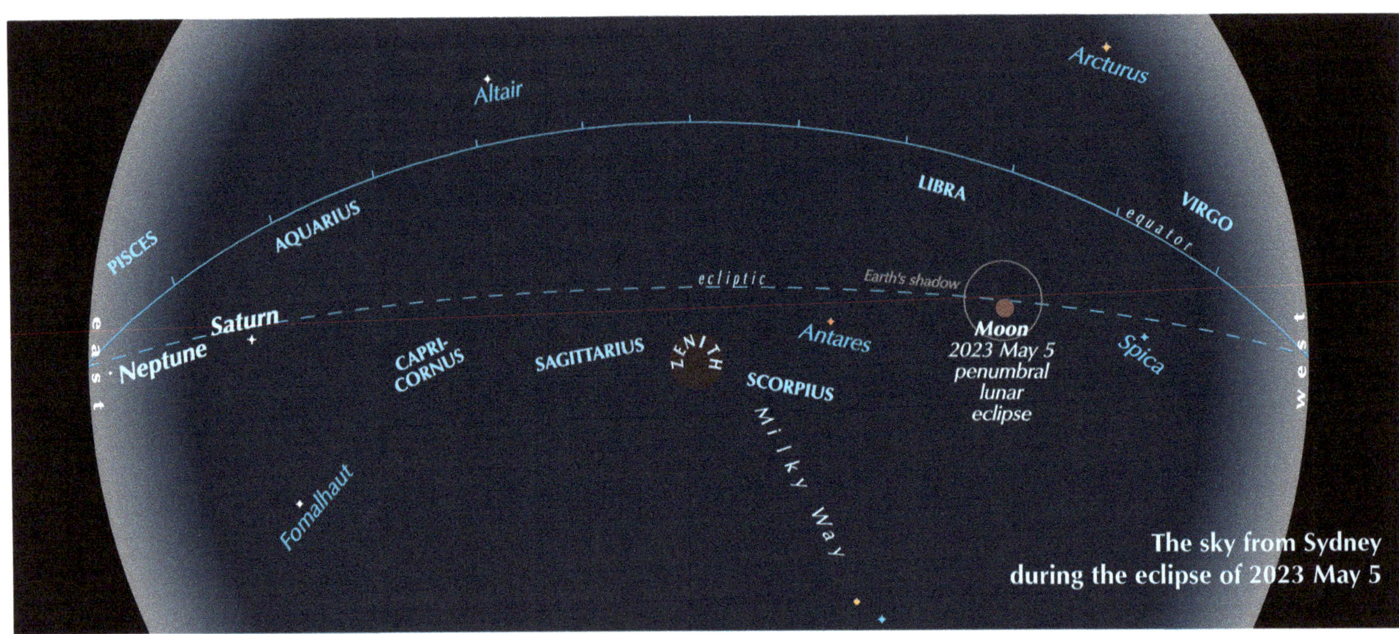

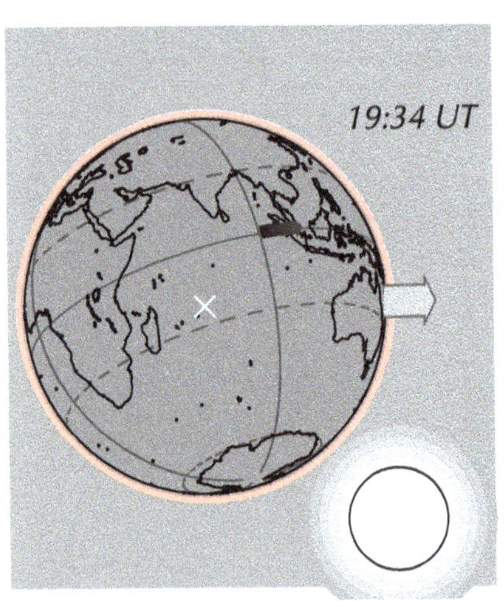

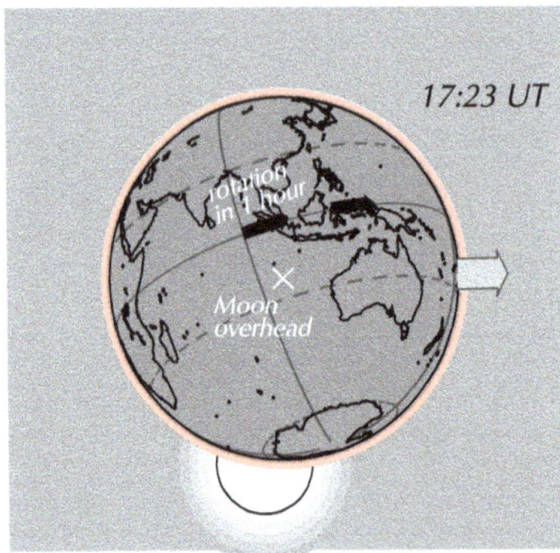

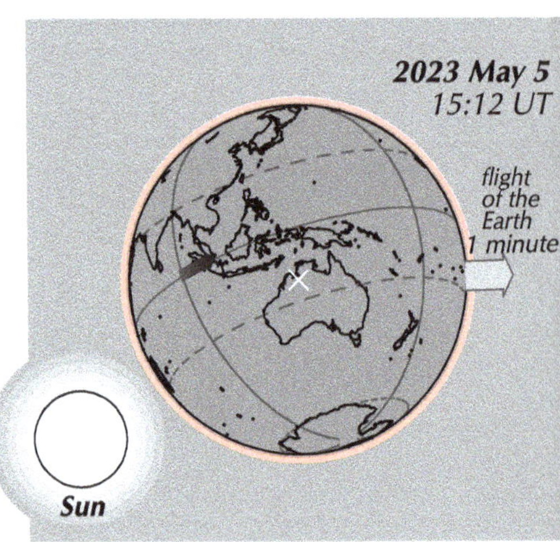

Views from the Moon toward Earth and Sun. The viewpoint is for an observer lying on his back on the midpoint of the Moon, with the Earth at the zenith. The event, which is for us an eclipse of the Moon, is for this man-on-the-Moon an eclipse of the Sun. If he were to move farther north, the Earth would appear farther south. The orientation of the pictures is with the ecliptic (the plane of Earth's motion) horizontal. Thus the northern hemisphere of the Earth tilts backward in (northern) spring. The Moon-inhabitant sees the Earth apparently moving backward (to the left or east): it is really traveling forward, but the Moon, at the full stage of its orbit, is overtaking the Earth on the outside. He also sees the Sun's path slanting to the ecliptic at about 5° because that is the angle at which the Moon itself is cutting across the ecliptic. Around the Earth appears a thin ring of light refracted and reddened by the atmosphere.

The practical purpose of the diagrams is to show which parts of the Earth can see the eclipse: all those on the side facing the Moon. In each diagram, lands on the right are about to move out of sight: for them the Moon is setting and the Sun is about to rise at the same time. Lands on the left have just come into view of the Moon at the end of their day. Places that appear in all three pictures see the whole course of the eclipse.

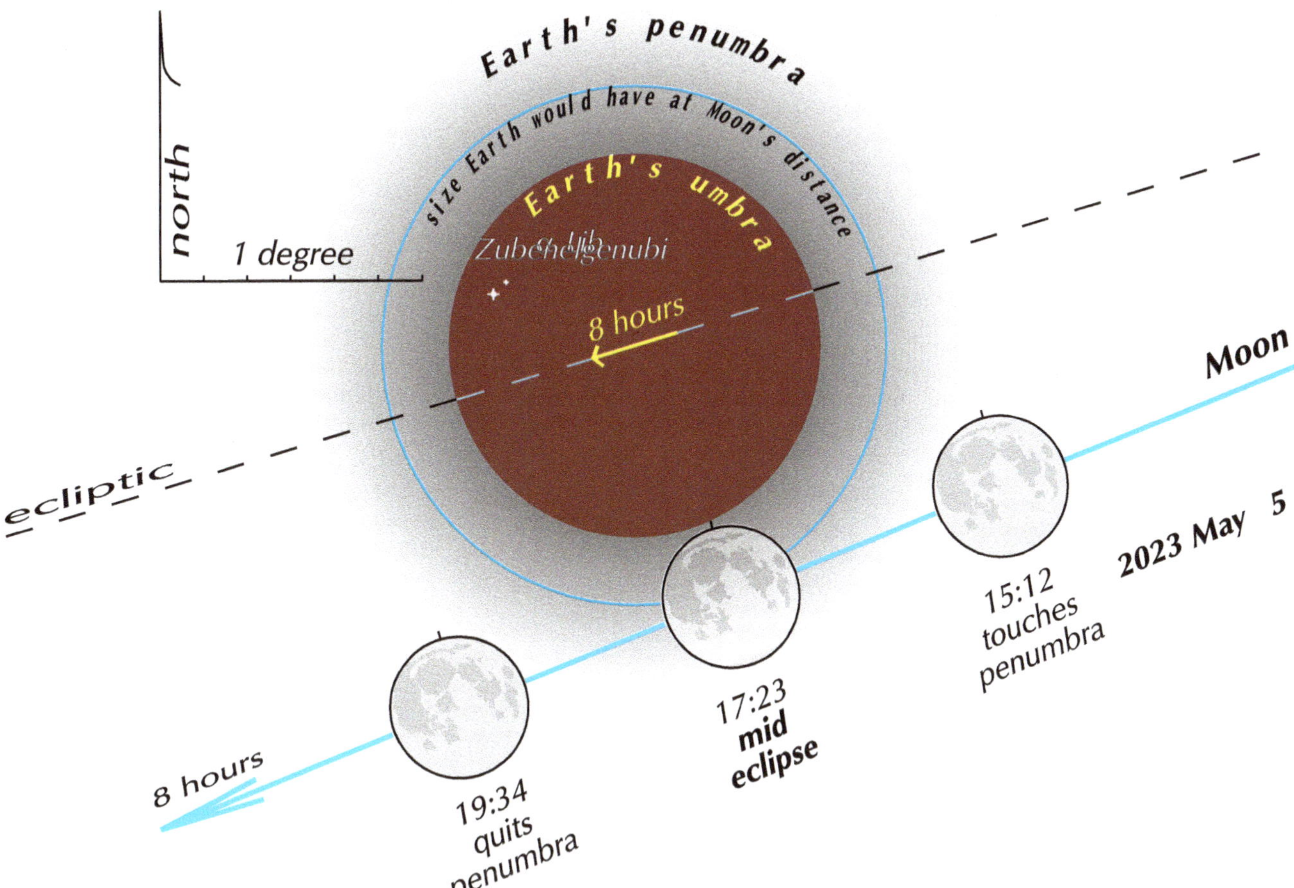

Views toward the Moon at successive stages of its encounter with the Earth's shadow. This is what can be seen from almost everywhere on the night side of the Earth. The umbra and penumbra are represented by cross-sections through them at the distance where the Moon is. They are visible only where they fall on the Moon. The umbra has a fairly abrupt edge; its darkness and color vary with atmospheric conditions around the Earth. The penumbra is imperceptible except in its inner part. The umbra gets narrower as it goes farther away; the penumbra, wider. A circle between them represents the size of the body casting the shadows: the Earth we are standing on. Arrows show the motion of the shadow and the Moon over a span of 8 hours. The Moon moves faster because it takes only a month to go around the sky, while the shadow (like the Sun opposite to it) takes a year. However, since the shadow does move along somewhat during the eclipse, the diagram, representing the relation of the Moon to the circular umbra and penumbra, cannot be exactly true in all respects: the Moons ought to be slightly wider apart. Any star or planet shown in the field is plotted in relation to the Moon and shadow at the middle moment of the eclipse, and as seen from the center of the Earth.

II: penumbral eclipse of the Moon, May 5

May 4 21:58--Moon's center reaches descending node through the ecliptic

May 5 15:12--first contact of Moon with Earth's penumbra

May 5 17:23--middle of eclipse: Moon nearest to center of Earth's shadow. The penumbral magnitude of the eclipse is 0.989; the penumbra reaches across this fraction of the Moon's diameter

May 5 17:36--full Moon (Moon at opposition to Sun in ecliptic longitude). Moon's center is north of the center of Earth's shadow, as measured perpendicularly to the ecliptic

May 5 18:10--Moon at opposition to Sun in right ascension; its center is north of the center of Earth's shadow, as measured perpendicularly to the equator

May 5 19:34--last contact of Moon with Earth's penumbra

May 11 05:07--Moon at perigee

The October eclipse season

III — Annular eclipse of the Sun, October 14

About half a year later, Earth and Moon have come around to the opposite side of the Sun, so that the Moon's new and full positions again roughly coincide with the orbit's line of nodes.

It arrives between us and the Sun more than 7 hours before reaching the descending node, so its shadow sweeps our northern hemisphere. In the northern-autumn part of Earth's orbit, the north pole is leaning forward, hence the shadow's track slants somewhat into the geographical southern hemisphere.

The Moon is in the distant part of its orbit; it passed the apogee only 4 days ago, and is at 63 Earth-radii (from Earth's center). So the umbra, the cone of total shadow, ends about 3.9 Earth-radii short of reaching Earth's central plane. What reaches Earth's surface is the antumbra, the imaginary extension. Anyone standing inside this and looking up sees the dark Moon surrounded by an annulus or ring of Sun. The antumbra gets broader as it goes away, and makes a broad track on Earth's surface.

Though exciting to watch, an annular eclipse falls far short of a total one. The sky darkens like twilight but not like night, so stars and planets less bright than Venus are unlikely to become noticeable. The Moon itself is likely to appear blue like the sky around it, rather than black, and details on its surface are unlikely to be discernible. However, the bright and extremely slender annulus may get broken into beads by mountains on the Moon's limb, if you are just inside the northern or, especially, the southern limit of the path.

The penumbra first touches the sunrise horizon at about 15h Universal Time, in the ocean west of the Oregon-California border. Within minutes it reaches the coast, where watchers (using eye protection) may detect a slight dent in the Sun's western edge. By 16 UT, the partial shadow spreads over almost the whole of North America. The bite taken out of the Sun by the invisible Moon is displaced southward if you are in northern Canada or near the Great Lakes, northward if you are Nexico.

At 15:12 UT, the central axis of the Moon's shadow meets Earth, much farther back west and north (west of Vancouver Island, south from Alaska) because the Earth has been rotating eastward. As seen from a ship out here, the Sun rises as a ring. This annular eclipse reaches the southern Oregon coast at 16:16 UT (9:16 PDT, really 8:16 standard time),

The spectacle is seen across parts of nine states, higher in the sky and later into the day. Albuquerque, New Mexico, is on the center line, and at 16:37 UT (10:16 MDT) sees the annular eclipse last for 4 minutes and about 40 seconds, 37° above the southern horizon. San Antonio, Texas, is in the northern half of the path

The longest duration of annular eclipse, 5m 12s, comes over the water north of Costa Rica at 18:15 UT, with the Sun 66° high. The path of annularity passes across across the Panama isthmus, with the canal just outside the northern edge.

Oct 10 03:55--Moon at apogee

III: annular eclipse of the Sun, Oct. 14

Oct 14 15:04--first contact of Moon's penumbra with Earth, at local sunrise
Oct 14 16:12--central eclipse begins: first contact of axis of Moon's shadow with Earth, at local sunrise
Oct 14 17:37--mid eclipse; Moon's center north of Sun's as measured perpendicularly to Earth's equator
Oct 14 17:54--new Moon (conjunction of Moon with Sun in ecliptic longitude): Moon's center is north of Sun's as measured perpendicularly to the ecliptic
Oct 14 19:47--central eclipse ends: last contact of axis of Moon's shadow with Earth, at local sunset
Oct 14 20:55--last contact of penumbra with Earth, at local sunset

Oct 15 01:10--Moon's center reaches descending node through the ecliptic
Oct 18 11--middle of eclipse season: Sun at same longitude as Moon's descending node

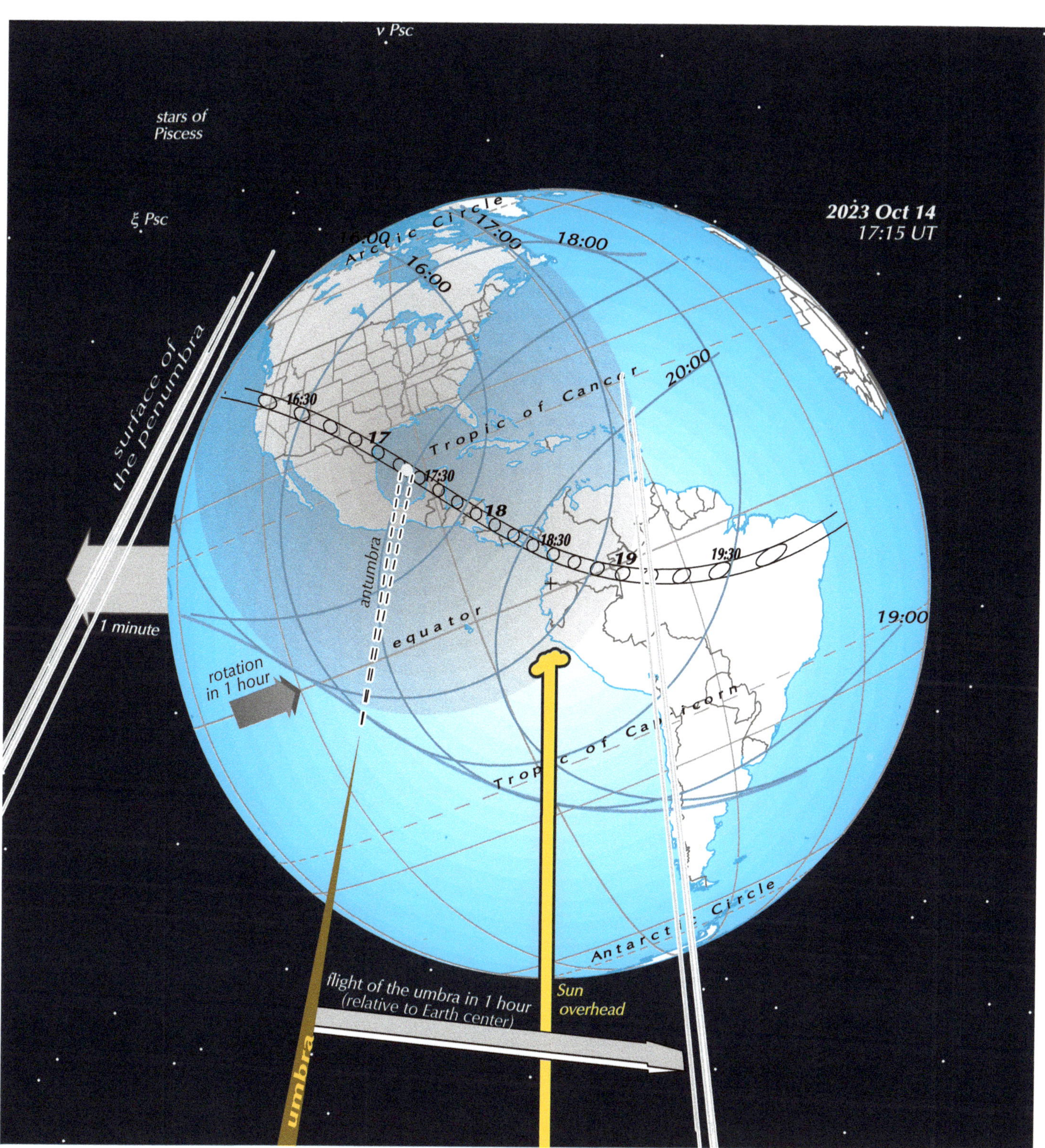
ν Psc
stars of Piscess
ξ Psc
2023 Oct 14
17:15 UT
Arctic Circle
16:00
17:00
18:00
16:00
surface of the penumbra
16:30
17
Tropic of Cancer
20:00
17:30
18
18:30
19
19:30
antumbra
equator
19:00
1 minute
rotation in 1 hour
Tropic of Capricorn
Antarctic Circle
flight of the umbra in 1 hour (relative to Earth center)
Sun overhead
umbra

IV — partial eclipse of the Moon, October 28

The Moon ascends through the ecliptic plane as much as 17 hours before coming to its full position. So it traverses the northern fraction of Earth's shadow, dipping into the umbra, the core of total shadow, for only an hour and 19 minutes. The umbral magnitude of the eclipse is 0.127, meaning that at maximum eclipse the umbra reaches across this fraction of the Moon's diameter.

When the imperceptible first stage of the eclipse is beginning, the full Moon is setting for the western Pacific, high in the sky for Asia, climbing in the east for Europe and Africa. At the exciting moment when the umbra, the sharp-edged central shadow, possibly reddish-black, touches the southeastern limb of the Moon, it is setting for Japan, mid Indonesia, western Australia, high in the sky for the rest of Asia and for Europe and most of Africa, and rising into view for Greenland and Northeastern North America and northeastern South America then comes into view of the Moon as it passes out of the umbra and then the penumbra. All stages of the eclipse have ended by the time the Moon rises into view for central and western Canada, most of the USA, and western South America.

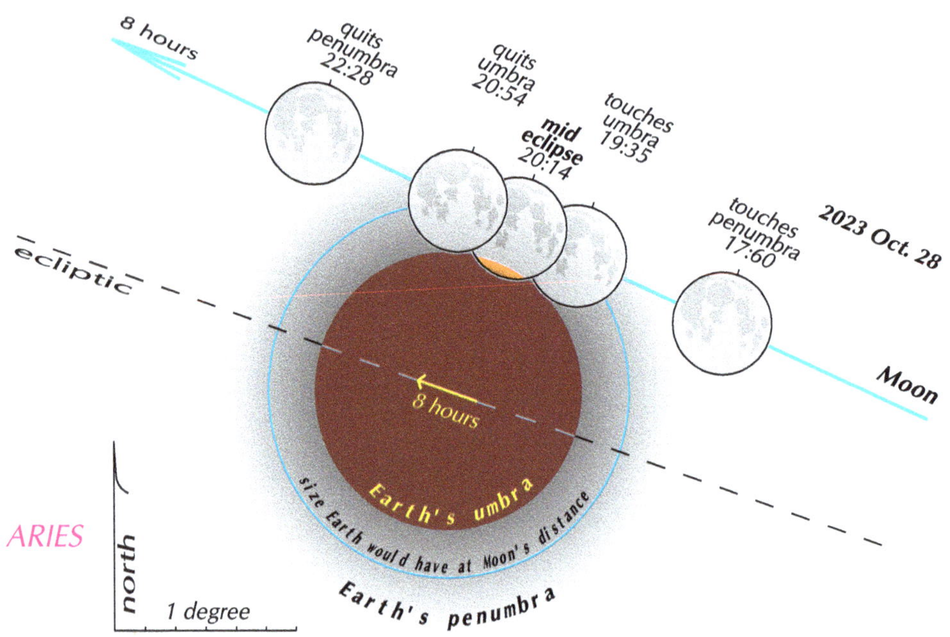

Oct 26 03:03--Moon at perigee

IV: partial eclipse of the Moon, Oct. 28

Oct 28 03:16--Moon at ascending node through the ecliptic
Oct 28 17:60--first contact of Moon with Earth's penumbra
Oct 28 19:35--first contact of Moon with Earth's umbra
Oct 28 20:14--middle of eclipse: Moon nearest to center of Earth's shadow. The penumbral magnitude of the eclipse is 0.127; the umbra reaches across this fraction of the Moon's diameter
Oct 28 20:23--full Moon (Moon at opposition to Sun in ecliptic longitude). Moon's center is north of the center of Earth's shadow, as measured perpendicularly to the ecliptic
Oct 28 20:54--last contact of Moon with Earth's umbra
Oct 28 21:02--Moon at opposition to Sun in right ascension; its center is north of the center of Earth's shadow, as measured perpendicularly to the equator
Oct 28 22:28--last contact of Moon with Earth's penumbra

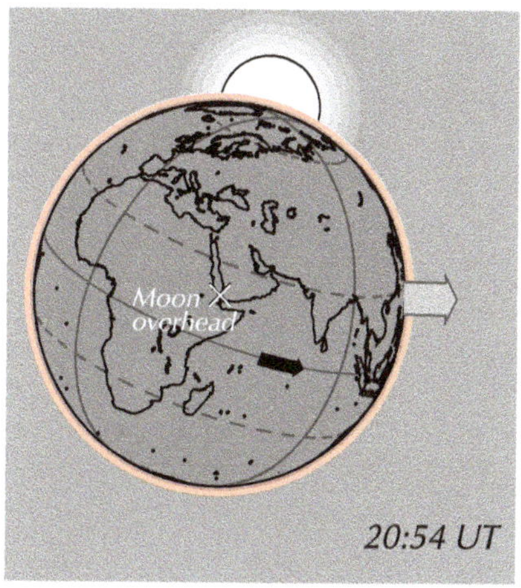

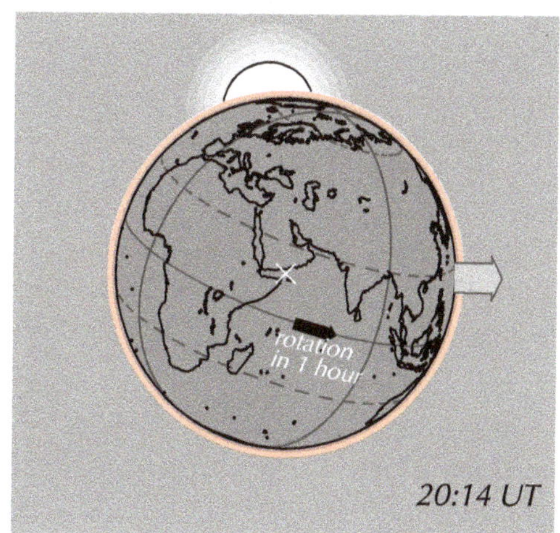

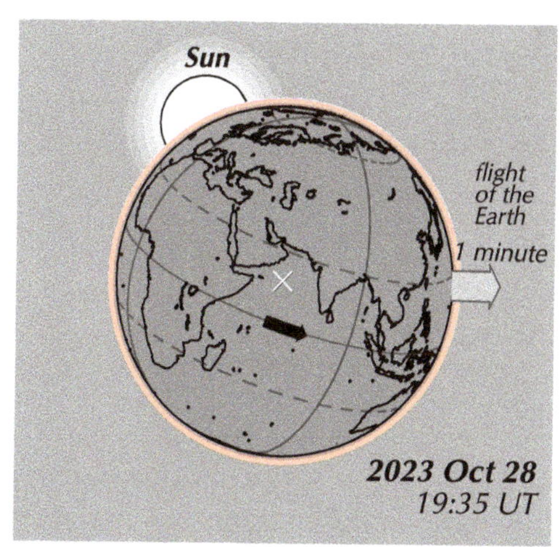

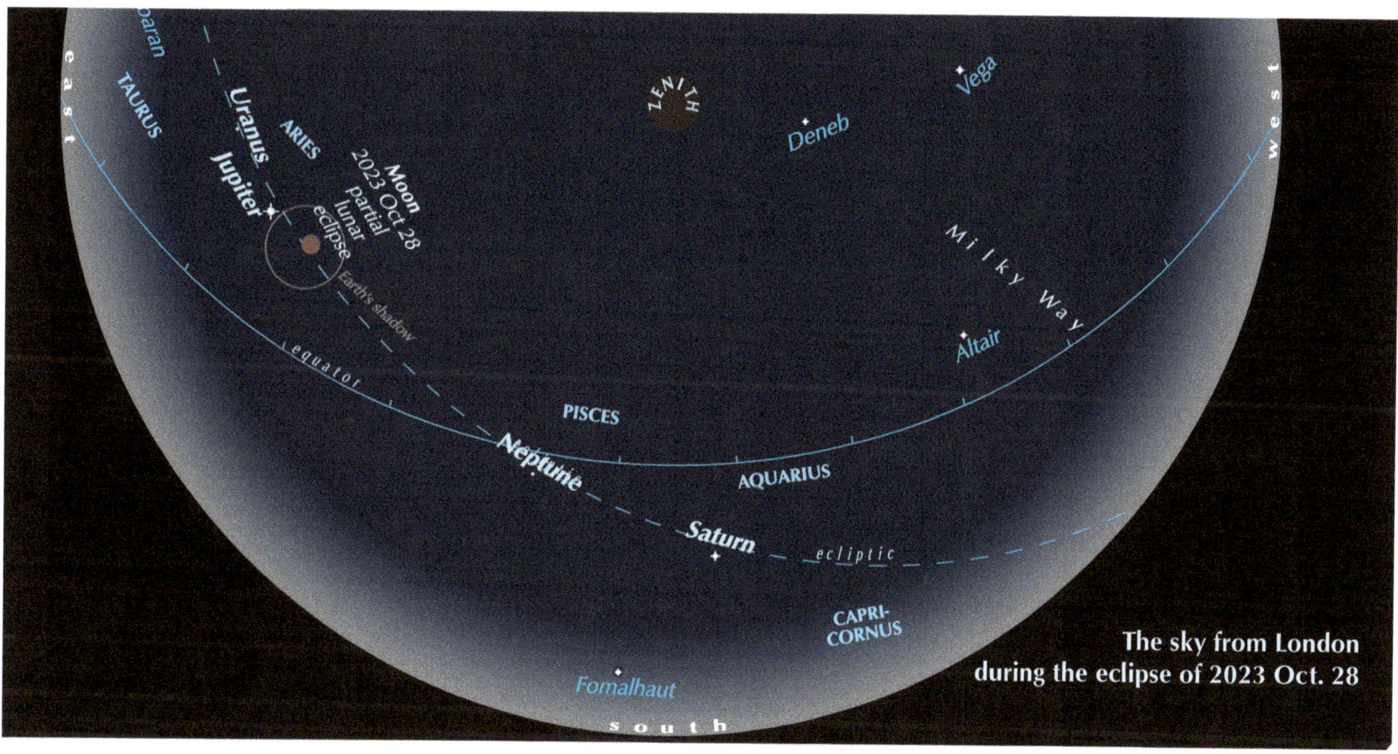

The sky from London
during the eclipse of 2023 Oct. 28

OCCULTATIONS of major planets by the Moon

Moving bodies can occult (hide) more distant objects. Planets and asteroids occult stars and, rarely, each other. Because of the Moon's nearness and angular size, occultations by it are abundant and fairly easy to observe. Timings of occultations give information on, for instance, the Moon's exact shape, rings around planets, and the presence of companions to asteroids or stars.

Each picture is a view from the occulted body to the Earth, with the Moon passing between. The Moon's outline is drawn at three moments: the hour nearest to the middle of the occultation, and an hour before and after. The outlines are the "shadow" of the Moon cast by the occulted body on the Earth; they are the regions from within which the occultation is seen. An arrow on the equator shows how much the Earth rotates in two hours. The outlines and their track on the surface are inexact, at left and right, because of the turning of the Earth during the two hours. And the most interesting and sensitive observations are of "grazing" occultations, seen from the track's north or south boundaries, where the planet or star may reappear several times between mountains at the Moon's limb (edge).

Side-diagrams show the phase of the Moon, and the occulted body passing behind it at intervals of 10 minutes over the same 2-hour span, as seen from the center of the Earth. These diagrams have **celestial north** at the top, to suit the view through telescopes. From places north of the Earth's center, the Moon will appear farther south, and vice versa.

Occultations have to be observed in a night sky, so observing opportunities are where the occultation track crosses a darkened part of the Earth. When the Moon is nearly full, the night area crossed by the occultation track is larger, but the star or planet has to be found in bright moonlight,

The disappearance of a body, at the Moon's leading limb, is the easier event to watch for. To be ready to catch the reappearance at the following limb, you need to know more exactly the predicted time and point. A star disappearing or reappearing at the Moon's bright limb is liable to be overwhelmed by glare. Yet the Moon's surface is far darker than that of Venus, whose tiny burning crescent, close to and far beyond the nearer, vaster, duller corresponding crescent of the Moon, is an amazing sight.

The time given is the Universal Time of the middle of the occultation, to the nearest hour. "Mag." is the magnitude (brightness) of the occulted body. "Elong." is its elongation or angular distance from the Sun. Thus "elong. 8° E" would mean it is in the evening sky (east of the Sun—left of it as seen from our northern hemisphere) but very close to the Sun; "elong. 80° W" would mean it is well away from the Sun, but in the morning sky.

Four first-magnitude stars are near enough to the ecliptic that they can be occulted by the Moon: Aldebaran, Regulus, Spica, and Antares. Last year the Moon's path missed all of them. This year it begins to touch Anares. 2024 will bring occultations of Antares and Spica.

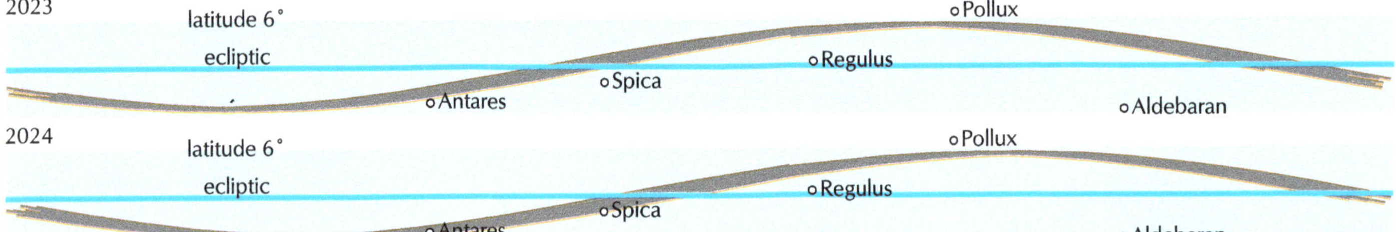

The Moon's path around the sky throughout this year, and the 1st-magnitude stars near the ecliptic. The horizontal scale is ecliptic longitude, and the vertical scale is ecliptic latitude exaggerated by 2. Gray lines trace the path of the Moon's center; the colored band includes its width of about half a degree. But the band of possible occultation is wider because of parallax (the shifted view from places north and south on the Earth). The path keeps shifting; its ascending and descending nodes slide backward (westward, rightward) along the ecliptic, going all around in 18.6 years. Pollux is far enough north that it is never occulted.

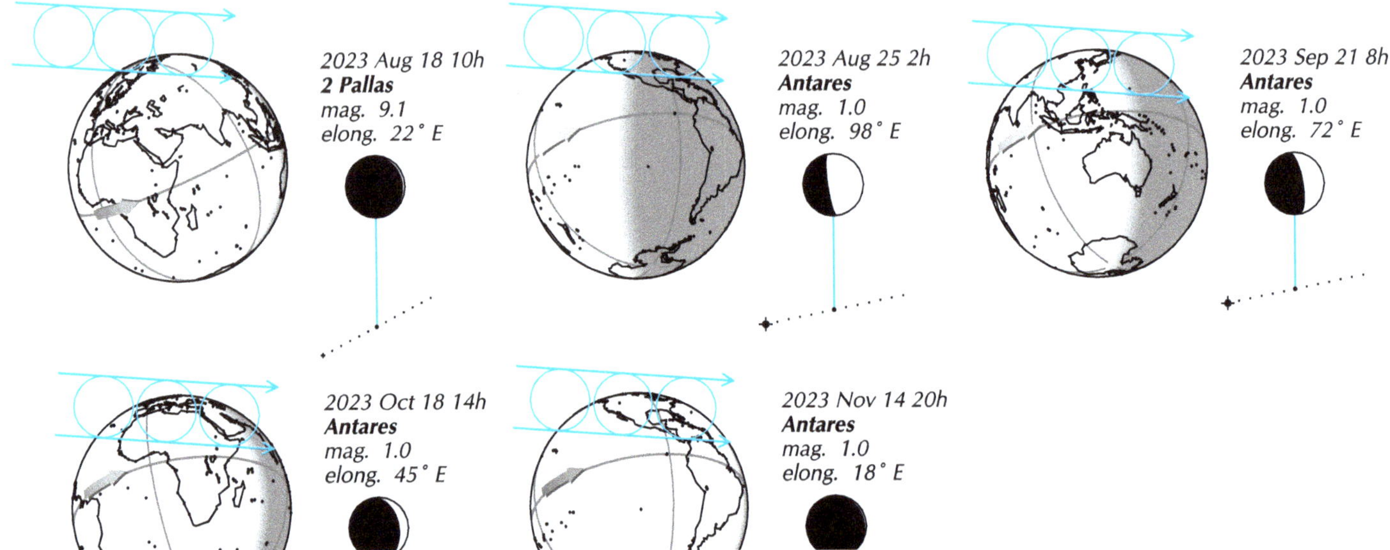

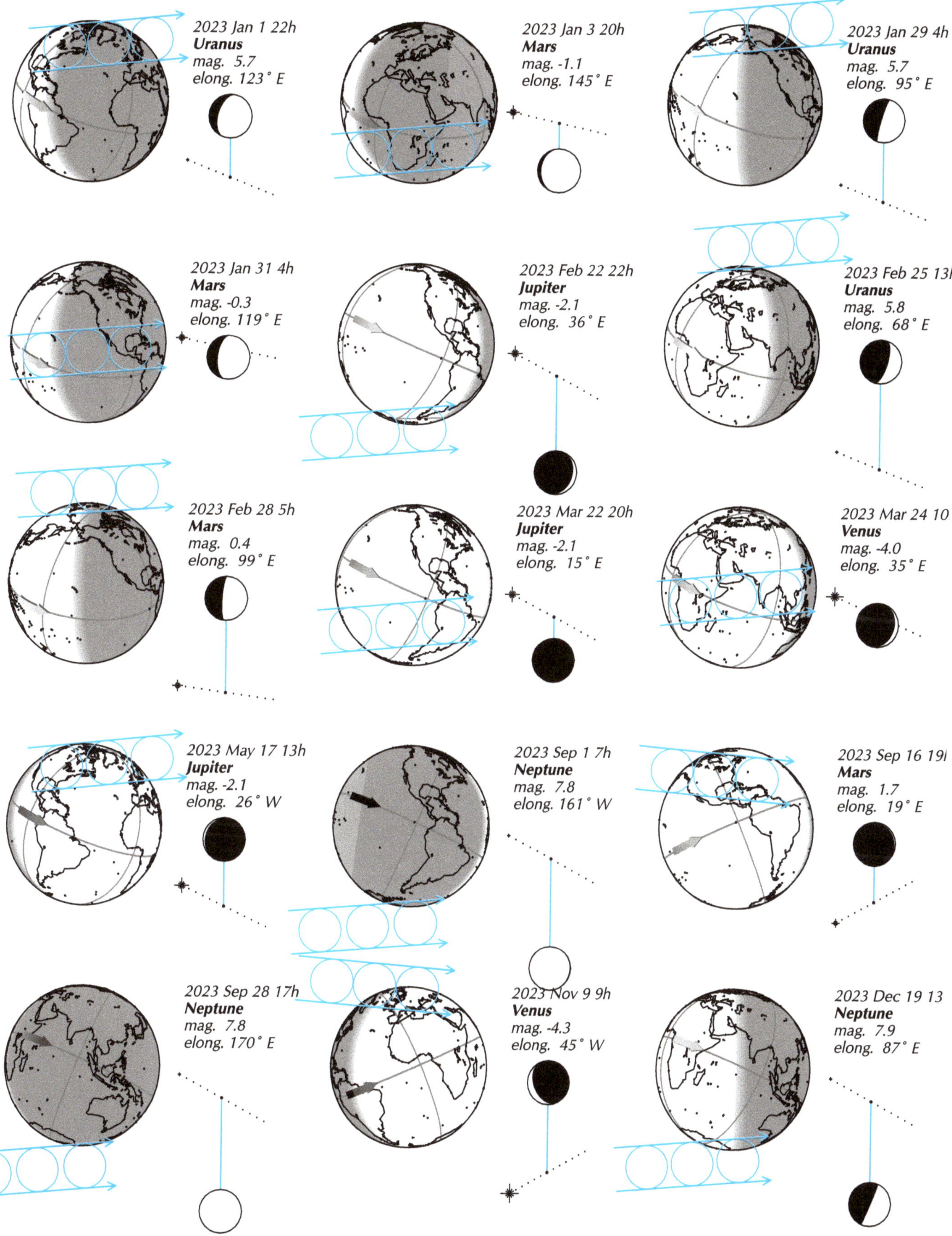

2023 Jan 1 22h
Uranus
mag. 5.7
elong. 123° E
2023 Jan 3 20h
Mars
mag. -1.1
elong. 145° E
2023 Jan 29 4h
Uranus
mag. 5.7
elong. 95° E
2023 Jan 31 4h
Mars
mag. -0.3
elong. 119° E
2023 Feb 22 22h
Jupiter
mag. -2.1
elong. 36° E
2023 Feb 25 13h
Uranus
mag. 5.8
elong. 68° E
2023 Feb 28 5h
Mars
mag. 0.4
elong. 99° E
2023 Mar 22 20h
Jupiter
mag. -2.1
elong. 15° E
2023 Mar 24 10
Venus
mag. -4.0
elong. 35° E
2023 May 17 13h
Jupiter
mag. -2.1
elong. 26° W
2023 Sep 1 7h
Neptune
mag. 7.8
elong. 161° W
2023 Sep 16 19h
Mars
mag. 1.7
elong. 19° E
2023 Sep 28 17h
Neptune
mag. 7.8
elong. 170° E
2023 Nov 9 9h
Venus
mag. -4.3
elong. 45° W
2023 Dec 19 13
Neptune
mag. 7.9
elong. 87° E

INNER PLANETS

Mercury and Venus are the planets lower (Latin *inferior*) than Earth in the gravitational field of the Sun.

TAURUS
ARIES
PISCES
GEMINI
CANCER
AQUARIUS
CAPRICORNUS
SAGITTARIUS
SCORPIUS
LIBRA
VIRGO
ecliptic
equator
vernal equinox direction
ECLIPTIC PLANE
Venus
Sun
Mercury
Earth
Jan
Feb
Mar
Jan 30 elong west

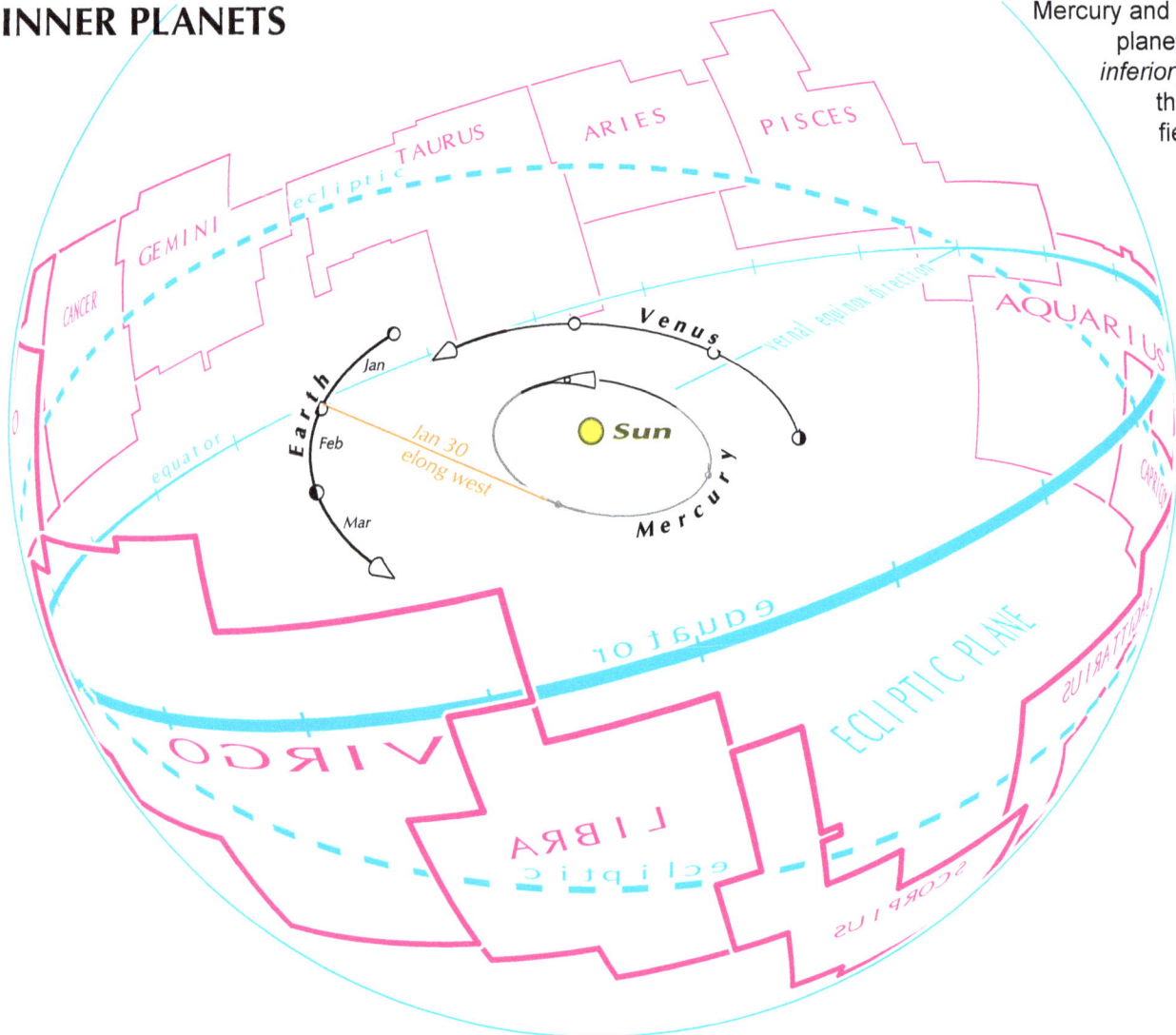

Orbits of the three inner planets in the first quarter of the year, seen from a viewpoint 6 astronomical units (Sun-Earth distances) from the Sun. On an imaginary sphere, 2 AU out, are shown the planes of the equator and ecliptic, and the boundaries of the zodiacal constellations. The planets move nearly in the ecliptic plane, so as seen from the Earth they appear against the background of these constellations (except that part of the ecliptic, in the foreground, lies in Ophiuchus rather than Scorpius). Globes represent the planets at the start of each month. Their size is exaggerated 500 times, the Sun's only 5 times. When a planet is in or north of the ecliptic plane, its path is drawn with a thicker line. When it is in the morning sky (west of the Sun) as seen from the Earth, its course is shown in gray.

Mercury	
mean distance from Sun	0.39 AU
sidereal period 0.24 year =	88 days
synodic period	116 days
eccentricity	0.206
inclination	7°
diameter	4,880 km

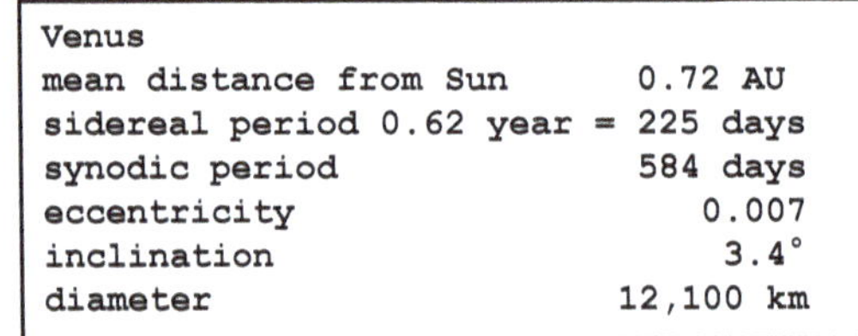

Venus	
mean distance from Sun	0.72 AU
sidereal period 0.62 year =	225 days
synodic period	584 days
eccentricity	0.007
inclination	3.4°
diameter	12,100 km

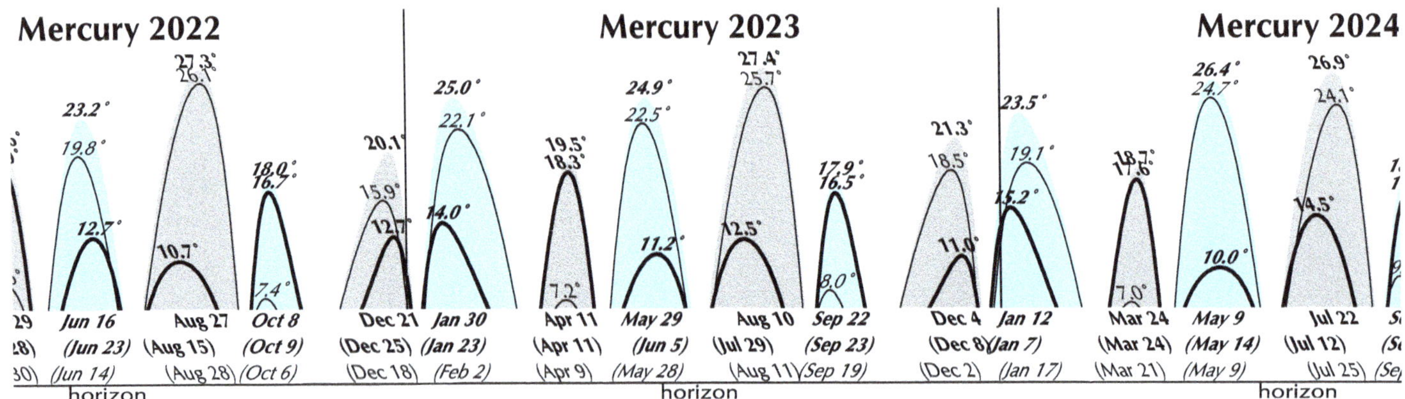

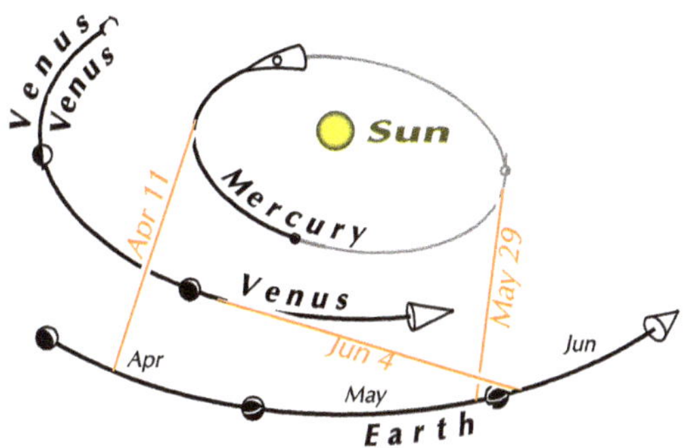

Continuations of the motions of Mercury, Venus, and Earth in the other quarters of the year. The Earth goes around the Sun once in the year, but Venus 1.625 times and Mercury 4.15 times.

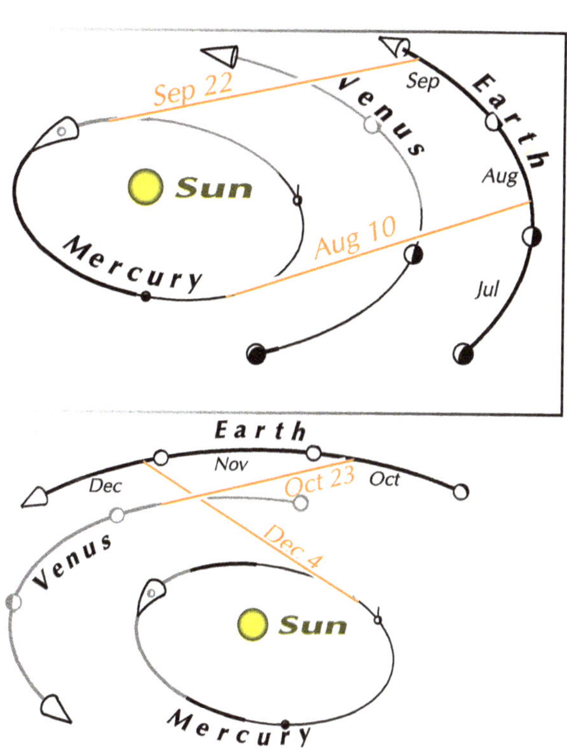

George III made drawings and calculations of Venus's 1769 transit of the Sun, and watched it from the King's Observatory he had founded that year between Richmond and Kew. He accurately predicted the transits of 1874 and 2004. A mass of his papers made public in 2016 revealed that the "mad" king who "lost America," considered dull and called by Thomas Paine "the royal brute," was compassionate, intelligent, interested in science, and bi-polar—all to an extreme degree.

Apparitions compared. Gray areas represent evening apparitions (eastward elongation); blue, morning apparitions (westward). The top figures are the maximum elongations, reached at the top dates shown beneath.

Curves show the altitude of the planet above the horizon at sunrise or sunset, for latitude 40° north (thick line) and 35° south (thin), with maxima reached at the parenthesized dates below (40° north bold).

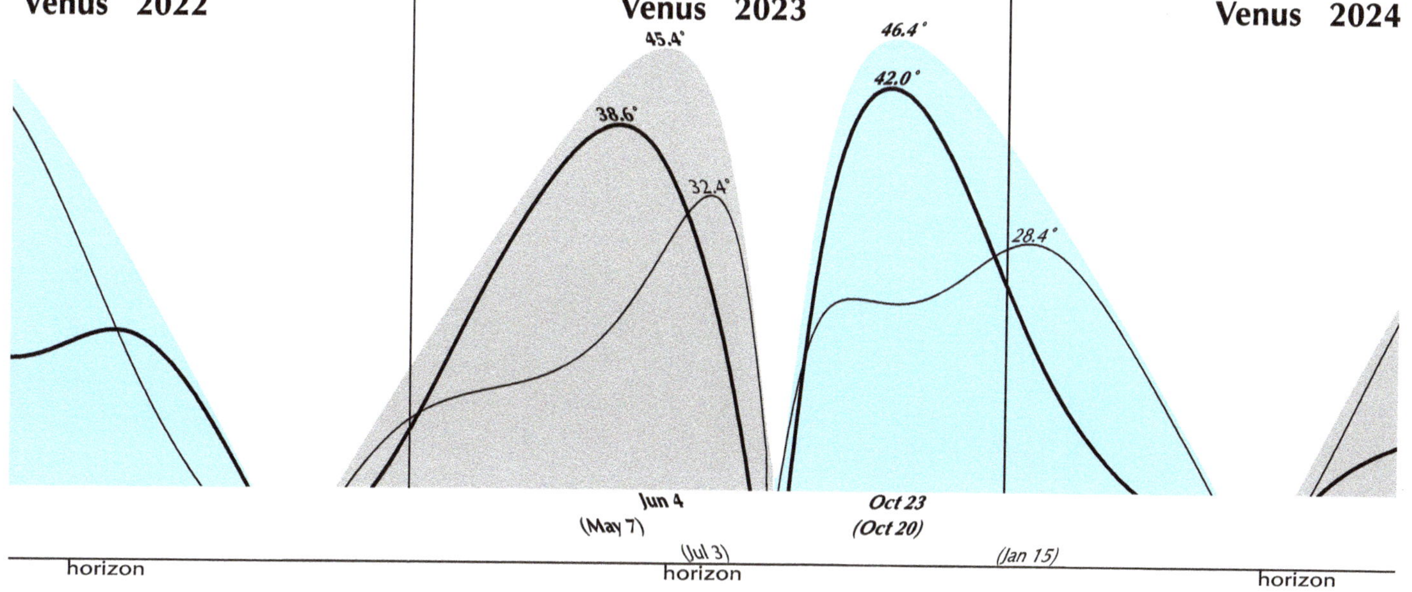

MERCURY

The solar system's innermost and fastest-moving major planet is also the smallest: only 1.4 times wider than the Moon. It orbits the Sun 4.15 times a year, but from our moving viewpoint it appears to go around 3.15 times. Its orbit is the most tilted (inclined 7° to the ecliptic plane) and most elliptical (eccentricity 0.206, where 0 means circular). (Pluto is smaller, and more tilted and eccentric in orbit, but is no longer considered a major planet. A planet nearer to the Sun, Vulcan, was once suspected but never found.)

Each year, Mercury makes three apparitions in the evening sky, often part of a fourth, and similarly in the morning sky. These excursions are, because of the eccentric orbit, unequal. The angular distance (elongation) to which Mercury goes out from the Sun is greater if it is near aphelion (greatest physical distance)

Earth's southern hemisphere is favored, because aphelion is in the direction where Mercury is south of the ecliptic. Often, when Mercury is at great elongation, it gets as high as 26° above the sunset or sunrise horizon for our southern hemisphere, but for the northern it gets less than half as high, going out at a low angle to the horizon.

Around inferior conjunction, when Mercury rushes westward between us and the Sun, its retrograde path traces loops well north or south of the ecliptic, because of the orbit's inclination.

This year has three performances on each side of the Sun.

For our northern hemisphere, the April evening apparition is as good as it could be, with Mercury about vertically above the setting Sun. The best morning apparition is September's. The others are middling to low.

For south-hemisphere observers, the January and May morning apparitions, and especially the August evening one, are higher than anything the north ever sees. But the April evening and September morning apparitions go to the opposite extreme of lowness.

Maps of the planets' paths, plotted in ecliptic latitude and longitude to save vertical space. The more familiar grid of equatorial coordinates (lines of declination and right ascension) is also shown, slanting and curving in relation to the ecliptic system. At day 1 of each month, the disk of the planet is drawn, exaggerated 600 times in size. Ticks are at days 11 and 21. Dashes along the ecliptic are 2° long. When the planet is in the morning sky its course is shown in gray. Transition from black to gray is at inferior conjunction, when the planet passes in front of the Sun; transition from gray to black is at superior conjunction. Mercury and Venus stay within 28° and 48° of the Sun, so they start and end each calendar year in the south.

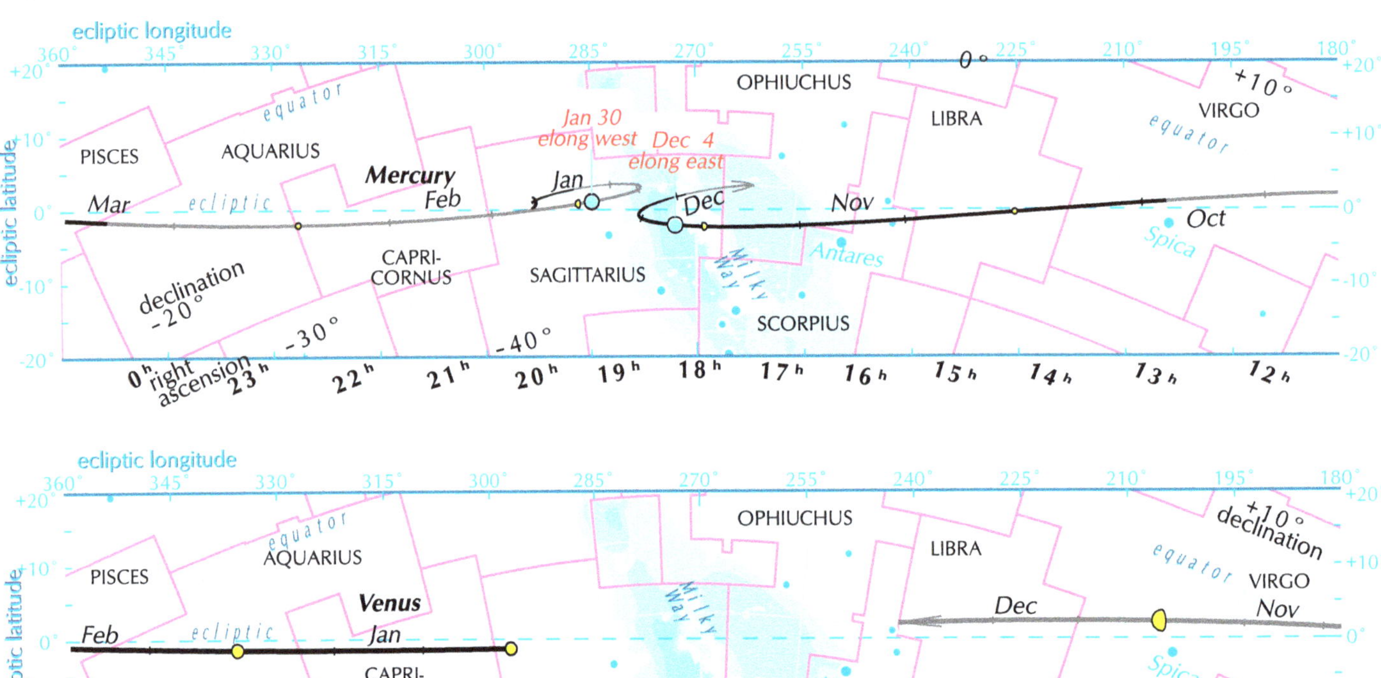

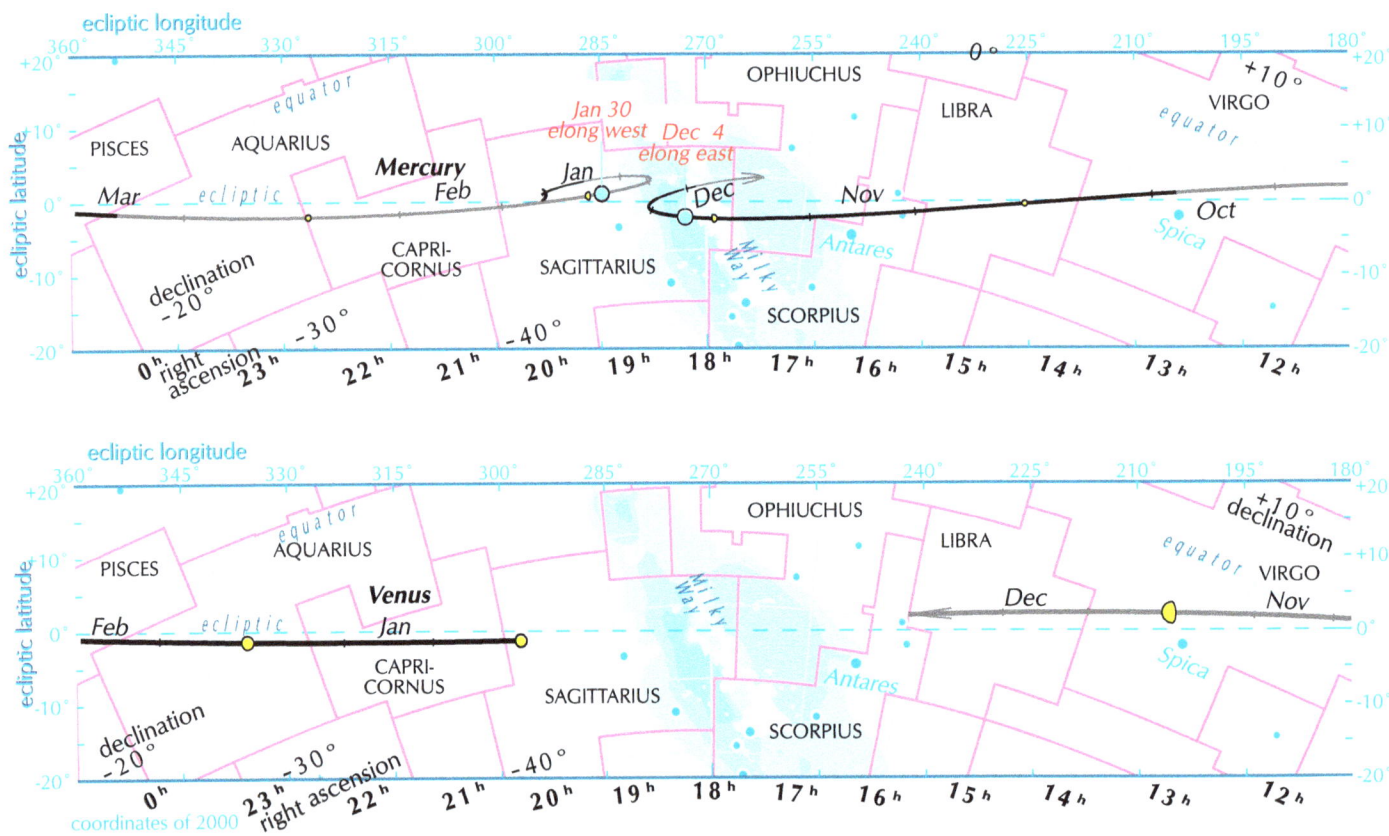

VENUS

Earth's "sister" is only slightly smaller (diameter 12,100 km versus Earth's 12,756) but fiercely different. Under a complete blanket of thick cloud is a dense hot carbon dioxide atmosphere. Venus rotates backward (clockwise as seen from the north) and slowly, in 243 Earth-days. The rotation is upright, without inclination to the orbit; and the planet is almost perfectly spherical; so there is almost no variation by season.

Venus is the major planet to which we come nearest, though Mars can be almost as near, and Mercury is nearest for more of the time.

Nearness, size, and surface of high albedo (whiteness, reflectivity) make Venus at almost all times the brightest celestial body after the Sun and Moon. It can be glimpsed in daylight, is typically the first "star" noticed as the sky dims toward a solar eclipse, and can be seen up to several hours after sunset or before sunrise. Early cultures saw it as separate Morning and Evening Stars, and many associated it with a great goddess of love and generation.

Venus travels in a nearly perfect circle, 13 times around the Sun in 8 Earth-years, so that as seen from Earth it appears to make 5 circuits around the sky. (5, 8, 13 are consecutive numbers in the Fibonacci series, which shows up in many details of nature.) This 8-year cycle was known to the Babylonians and the Maya. Each year of the cycle has a distinctive pattern. The phenomena of 2014 repeat, 2 or 3 days earlier, in 2022.

This year of the cycle is divided by the August inferior conjunction into a long evening apparition and the first months of a morning one, both favorable for our northern hemisphere.

In the early months, Venus is southerly in Capricornus and Aquarius, setting less than an hour after the Sun. Through (northern) spring, it slopes north, ascending across the ecliptic on March 14, passing 2.5° south of the Pleiades on April 11. On May it is at its northernmost both in latitude above the ecliptic and in declination, 2.6° north of the ecliptic's northernmost point at the beginning of Gemini. This is not quite the northernmost Venus can reach; see our chart of "northness and southness" on page 38 of *Venus*.

All this means that for most of the south-hemisphere autumn Venus is low in evening skies.

Venus is overtaking us on the inside, growing in apparent size, and on June 4 reaches its easternmost elongation of 45.4° from the Sun. The moment of dichotomy, when the terminator or boundary of the sunlit half looks in your telescope exactly straight, should be close to maximum elongation but, because of the soft cloudy surface, may come up to 10 days sooner. Another climax, the moment of greatest illuminated extent, when the sunlit area appears largest, is not till July 7, and the peak of total light, at magnitude -4.5, on July 9.

By this time Venus has already descended southward across the ecliptic (on July 4). Its sunlit crescent grows spectacularly in size and thinness as it whirls between us and the Sun.

Shooting out into the morning sky in late August, the reversed crescent dwindles and thickens as Venus climbs rapidly and races away from us toward the other side of the Sun. The phenomena comes in reverse order: brightest on Sep. 18; greatest illuminated extent on Sep. 19; greatest elongation on Oct. 23; apparent dichotomy up to 6 days later. As Venus travels south into Virgo, it becomes somewhat lower in morning skies for the northern hemisphere and somewhat higher for the southern.

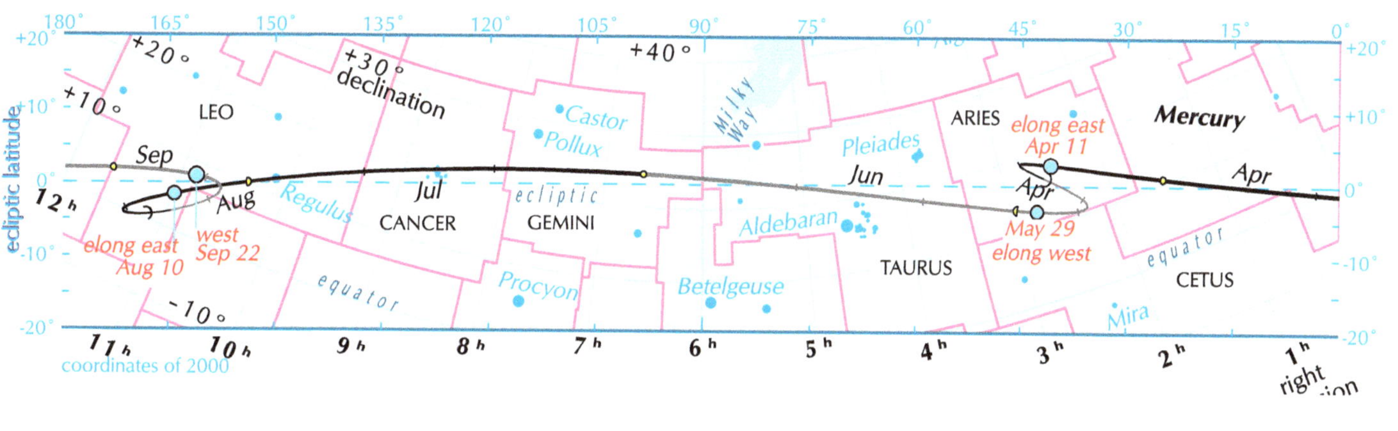

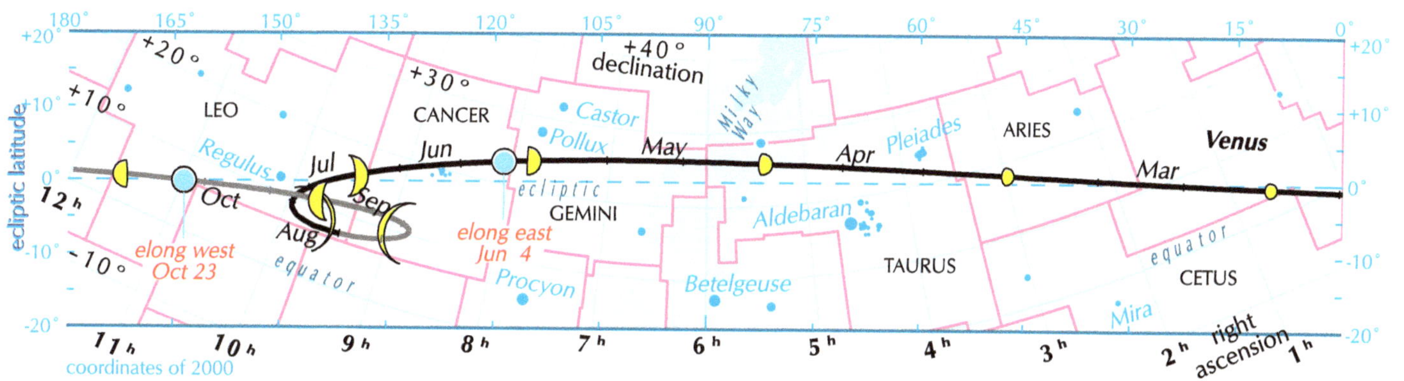

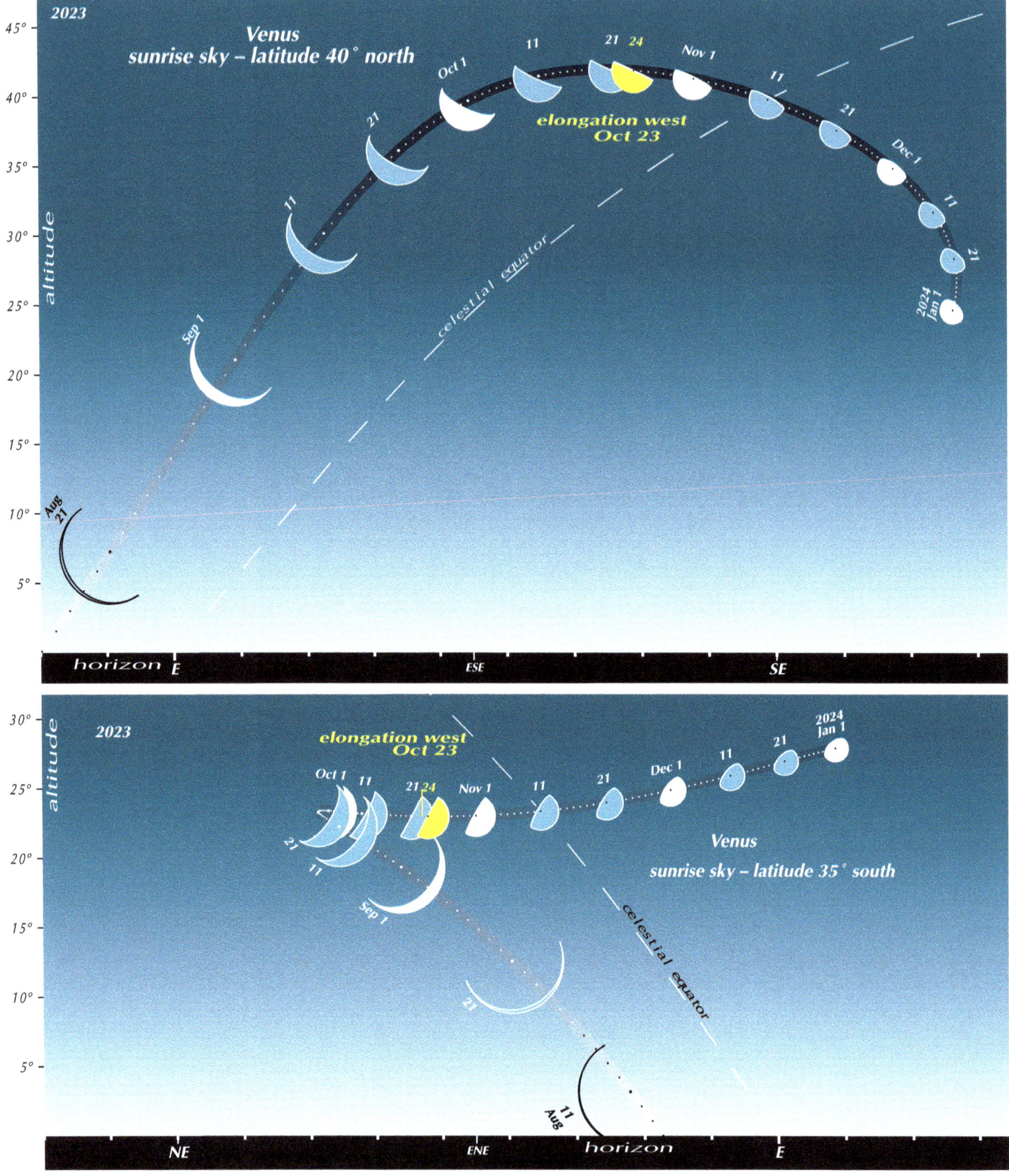

These are the visits of Mercury and Venus to the morning and evening skies, as they appear at sunset and sunrise for latitudes 40° north and 35° south.

You can imagine the changed situation at, say, an hour after sunset by mentally raising the horizon-line about 12°. The sky will be darker, but the planets will be lower.

Longitude makes little difference, because planets do not change their positions as fast as the Moon—even Venus never moves more than about a degree a day.

Latitude makes a great difference. If you are farther north, you must imagine the horizon tilted up on the south like a seesaw, around the Sun's position as a pivot. Or, which is the same thing, you will find the equator and also the trajectories of Mercury and Venus lying flatter. If you are at the north pole, the equator is the horizon, and the planets' travels south of the equator are below the horizon. Conversely as you move south, the horizon tilts down at its south end; in Ecuador or Uganda on the terrestrial equator, the celestial equator is vertical, and the planets' sallies, too, are roughly vertical. For South Africa or New Zealand, the celestial

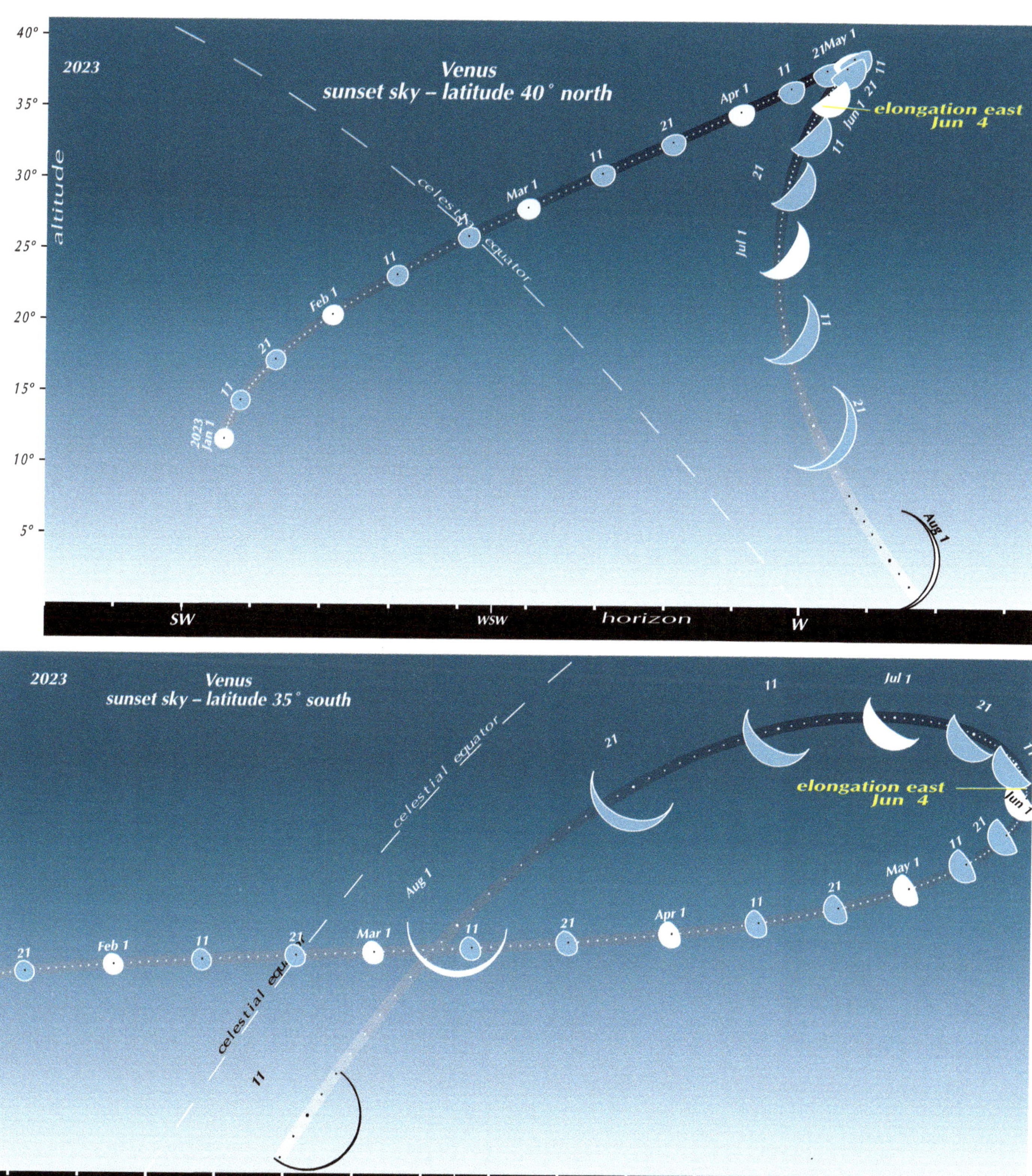

equator slopes in the opposite direction and so do the planets, leaping from the Sun generally rightward into the evening sky and leftward into the morning.

The coordinate system used is simple altazimuth. Ticks along the horizon are 5° apart. The scale is 2.5 mm to 1°. The planets are exaggerated 480 times in size. Each image of Mercury or Venus is like a view of this spot in the sky through a powerful telescope.

The planet images are at the 1st, 11th, and 21st of each month. Dots show the actual positions of the planets for every day. These dots give a truer idea of the planets' actual sizes—though even they can be up to 9 times too large in Mercury's case. Mercury is cut off when it is below 5° of altitude; Venus is followed all the way down to the horizon.

Dashes on the celestial equator are 2° long, but do not represent fixed points on the equator, since the horizon is always moving in relation to it. The ecliptic cannot be shown, since it is in a slightly different altazimuth position each day. It always runs through the Sun, toward which the planets' illuminated crescents face.

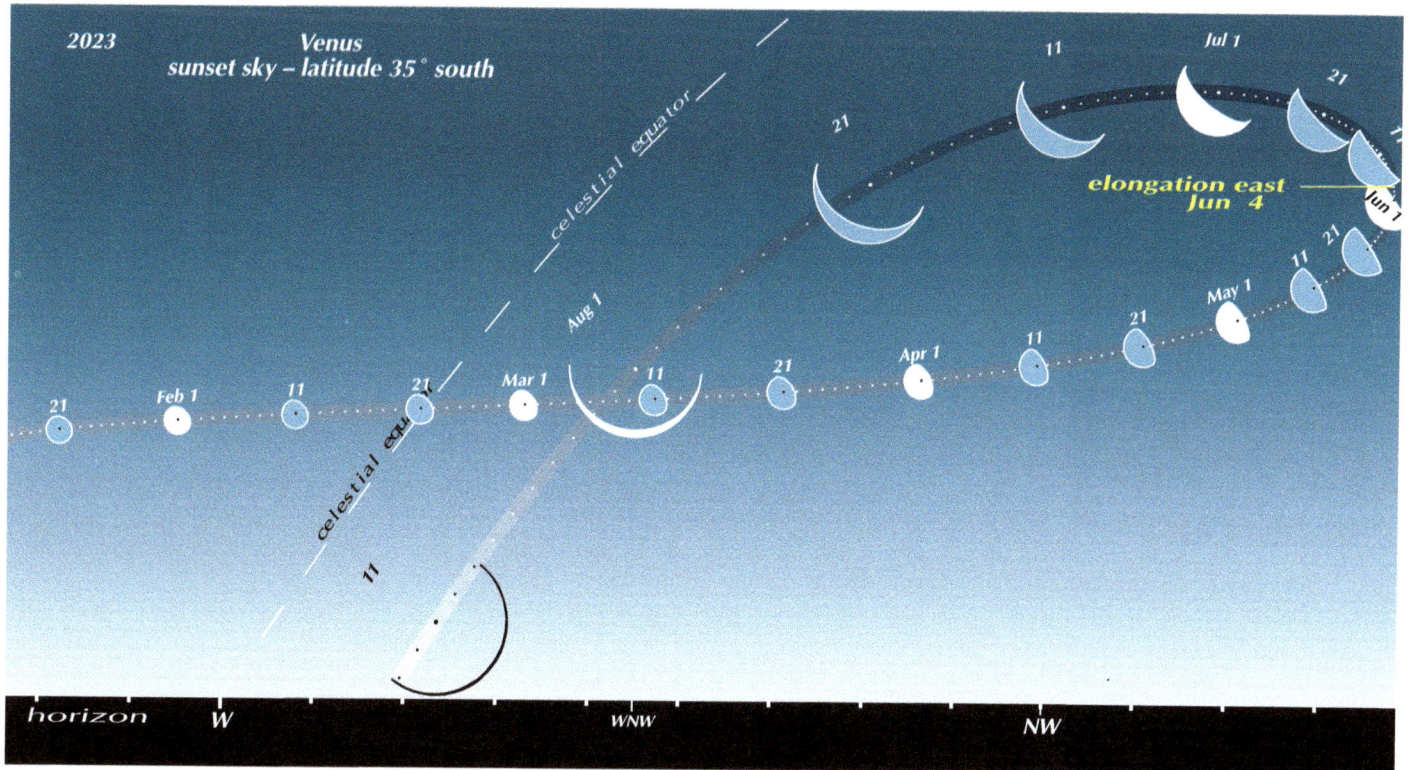

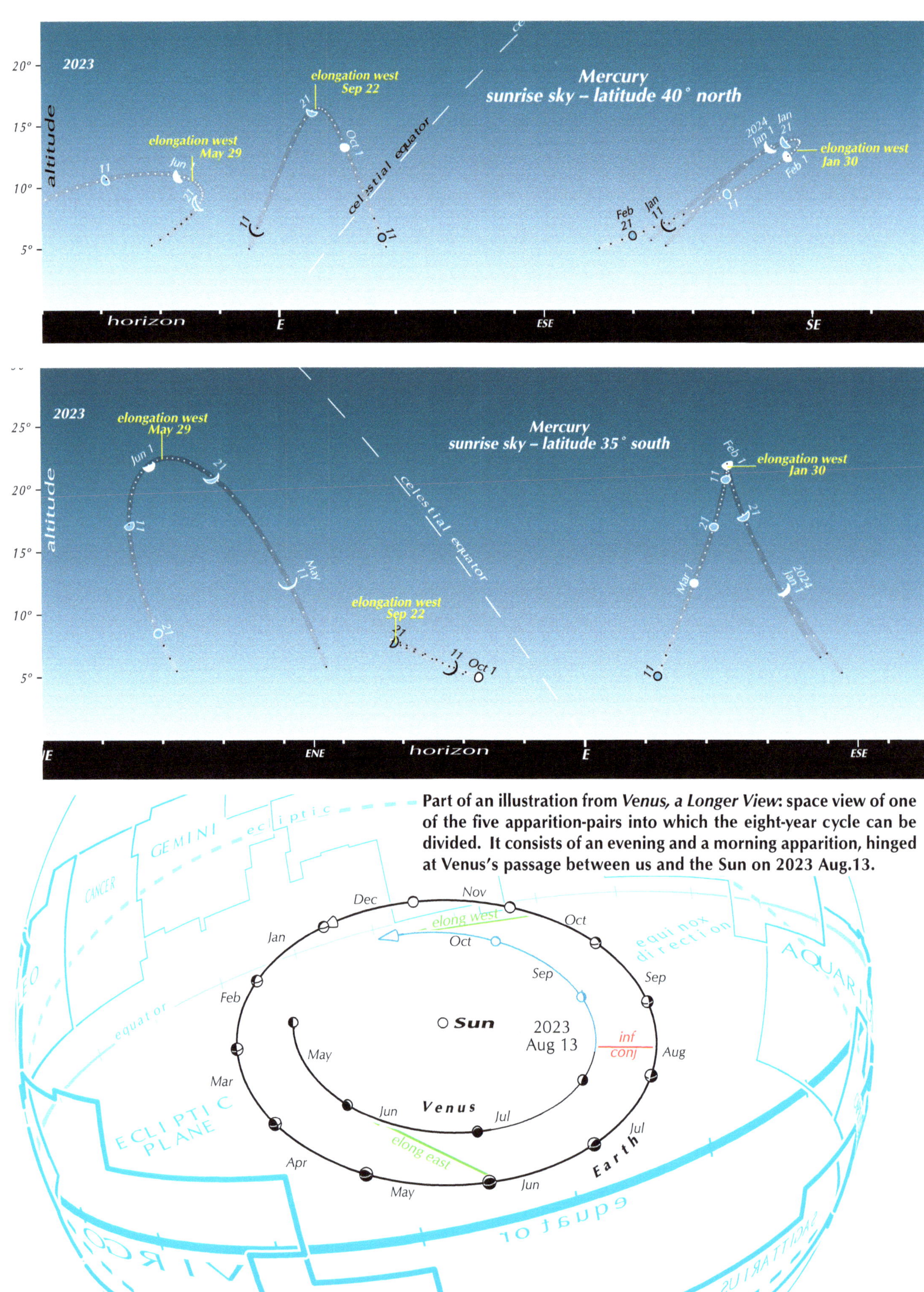

Part of an illustration from *Venus, a Longer View*: space view of one of the five apparition-pairs into which the eight-year cycle can be divided. It consists of an evening and a morning apparition, hinged at Venus's passage between us and the Sun on 2023 Aug.13.

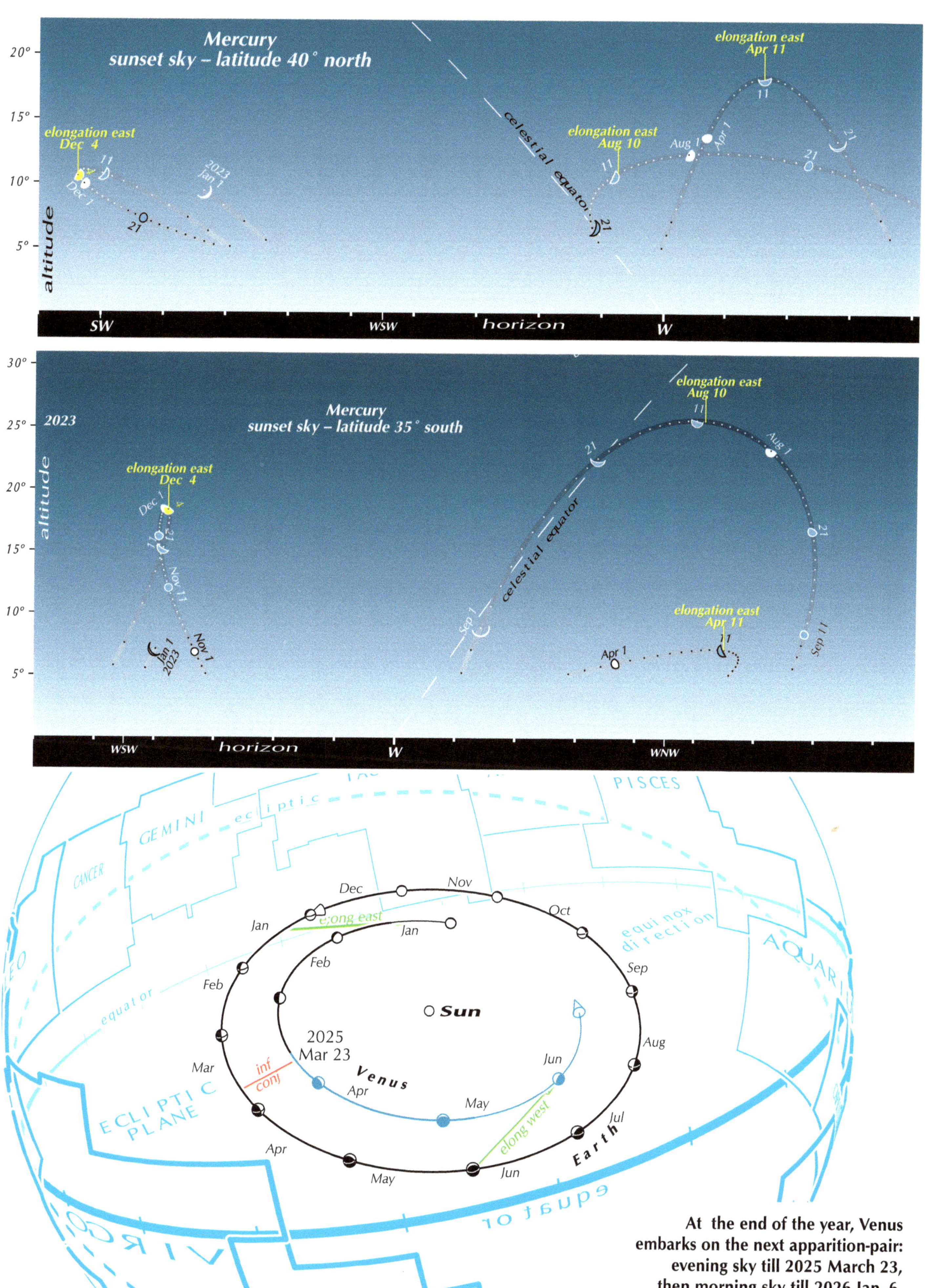

At the end of the year, Venus embarks on the next apparition-pair: evening sky till 2025 March 23, then morning sky till 2026 Jan. 6.

MARS

Phenomena. Columns: right ascension (hours, minutes, seconds) and declination (degrees, minutes), for epoch 2023; distance from Sun and Earth, in astronomical units; elongation from Sun (degrees; negative = westward); magnitude.

Mars 2023		RA(2023)	dec	hedis	gedis	elo	mag
Jan 12 20	stat.in r.a.>dir.	4 24	24 28	1.578	0.717	136	-0.8
Jan 12 21	stat.in long>dir.	4 24	24 28	1.578	0.717	136	-0.8
Jan 14 11	max.declin.south	4 24	24 28	1.580	0.729	134	-0.8
Mar 16 18	east quadrature	5 42	25 36	1.638	1.302	90	0.7
Mar 19 0	max.declin.north	5 46	25 36	1.640	1.325	89	0.8
Apr 22 17	max.lat.north	7 4	24 29	1.659	1.662	72	1.2
May 30 21	aphelion	8 35	20 14	1.666	1.991	57	1.6
Aug 29 22	on equat.,to sou.	12 7	0 0	1.626	2.482	25	1.8
Nov 6 15	descending node	15 0	-17 3	1.552	2.540	4	1.5
Nov 18 6	conjunc.with Sun	15 33	-19 17	1.537	2.526	0	1.4

The fourth planet is a little more than half as wide as Earth. It is one and a half times farther out from the Sun, and takes nearly twice as long to travel around its orbit. (Kepler's third law of planetary motion: period squared equals average distance cubed; in which period is in Earth years and distance is in astronomical units or Sun-Earth distances.)

The result is that Mars has "good" and "poor" years — opposition and non-opposition — in rough alternation. We overtake it at opposition, where it appears directly outward from the Sun, therefore on the meridian at midnight and nearest, largest, and brightest. We next do so 2.13 years later on average.

So the oppositions come in each second year but progressively later: 2010 Jan. 29, 2012 Mar. 3, 2014 Apr. 8, 2016 May 22, 2018 July 27, 2020 Oct. 13, 2022 Dec. 8. After 7 oppositions, two years are skipped; the next opposition is 2025 Jan. 16.

The oppositions are spaced around the orbit. The nearest ones fall in August, in the rather southerly Aquarius part of the ecliptic zone, because that is the direction of the perihelion, the near-in point of Mars's fairly eccentric orbit. 2003 Aug. 28 fell only 1¾ days before perihelion, and was the nearest opposition in several thousand years.

Around opposition, when we are overtaking it on the inside, a planet appears to make a retrograde track; so its total track around the sky is shorter. In an oppositionless year, the track is long and distant and dimmer. We now have the first of two such years, centered around the Sun-conjunction of Nov. 18. From the opposition of 2022 Dec. 8, we pull away ahead of Mars, pass around the Sun, and begin catching up toward the opposition of 2025 Jan. 16. The long straight course this year is all the way from Taurus along the ecliptic to Sagittarius.

At the start of the year, Mars is still high in the midnight sky and, at magnitude about -1, brighter than almost all the stars. For the first four months it is still the most favorably placed planet, though sinking lower in the evening sky. At the end of August it descends south of the equator and is essentially lost to northern viewers.

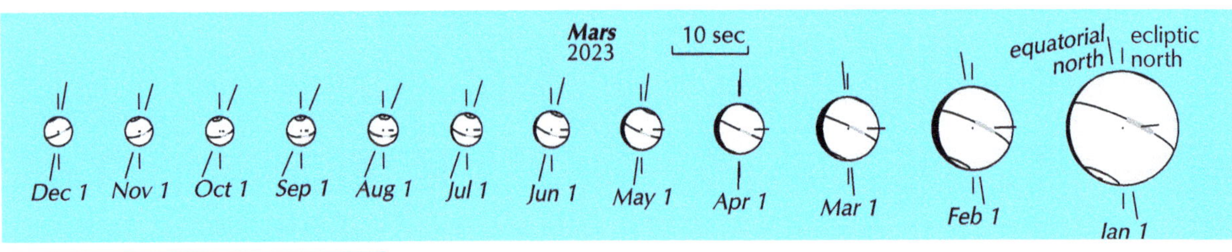

The disk of Mars. Arrow on the equator shows rotation in 2 hours.

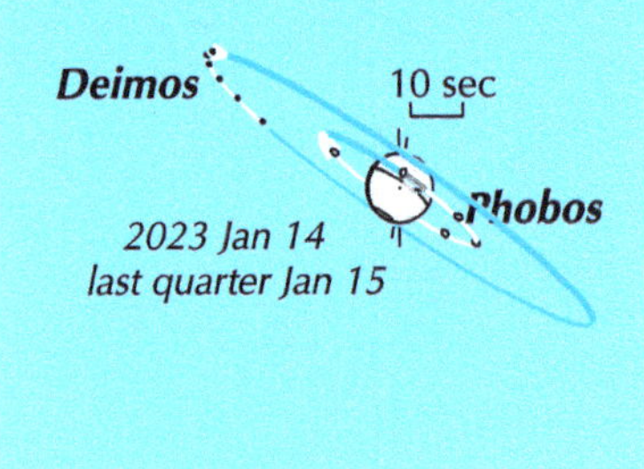

The satellites move in almost circular orbits, in planes slightly varying from Mars's equator. Their tracks are shown in white for 6 hours, starting at 0h UT.

Phobos orbits in only 7.65 hours, Deimos in 30.3 hours. Since Mars rotates in 24½ hours, Phobos travels more than three times faster than the planet's surface: seen by a Martian, Deimos goes over slowly from east to west (more than 2 days from rising to setting); Phobos goes in the opposite direction, rising in the west and setting in the east, twice a day! The satellites are exaggerated 30 times

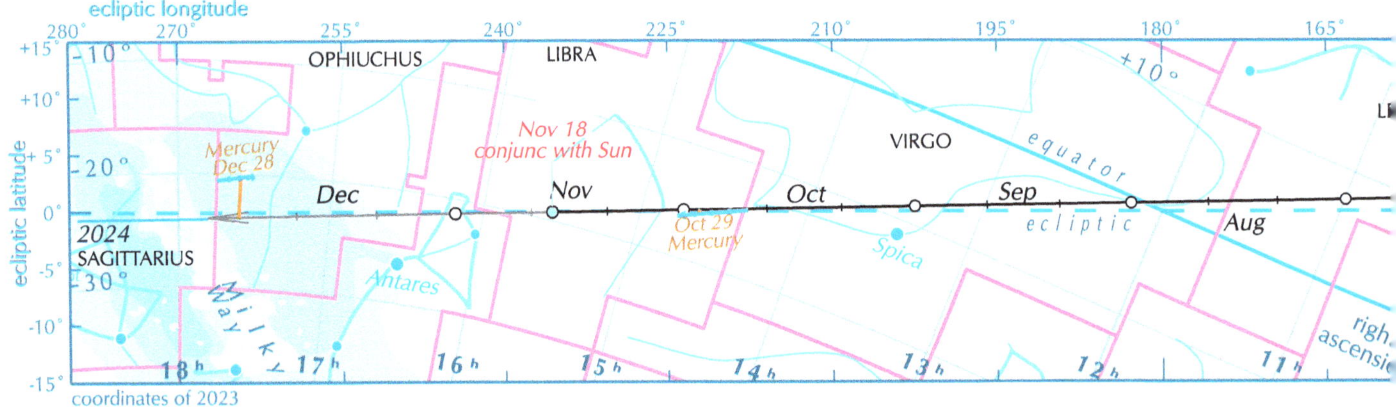

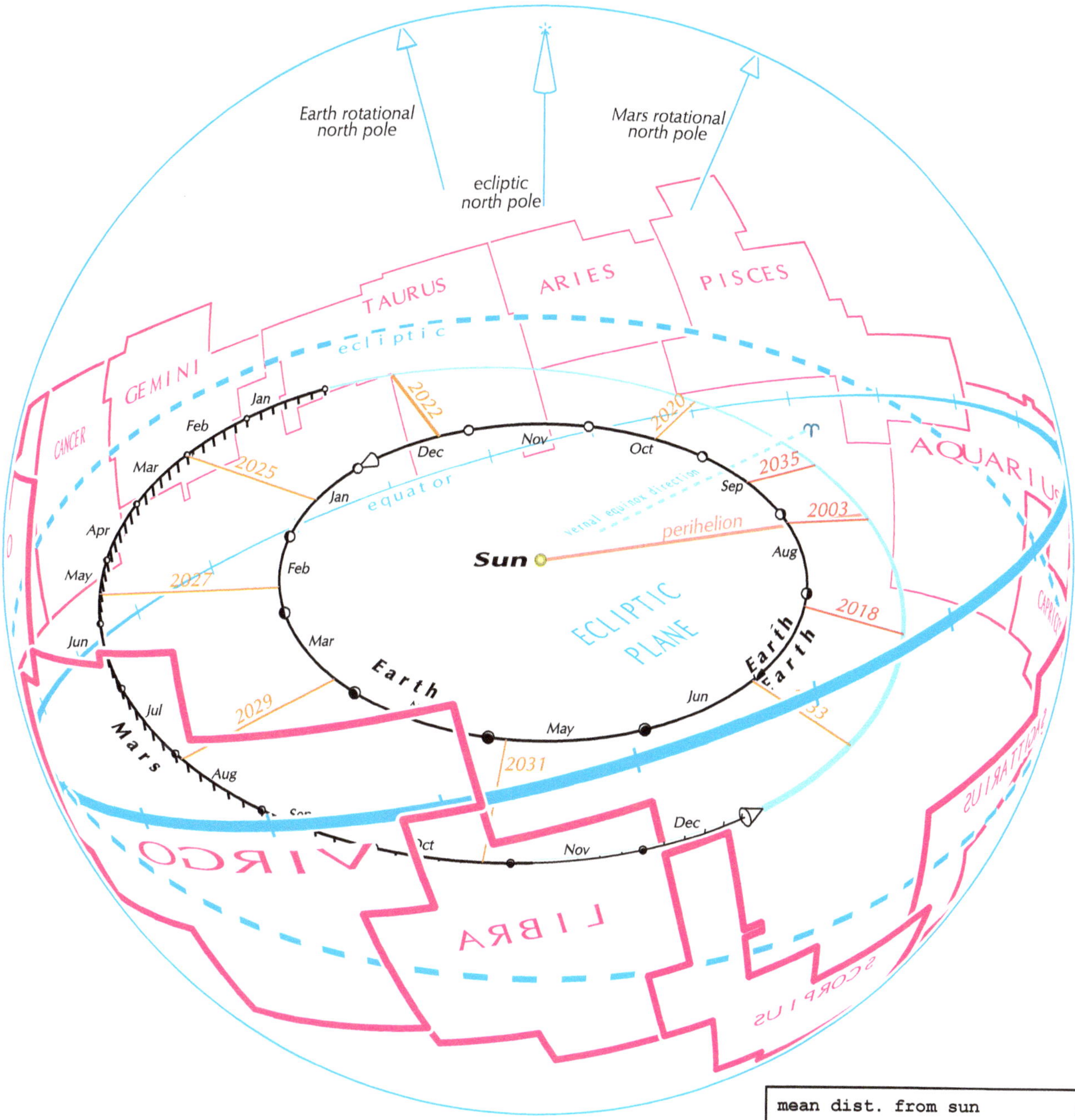

mean dist. from sun	1.52 AU
sidereal period 1.88 years =	687 days
synodic period 2.13 years =	780 days
eccentricity	.093
inclination	1.9°
diameter	6,790 km
satellites	2

in size. Both are elongated: dimensions of Phobos are 27x22x19 kilometers, and Deimos 15x12x11 (as against the 6,800 km diameter of Mars). So they are shown as ellipses. They rotate synchronously, keeping the same face to Mars. Phobos and Deimos are about 13 and 14 magnitudes fainter than Mars.

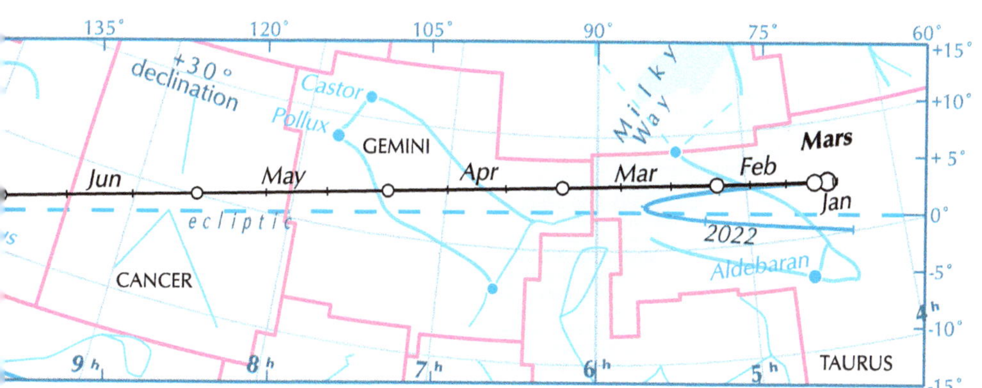

Mars's track against the star-
round, ecliptic-based like the
and Venus maps. The scale
m to 1°. The track is drawn
vhen Mars is in the morning
r conjunction with the Sun).
the tracks for the neighbor-
are included (in blue). Short
es connect Mars to other
vhen they appear closest.

OUTER PLANETS

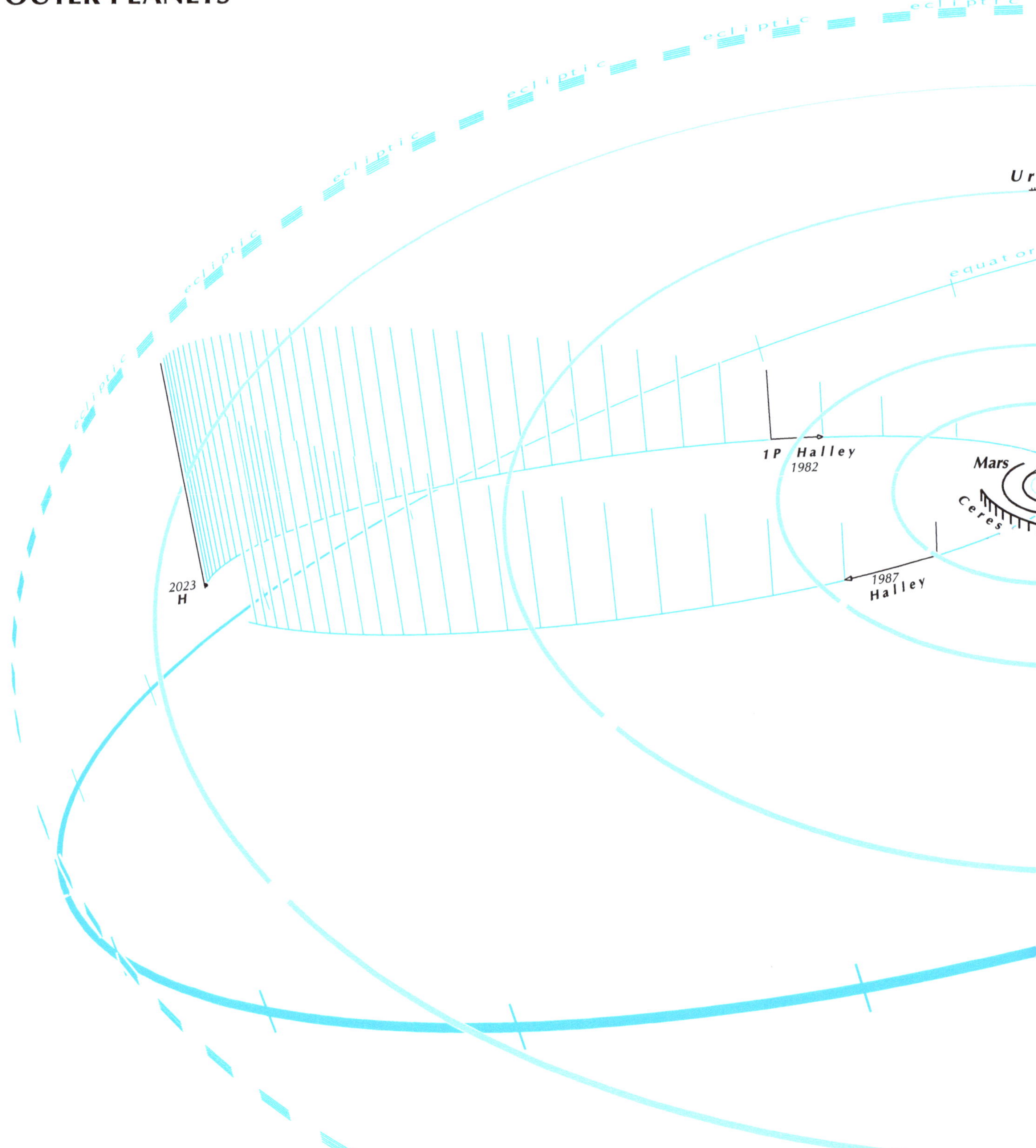

Space view of the planetary system. The major planets' orbits are shown in blue, their paths for this year (omitting Mercury and Venus) in black, with stalks to the ecliptic plane at monthly intervals. The lines from Earth to the planets are at their oppositions. Also shown are a few minor bodies (of which there could be thousands in the picture): dwarf planets Ceres (the largest of the Main Belt asteroids, between Mars and Jupiter) and Pluto (the most prominent of the trans-Neptunian objects; it crossed outward over Neptune's orbit in 1999); and Comet 1P Halley, which at its last visit was first observed in 1982, was at perihelion in 1986, and is now 35.3 AU from the Sun. The viewpoint is 100 AU from the Sun. The equatorial and ecliptic planes are represented by circles around the sky at 35 AU from the Sun.

The four giant planets are in a narrowing quadrant of the sky. The longitude from Saturn to Uranus shrinks from about 81° to 74°. Jupiter overtook Neptune on 2022 June 6, and will overtake Uranus on 2024 March 14.

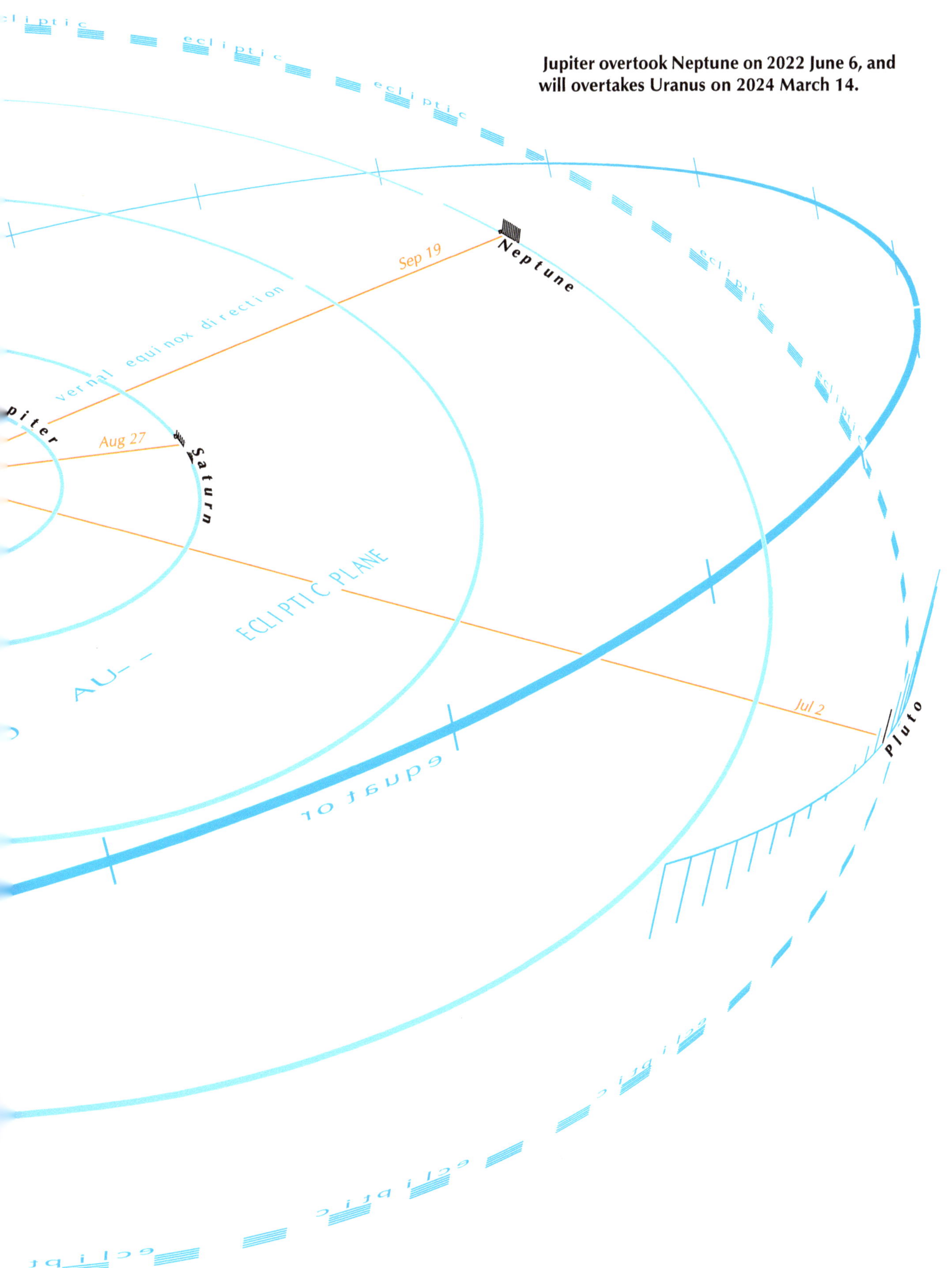

Jupiter overtook Neptune on 2022 June 6, and will overtakes Uranus on 2024 March 14.

JUPITER

mean dist. from sun	5.2 AU	inclination	1.3°
sidereal period	11.86 years	diameter	143,000 km
synodic period	399 days	satellites	67
eccentricity	.048		

Jupiter 2023		RA	.... dec	hedis	gedis	elo	mag
Jan 13	5 on equat.,to nor.	0 12	0 0	4.951	5.198	70	-2.3
Jan 20	9 perihelion	0 17	0 29	4.951	5.303	64	-2.3
Apr 11	22 conjunc.with Sun	1 22	7 30	4.953	5.955	1	-2.1
Aug 7	0 west quadrature	2 49	14 56	4.962	4.858	-90	-2.4
Sep 4	14 stat.in long>retr	2 54	15 14	4.966	4.438	-116	-2.6
Sep 4	20 stat.in r.a.>retr	2 54	15 14	4.966	4.434	-116	-2.6
Nov 3	5 opposition	2 34	13 38	4.975	3.983	-179	-2.9
Dec 31	2 stat.in long>dir.	2 15	12 16	4.985	4.468	116	-2.6

The giant planet has nearly 3 times the mass of all the others together (though less than 1/1000 of the Sun's). It takes nearly 12 years to go around the Sun. Each year, it advances about 30°, spending roughly a year in each zodiacal constellation.

So the Earth takes a bit more than 13 months to catch up and again pass Jupiter at opposition. (This is Jupiter's "synodic" or seen-from-Earth period.) The oppositions advance from January to February and so on, also moving later in each month, so that occasionally a month is skipped; and, after 11 oppositions, a year is skipped, as 2013 and 2025.

The cycle of oppositions has to start with Jupiter in Gemini, to which Earth faces outward in January. Now comes the 10th opposition of the cycle, in November, and it is in Arites that Jupiter performs its retrograde loop, going into apparent backward motion as we pass it. The retrograde loops run along the planet's celestial track like kinks along a flicked rope; next year's will be in Taurus.

At the start of this year, Jupiter has just come out of the retrograde loop of which the last opposition was the center. When this happens in Pisces or in Virgo, there can be, as now, a triple crossing of the celestial equator: northward on 2022 May 25, southward 2022 Sep. 25, and now again northward on Jan. 13. Jupiter remains in the north celestial hemisphere till 2028 Sep. 5 — a single crossing, as is much more common.

Jupiter moves down closer to the setting Sun, in April disappears behind it, and in May appears above Mercury in the morning sky, which it dominates until Venus appears in August. In September the retrograde loop begins: Jupiter appears to swirl back toward us as we overtake it.

At opposition on Nov. 3, Jupiter is fairly well north. Being now 10 months past perihelion, it is not quite as near as at the previous opposition. Its rather oblate (flattened) globe appears 49.4" wide, and shines at magnitude -2.9, about as bright as it can be.

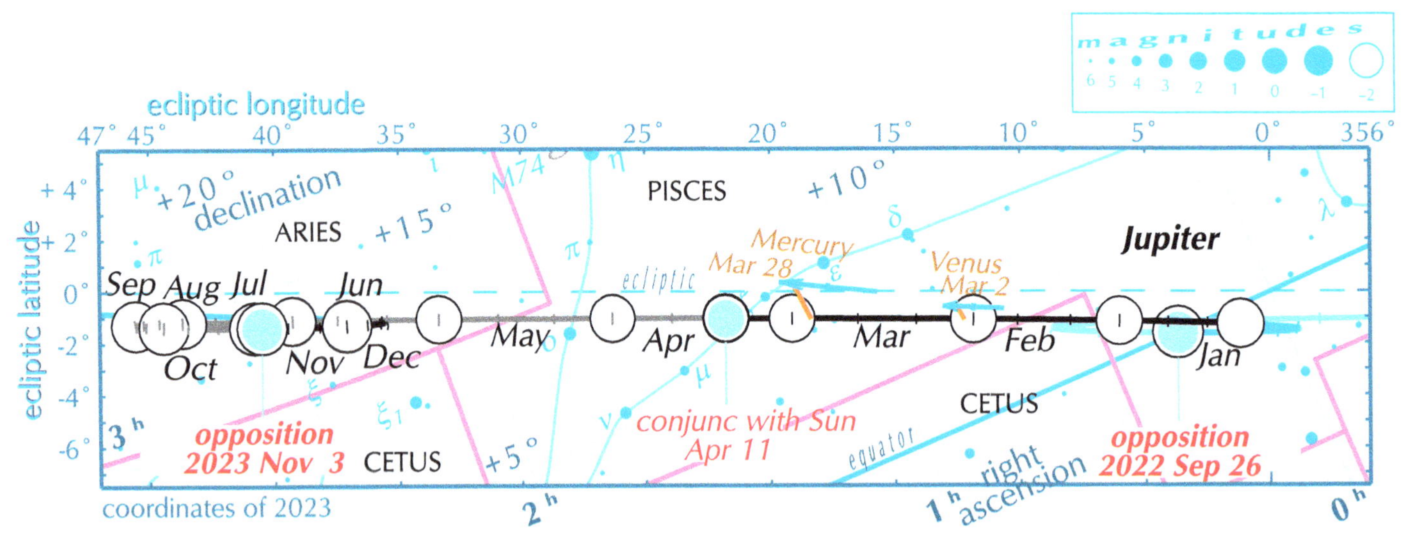

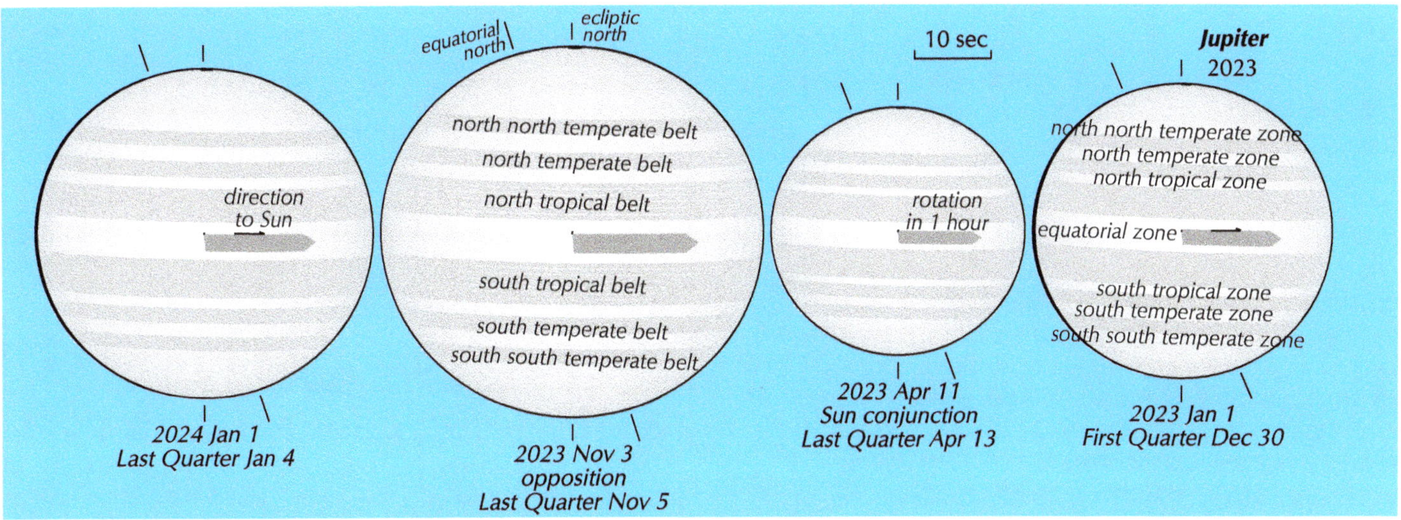

Jupiter at start and end of the year and when farthest (Sun-conjunction) and nearest (opposition). Shown are the planet's light "zones" and darker "belts." Ecliptic north is up, and a longer line shows the direction to equatorial north.

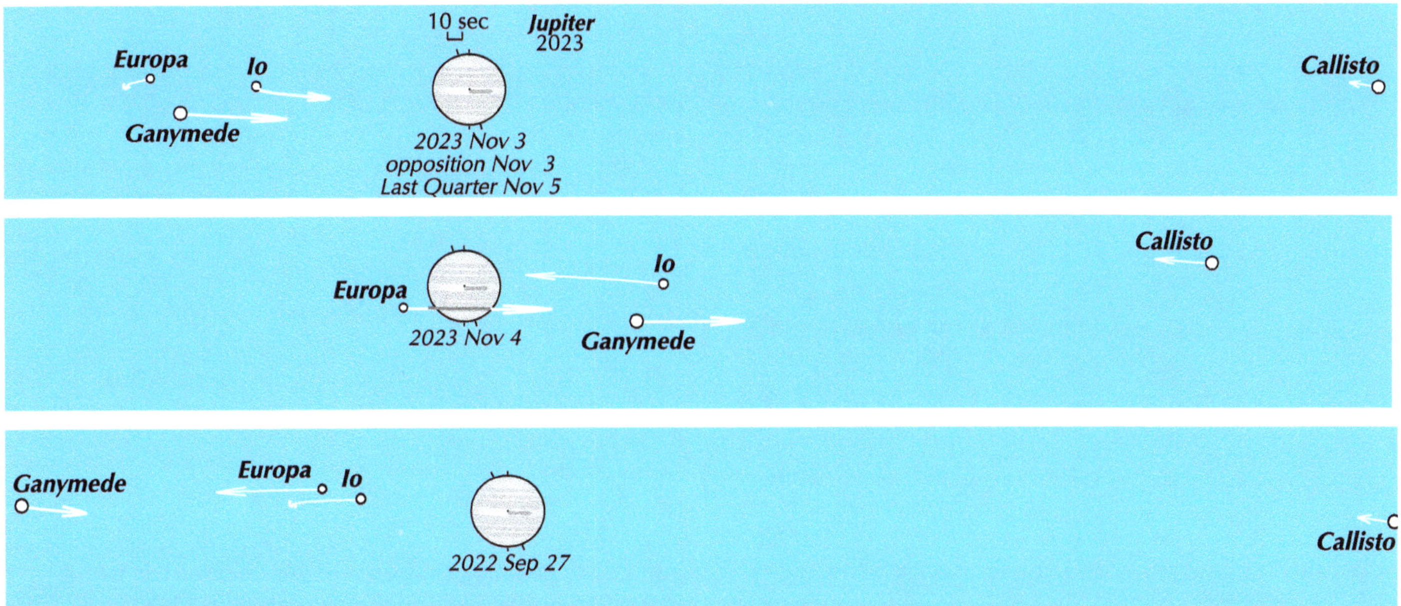

The four major or "Galilean" satellites of Jupiter (among Galileo's first discoveries with his telescope, in Jan. 1610), on the day of opposition and the next two days. Their movements are shown from 0h to 6h UT: by North American clocks, 5 or more hours earlier, so in the early night of the previous date. Ecliptic north is at top. The satellites are exaggerated 5 times in size. They move almost in circles, almost exactly in the plane of Jupiter's equator, which is tilted about 3° to the plane of its orbit. This plane pointed straight at us in 2021, so that the satellites appeared to move in straight lines. Since then the plane has passed south of us.

Jupiter-Saturn conjunctions

Jupiter overtakes Saturn roughly each 20 years and roughly 2/3 of the way onward around the circle. The latest of these "Great Conjunctions" was on 2020 Dec. 21. It was the closest since 1623. The two came to within a tenth of a degree apart, and could be seen, with their families of satellites, in one telescope field. It happened rather low in the evening sky, 30° from the Sun. But the 1623 event, shortly after the 1609 invention of the telescope, was less than half as far out, 13°, in the sunset sky. For a great deal more about Great Conjunctions, see www.universalworkshop.com/conjunctions-of-jupiter-and-saturn/

Jupiter is now ahead of Saturn by about 46° at the beginning of the year and 68° at the end.

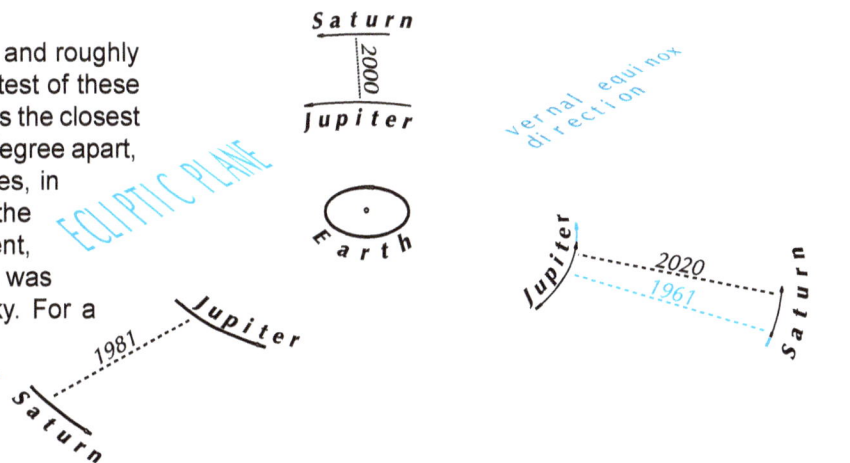

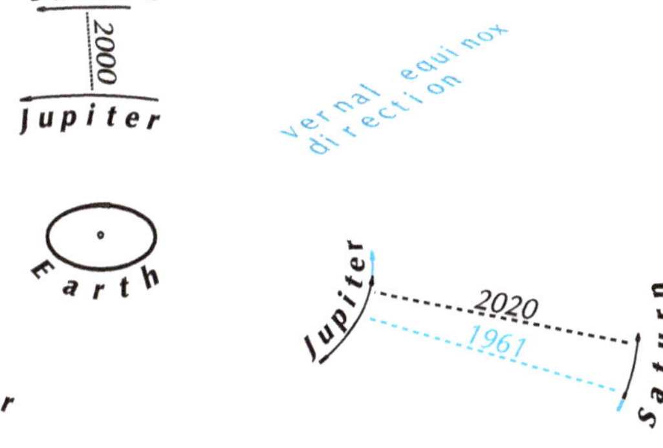

SATURN

mean dist. from sun	9.5 AU	inclination	2.5°
sidereal period	29.46 years	diameter	120,500 km
synodic period	378 days	satellites	62
eccentricity	.056		

Saturn 2023			RA .... dec		hedis	gedis	elo	mag
Feb 16	17	conjunc.with Sun	22 2-13	26	9.824	10.812	1	0.8
May 28	11	west quadrature	22 37-10	23	9.798	9.745	-90	1.0
Jun 17	16	stat.in long>retr	22 38-10	20	9.792	9.414	-109	0.9
Jun 18	13	stat.in r.a.>retr	22 38-10	20	9.792	9.400	-110	0.9
Aug 27	8	opposition	22 26-11	46	9.773	8.763	-178	0.4
Nov 4	6	stat.in long>dir.	22 13-12	55	9.754	9.387	109	0.8
Nov 4	16	stat.in r.a.>dir.	22 13-12	55	9.754	9.393	109	0.8
Nov 23	10	east quadrature	22 14-12	46	9.749	9.699	90	0.9

The sixth planet was the outermost—beyond it only the sphere of the stars—till the discovery of Uranus in 1781.

In everything except its superb rings, Saturn ranks second to Jupiter: 0.8 times as wide, 0.3 as massive, hence 0.5 as dense—lighter than water. It is 1.8 times farther from the Sun, thus appears on average 0.4 as wide and about 3 magnitudes dimmer. Moving through space 0.7 times as fast, it takes 2.5 times as long, or nearly 30 years, to finish an orbit. (5 revolutions of Jupiter equal about 2 of Saturn.) So whereas Jupiter spends about one year in each zodiacal constellation, Saturn averages 2½. Its events as seen from Earth, such as opposition, fall about 13 days later each year.

Saturn now advances from Capricornus into Aquarius, where it will spend two years.

At the beginning of this year, Saturn is low in the sunset sky, disappears behind the Sun in February, appears beside Mercury in the morning sky in March. It is still quite low for northern observers, being south of both the equator and the ecliptic. The retrograde phase, when we come nearest to Saturn, stretches from June to November.

Saturn is on the inward curve of its slightly elliptical orbit (eccentricity 0.057, slightly more than Jupiter's); it was at aphelion in 2018 and will reach perihelion in 2032. So at opposition on August 27, it is a few million kilometers nearer than last year. It is 9.77 AU from the Sun and 8.76 AU from Earth. The ball of the planet appears 18.9" wide; the rings are 2.29 times wider. Saturn's magnitude at opposition is 0.44; it can be as bright as -0.48, as in 2002, 2031, and 2032, or as dim as 0.6, as in 2024.

The rings, revolving in the plane of the plane's equator, were edge-on to Sun and Earth in 2009. The planet went on into the half of its orbit in which the northern face of the rings is sunlit and open toward us. A quarter of the way round, in 2017, it was most open, and it will be again edge-on in 2025.

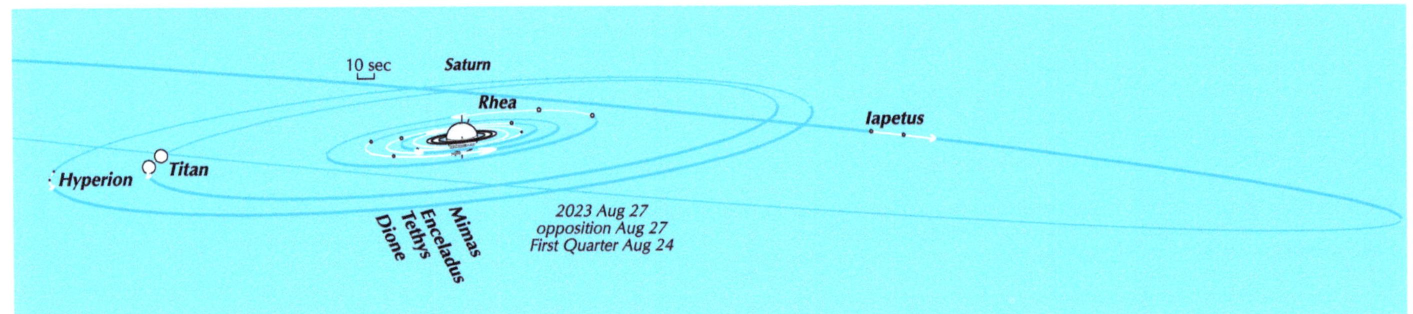

Major satellites of Saturn. They are drawn at 0h and 12h UT of the day of opposition, exaggerated 10 times in size. In contrast with the picture for Jupiter, it is more convenient to have equatorial north at the top. The shorter line from the planet points to ecliptic north. Saturn's equatorial plane, which is also the plane of the rings and inner satellites, is almost parallel to Earth's equatorial plane. The orbits of the inner satellites, like the rings, were most open toward us in 2017 and will be edge-on in 2025, so that already the inner three, Mimas Enceladus, Tethys, become occulted by the planet.

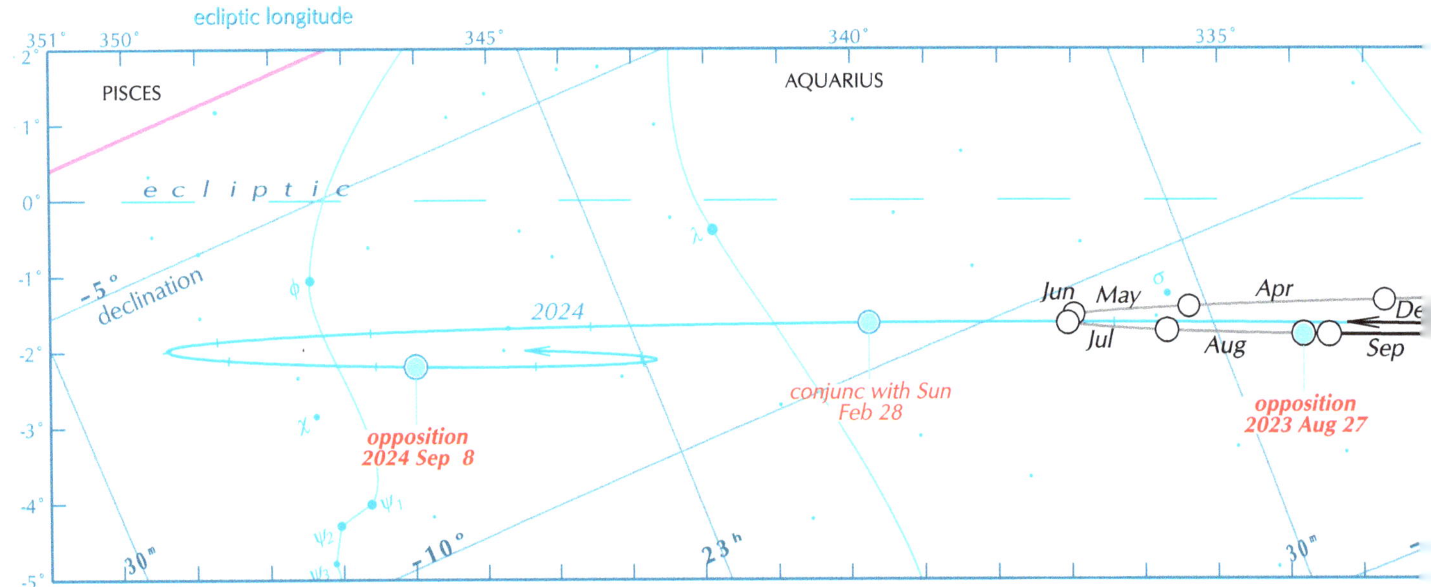

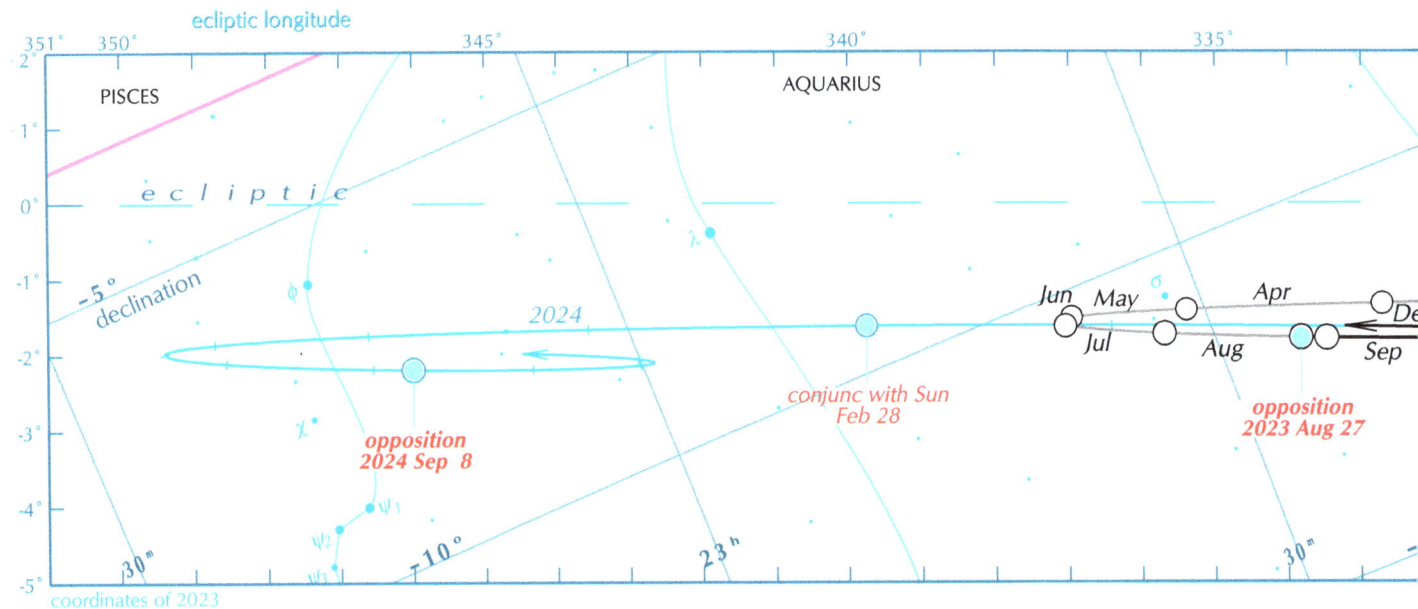

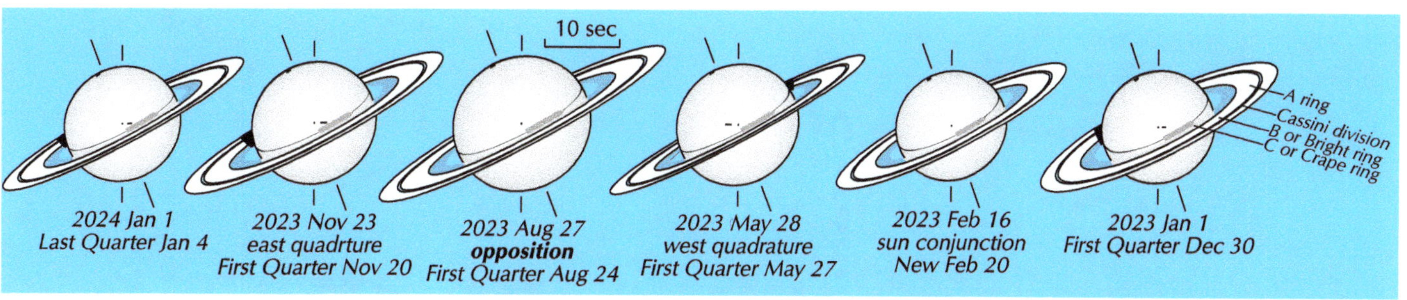

Saturn and rings this year. Times around east and west quadrature are better than opposition for seeing thee shadow on the rings, and eclipses of the satellites.

Latitude of Sun and Earth as seen from Saturn's equator and rings. For the Sun, two lines very close together represent its width of about 0.1° as seen from Saturn. The oscillation of Earth's view of Saturn is caused by its own circling around the Sun. Near dates of opposition and Sun-conjunction, Earth and Sun lie along one line to Saturn, so their latitudes as seen from Saturn are almost the same. The rings were edge-on to the Sun on 2009 Aug. 11, to Earth 2009 Sep. 4; were most open to the north in 2017; will be edge-on to Earth on 2025 Mar. 23, to the Sun 2025 May 6.

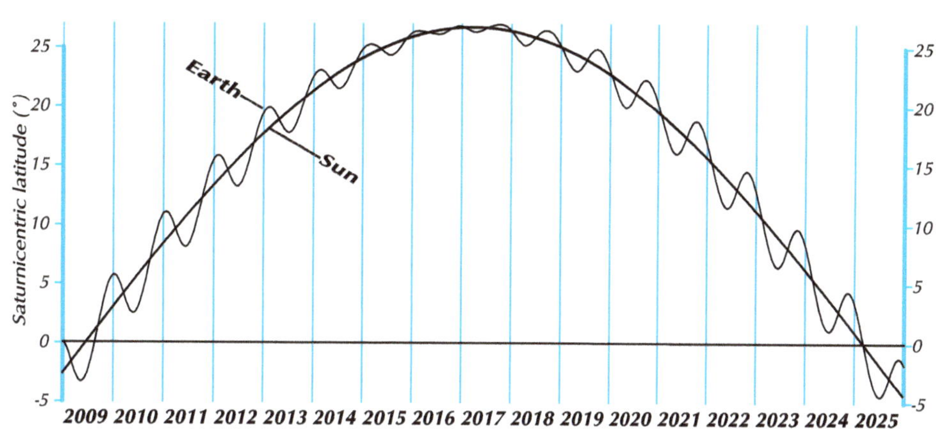

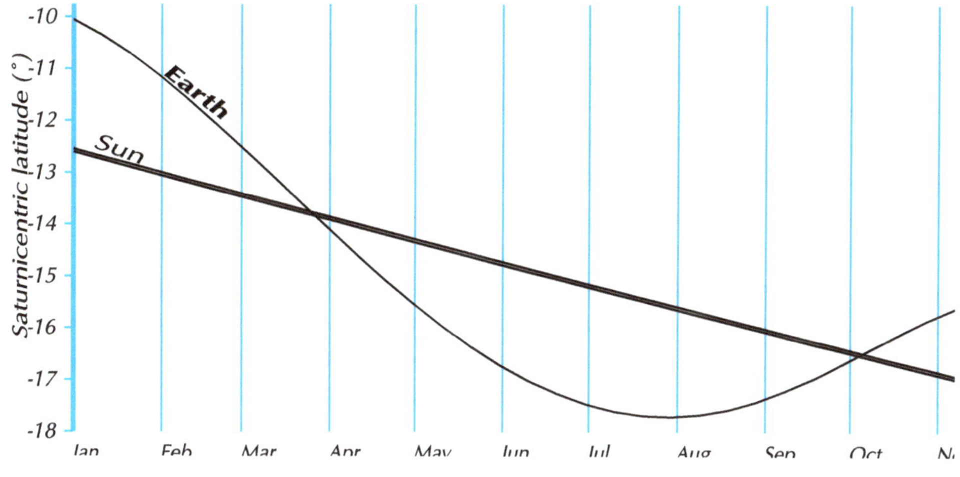

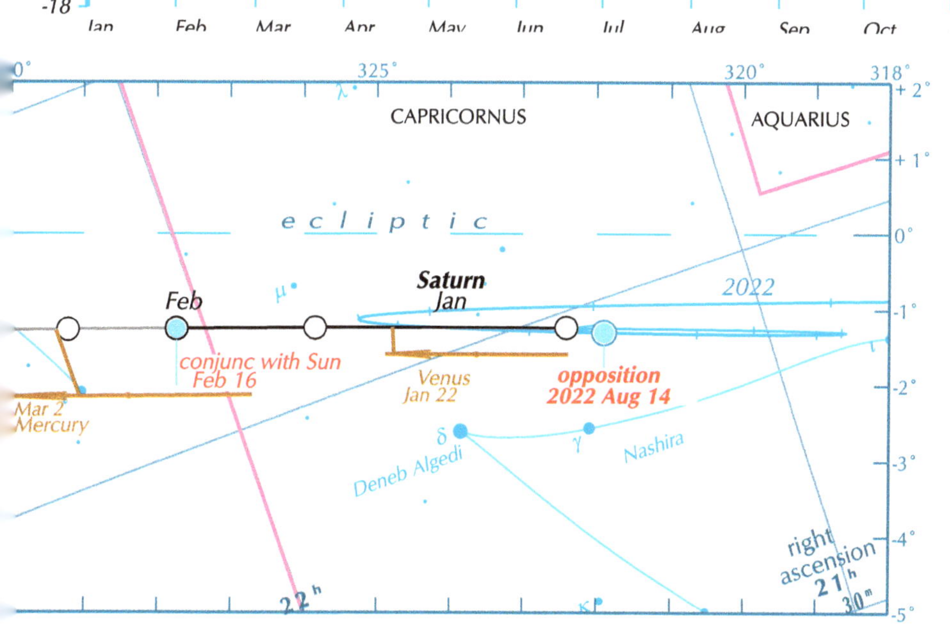

Ecliptic chart, as for Jupiter except that the scale is larger. Stars are plotted to magnitude 7.

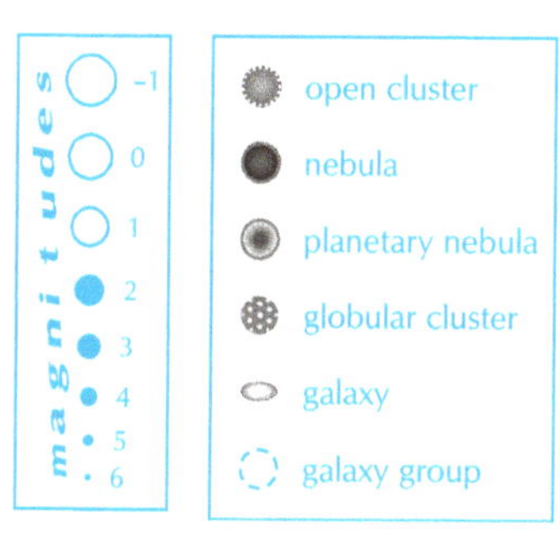

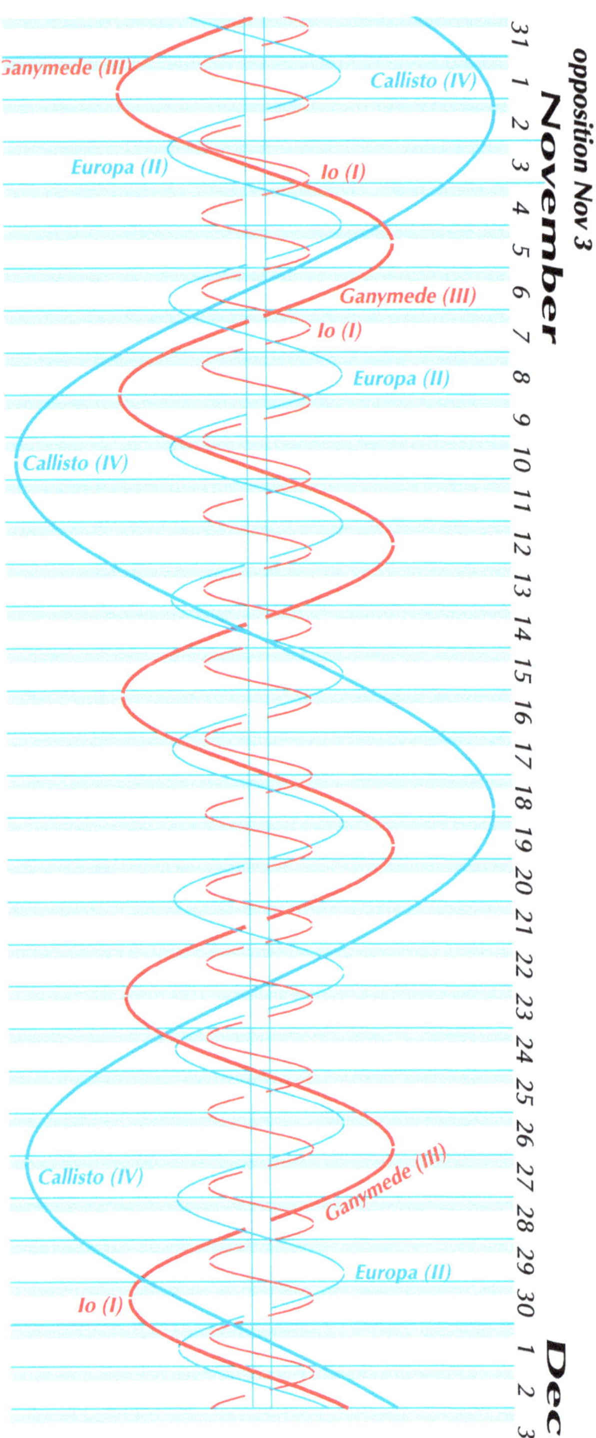

Corkscrew graphs of the major satellites of Jupiter and Saturn, for spans of 60 days around the planets' oppositions. The cross dimension is apparent distance from the planet's center. The scale bar at the top shows this distance in minutes of arc. The pair of lines down the middle represents the planet's width. For Saturn, an outer pair of lines shows the width of the rings. The curves are red for the odd-numbered satellites, and thick for the larger ones (Jupiter's over 4,000 km in diameter, Saturn's over 1,000 km).

Each cross line is at 0h UT. This is 7 PM EST (8 PM EDT). Thus the light blue bar roughly represents night in eastern America, starting in the evening of the previous date. In Europe this night-bar would be higher, centered on the cross-line; farther west in America, it should be slightly lower.

Jupiter is 143,000 km wide. The distances of the satellites from its center are: Io 422,000 km, Europa 671,000, Ganymede 1,070,000, Callisto 1,880,000. In apparent width, Jupiter varies from 33″ when farthest from us, at Sun-conjunction, to 49″ when nearest, at opposition. The apparent width of the satellite-system swells and shrinks likewise.

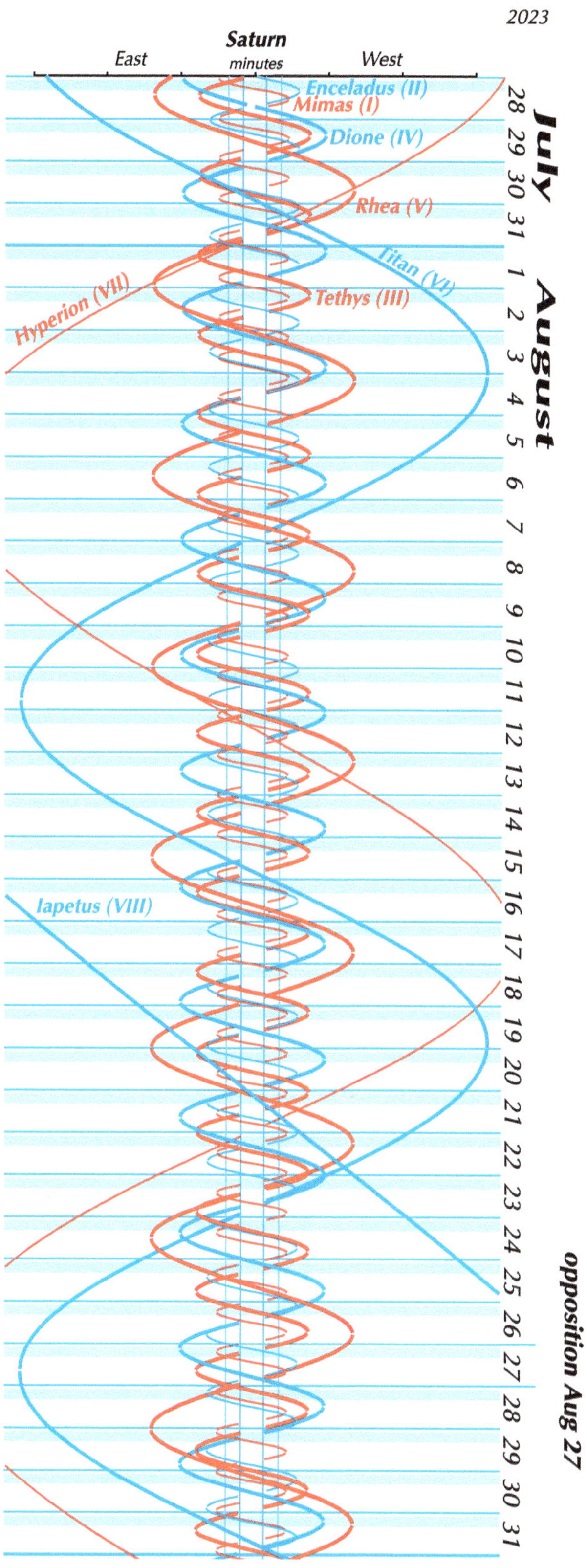

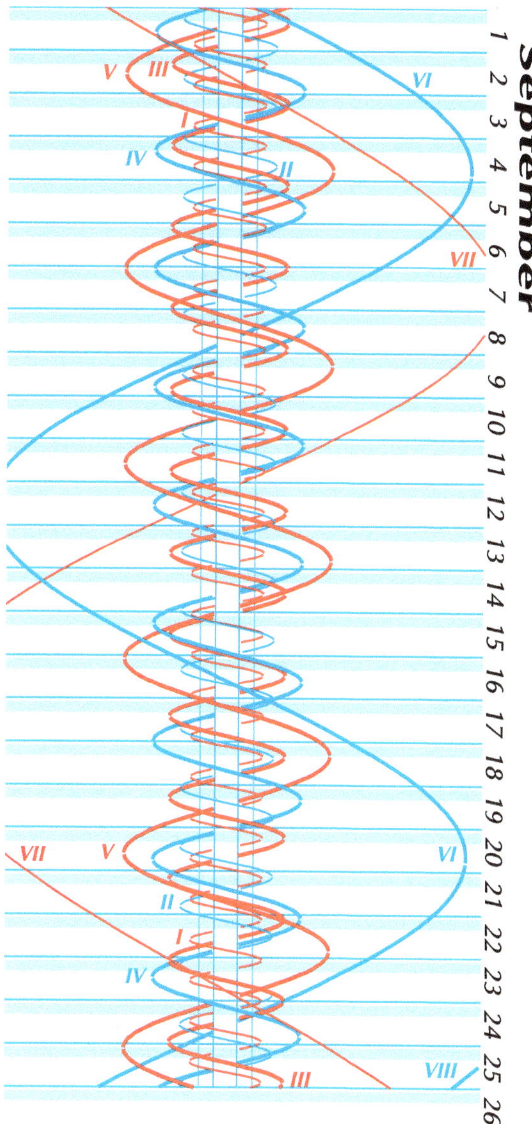

East is to the left, because north is to the top. If you're about to look in an inverting telescope, turn the picture around.

The satellites, orbiting their planet in the "direct" or typical solar-system way, pass on the near side when going from east to west (through their inferior conjunctions with the planet), so that (1) they transit the planet and (2) their shadows transit it. When going the other way they pass behind it, and then their curves are interrupted wherever they are (3) occulted behind the planet or (4) eclipsed in its shadow. Jupiter's major satellites go through these events at every revolution, except for the most distant, Callisto, which at present is passing too far north and south. Some of Saturn's satellites pass far north and south.

Near opposition, planet and satellites loom largest but get in front of their own shadows, so it is good also to look many days before or after, when the shadows are cast more sharply east or west.

URANUS AND NEPTUNE

URANUS	
mean dist. from sun	19.2 AU
sidereal period	84.0 years
synodic period	370 days
eccentricity	.047
inclination	.77°
diameter	51,118 km
satellites	27

NEPTUNE	
mean dist. from sun	30 AU
sidereal period	165 years
synodic period	367 days
eccentricity	.009
inclination	1.8°
diameter	49,568 km
satellites	14

The family of six known planets first increased on 1781 March 13, when William Herschel, a German musician living at Bath in England, discovered what he took for a comet: it was the planet soon named Uranus. It is just above the naked-eye brightness limit and had been at least 20 times plotted on maps as a star.

By 1844, small discrepancies in its motion suggested that there was a yet more remote planet, which Uranus would have passed in 1821. Urbain Leverrier in France calculated its position, as a result of which Galle and d'Arrest at Berlin telescopically found Neptune on the evening of 1846 Sep. 23. It used to be considered that John Couch Adams had equal credit. He had been earlier in working on the solution, but had bad luck with the British astronomy establishment, and his work did not lead to the discovery.

Uranus and Neptune creep along vast orbits, 19 and 30 times Earth's distance from the Sun, at 6.8 and 5.5 kilometers a second, advancing only about 4° and 2° a year. So their yearly events, such as oppositions, fall about 4 and 2 days later. After their unseen conjunction of 1821, Uranus pulled all around the sky to its first observed overtaking of Neptune, in 1993. In 2011 Neptune finished its first orbit since discovery. In 2022, Uranus is from about 53° to 55° ahead of Neptune.

For many decades both these planets were in the southern celestial hemisphere. Then Uranus crossed the equator northward in 2011-2012 (a triple event because of retrograding). Neptune will not follow across the equator till 2027.

Uranus and Neptune are now traveling 0.3° and 1.2° south of the ecliptic. Uranus was at southernmost latitude in 2007 and will be at ascending node in 2029. Neptune's orbit is inclined 1° more, but its apparent difference from the ecliptic is reduced by its distance. It descended through the ecliptic in 2003 and will not reach its 1.77° southernmost latitude till 2044. In years near the nodes, the planets' forward and backward paths become almost straight lines. From 2000 to 2007 Neptune remained near enough to the ecliptic that at each inferior conjunction it was occulted by the 1/2°-wide Sun (an event that of course we could not observe).

Uranus is at a distance such that the apparent retrograde parts of its path are just short of 5 months long. They almost kiss: the point where Uranus turned back last year is close to that to which it retreats this year. (And to that of last year's Sun-conjunction, halfway between them in time.)

Neptune's retrograde paths are a week longer, and overlap. This even more distant planet is further toward being like a "fixed" star. The overlapping means that Neptune goes five times past any adjacent star—a quintuple conjunction, as with the star mu Capricorni in 2009-2010.

Uranus was at aphelion in 2009 (20.099 AU from the Sun) and will be at perihelion in 2050, so it is still at greater than

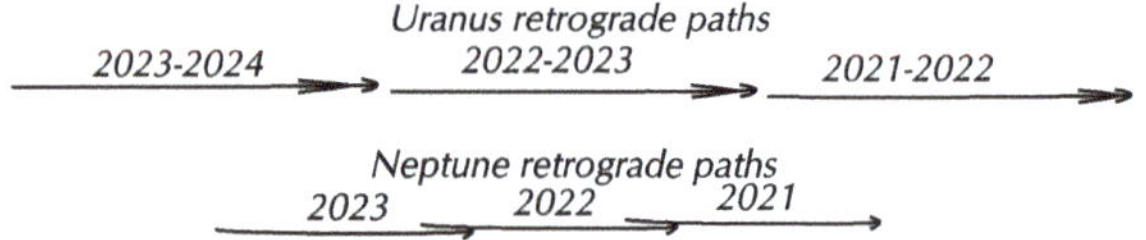

Retrograde paths of Uranus and Neptune in successive years, from each first stationary point back to the second.

average distance. Neptune was at aphelion in 1959 and will be at perihelion in 2042. But these are non-simple events, because Neptune's orbit, the nearest to circular after Venus's, has a ripple (explained in the "Neptune's circular yet wavy orbit" section of *Uranus, Neptune, Pluto*).

These giant planets are not much different in size, and despite Neptune's 50% greater distance their apparent widths at opposition are not much different 3.8" and 2.2". But the difference between their magnitudes of 5.6 and 7.9 means that the star-like image of Uranus is findable by a keen naked eye on a good night, whereas Neptune is 7 times fainter.

Close up, Uranus may be more greenly blue than Neptune, but in telescopes these colors are subtle. Uranus's

Ecliptic charts, at scale 2 cm per degree. Tracks are gray when the planets are in the morning sky. Lines to other planets are at dates of conjunction in right ascension

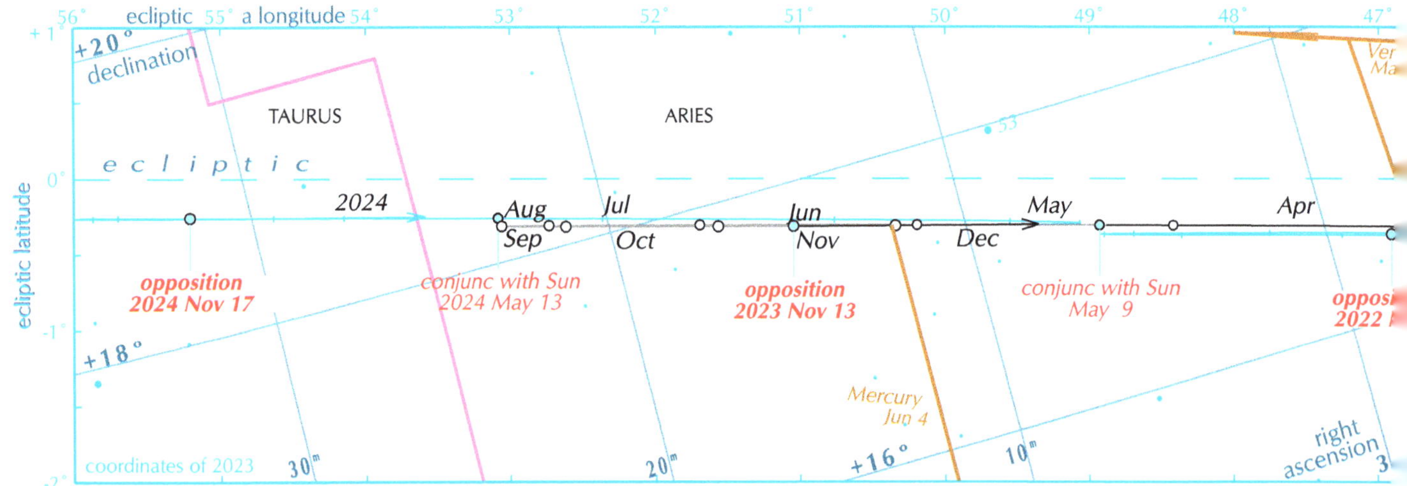

surface is bland, Neptune's shows more activity. Neptune is slightly smaller, but more massive, hence denser.

Uranus is the odd planet in that it rotates "on its side": almost perpendicularly to the plane of its orbit (and ours). (Which of its poles is really "north"?—for more on this problem, see "north" in our glossary *Albedo to Zodiac.*) Its equatorial plane becomes edge-on to the Sun each 42nd year; this Uranian equinox happened (for the 3rd time since discovery) on 2007 Dec. 16.

Uranus's first two satellites, Titania and Oberon, were also discovered by Herschel, in 1787; he suspected a ring in 1789, but the thin rings were really discovered in 1977, and 13 are now known. Rings and chief satellites revolve in the planet's equatorial plane. Transits of the satellites and their shadows across the planet were first observed in 2006, and were possible for about 3 years.

Uranus	RA	dec	hedis	gedis	elo	mag	dia"#
Jan 22 21 stat.in long>dir.	2 50	15 59	19.667	19.432	102	5.7	3.6
Jan 23 0 stat.in r.a.>dir.	2 50	15 59	19.667	19.435	102	5.7	3.6
Feb 4 3 east quadrature	2 51	16 1	19.665	19.640	90	5.7	3.6
May 9 20 conjunc.with Sun	3 6	17 9	19.650	20.660	0	5.9	3.4
Aug 16 2 west quadrature	3 23	18 13	19.635	19.609	-90	5.7	3.6
Aug 28 24 stat.in long>retr	3 23	18 14	19.633	19.393	-102	5.7	3.6
Aug 29 0 stat.in r.a.>retr	3 23	18 14	19.633	19.392	-102	5.7	3.6
Nov 13 17 opposition	3 15	17 43	19.621	18.632	-180	5.6	3.8

Neptune	RA	dec	hedis	gedis	elo	mag	dia"
Mar 15 24 conjunc.with Sun	23 44	-3 1	29.911	30.905	1	8.0	2.2
Jun 19 4 west quadrature	23 53	-2 3	29.908	29.891	-90	7.9	2.2
Jun 30 15 stat.in long>retr	23 53	-2 3	29.908	29.699	-101	7.9	2.3
Jul 1 7 stat.in r.a.>retr	23 53	-2 3	29.908	29.688	-102	7.9	2.3
Sep 19 11 opposition	23 48	-2 39	29.906	28.902	-179	7.8	2.3
Dec 6 9 stat.in long>dir.	23 43	-3 11	29.904	29.702	101	7.9	2.3
Dec 6 19 stat.in r.a.>dir.	23 43	-3 11	29.904	29.709	101	7.9	2.3
Dec 17 4 east quadrature	23 43	-3 10	29.904	29.888	90	7.9	2.2

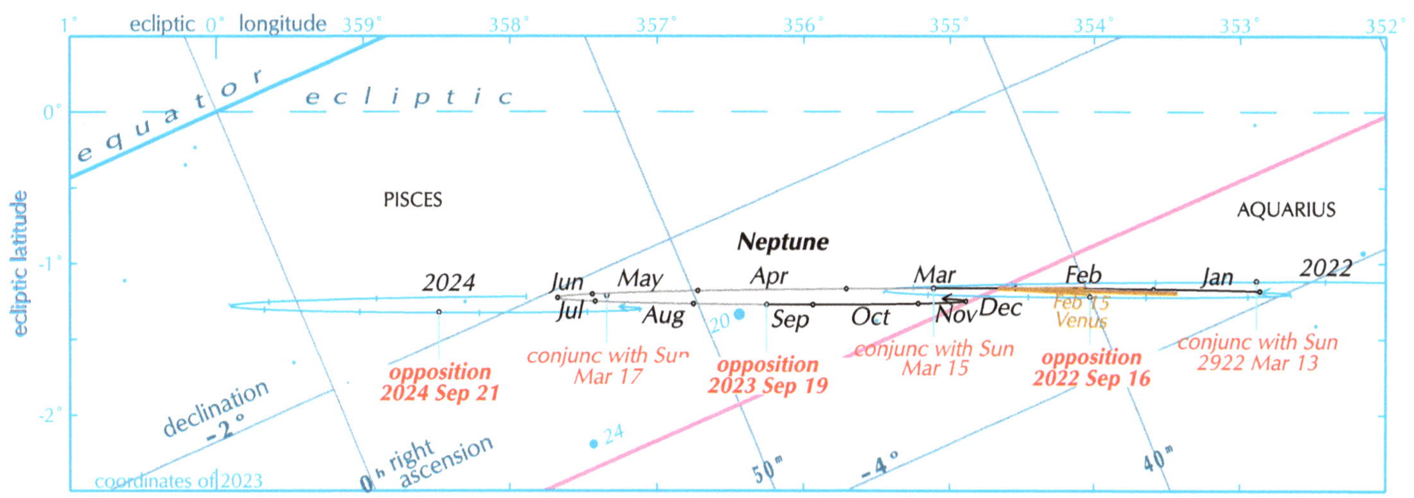

There is far more to be said about the discovery story of the outer planets, and the interlocking patterns of their motions. It's all in our book *Uranus, Neptune, Pluto: A Longer View.*

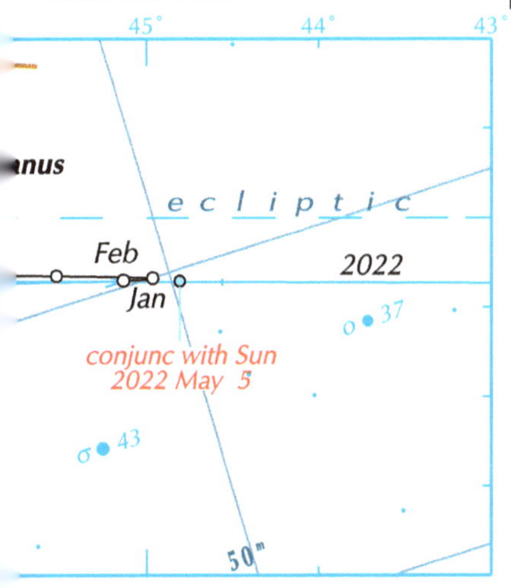

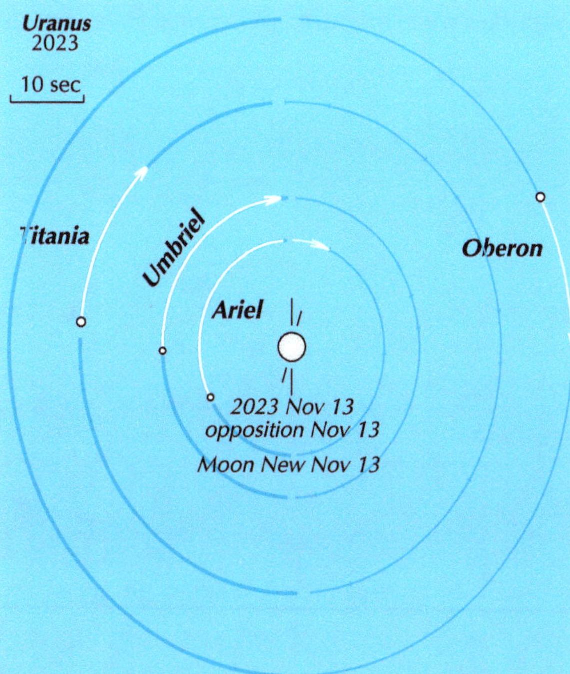

Major satellites. Their tracks are shown in white for one UT day. The satellites' sizes are exaggerated by 10. The longer pointers above the planets are to the north celestial pole (at top), the shorter to the north ecliptic pole.

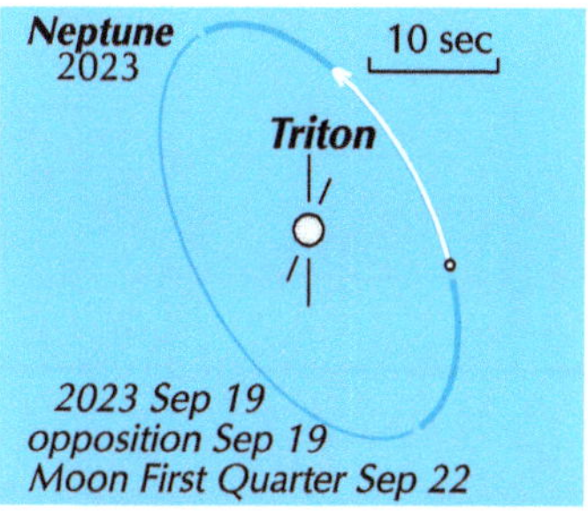

ASTEROIDS

Orbital and other facts.
q: perihelion distance.
a: mean distance.
Q: aphelion distance.
e: eccentricity.
P: period.
i: inclination.

name	discov.	diam. km	q AU	a AU	Q AU	e	P years	i °
1 Ceres	1801	939	2.56	2.77	2.98	0.07	4.61	11
2 Pallas	1802	511	2.13	2.77	3.41	0.23	4.62	35
3 Juno	1804	275	1.98	2.67	3.35	0.26	4.36	13
4 Vesta	1807	525	2.15	2.36	2.57	0.09	3.63	7

The age of the asteroids began on 1801 Jan. 1, when Giuseppe Piazzi, at Palermo in Sicily, discovered what he and the astronomers of Europe hoped was the "missing planet" in the wide gap between Mars and Jupiter. He named it Ceres, for the Roman goddess of the harvest and patroness of Sicily. Three more were discovered in the next few years, and also named for classical goddesses: Pallas, Juno, Vesta.

There was a pause till 1845, when the 5th was found. Then as these small bodies became numerous they were no longer called planets but *asteroids*, "star-like," because they did not show perceptible disks like the planets; later, more officially *minor planets*. Thousands are discovered each year, and there must be millions that are over a few meters in size. They receive temporary designations, then numbers when their orbits are securely known.

Most circulate in a Main Belt between Mars and Jupiter, but there are peculiar classes that go nearer in, some across Earth's orbit, or farther out. All are below naked-eye brightness except sometimes 4 Vesta.

The First Four are reliably observable each year, though 3 Juno is usually the dimmest and is surpassed by various others from lower down the list that happen to be at favorable oppositions. These Main Belt asteroids have periods of between 3 and 5 years; in each of their circuits of the sky they miss opposition in the year when they pass the direction outward from Earth's December-January position.

In 2023, 2 Pallas, 1 Ceres, and 4 Vesta pass through opposition (in January, March, and December).

1 Ceres is so much larger than the other asteroids that it has been re-classified as a dwarf planet, along with Pluto (formerly regarded as a major planet) and the trans-Neptunian bodies Eris, Haumea, Makemake, and possibly others—all larger than Ceres but far more distant.

Ceres was at perihelion on 2022 Dec. 7, 2.55 AU from the Sun. As we overtake it early in 2023, it goes into its apparent backward loop and through opposition in March, in Coma Berenices quite high north of the ecliptic plane. Because of the near-circularity of its orbit, its opposition magnitude does not vary much; it can be as bright as 6.8, as in 2012 Dec. and 2035 Dec., or as dim as 7.7, as in 2020 Aug. and 2029 Aug. This time it is slightly brighter than 7.0.

2 Pallas has an orbit of higher inclination (35°) than other large asteroids. The orbit's orientation, with descending node not far west of the vernal equinox point, puts Pallas usually nearer to our celestial equator than to the ecliptic.

Pallas avoided opposition in 2022, diving behind the Sun. It was farthest south (-32°) on Dec. 25, and at the beginning of 2023 climbs past the stars that form the feet of dog Canis Major. It is at opposition on Jan. 8, which means it is 180° from the Sun in longitude; but is still 31° south. Its magnitude is 7.6; this can range from 5.5 as in 2028 Mar. to 9.5 as in 2020 July. It repasses Sirius in late February; is at perihelion on Mar. 6; climbs northward in the evening sky, to cross the ecliptic plane in August, by which time it is 3 AU away from us and has dimmed to magnitude 9.

3 Juno, though among the first asteroids discovered, and though named for the queen of Roman goddesses, is small-

Asteroids' paths in the year, drawn thicker when brighter. Ticks are at start of each month.

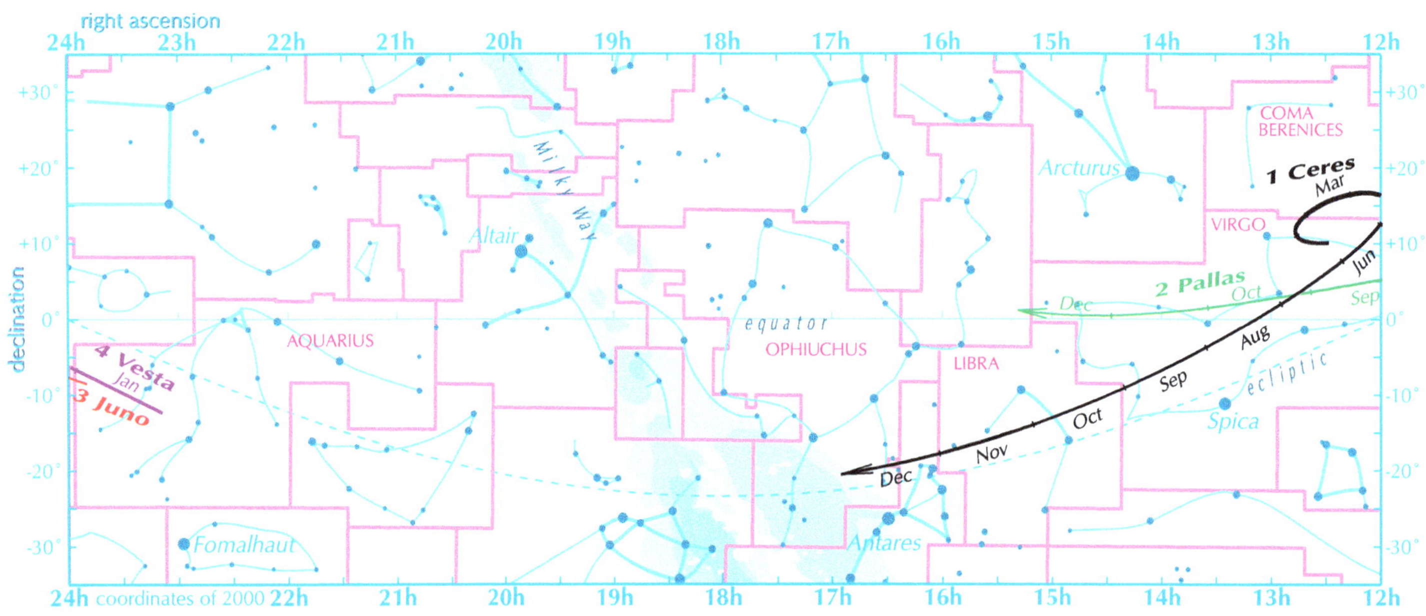

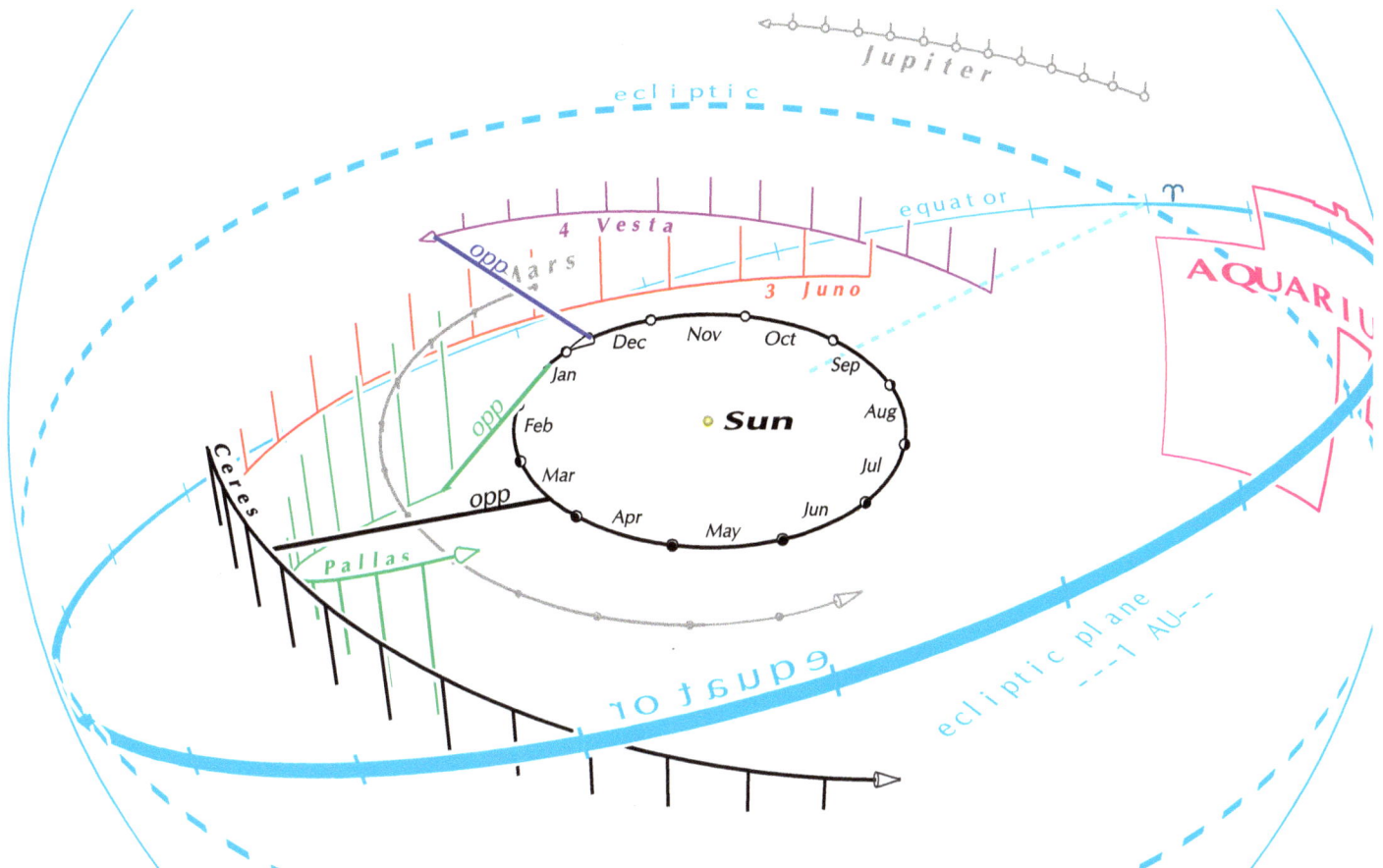

A sphere 3.5 AU in radius, seen from a viewpoint 10.5 AU from the Sun and 35° north of the ecliptic. Stalks to the ecliptic plane are at the start of each month. Lines connect Earth and asteroid at dates of oppositon.

er than the others of the First Four, and is in most years surpassed in brightness by at least one of those discovered later.

Last year was the closest and brightest for Juno since 2017, but 2023 is poor, with no opposition. Though at perihelion in April, Juno is on the far side of the Sun, passing behind it in the middle of the year. The best chances to see it are at the year's beginning and end, when it is fairly far out on the evening and mornin sides respectively.

4 Vesta, though only the 4th asteroid discovered, is the second largest, slightly wider than Pallas but considerably more massive. It is the brightest almost always (Ceres when brightest is 0.1 magnitude brighter than Vesta when faintest). This results from a combination of size; distance (it is usually nearer in than Ceres and Pallas, its aphelion being slightly farther out than Ceres's perihelion); and high "albedo" or reflectivity—that is, Vesta has a light-colored surface. It is aptly named for the Roman goddess of the hearth.

In 2023 Vesta continues to travel behind, south of, and

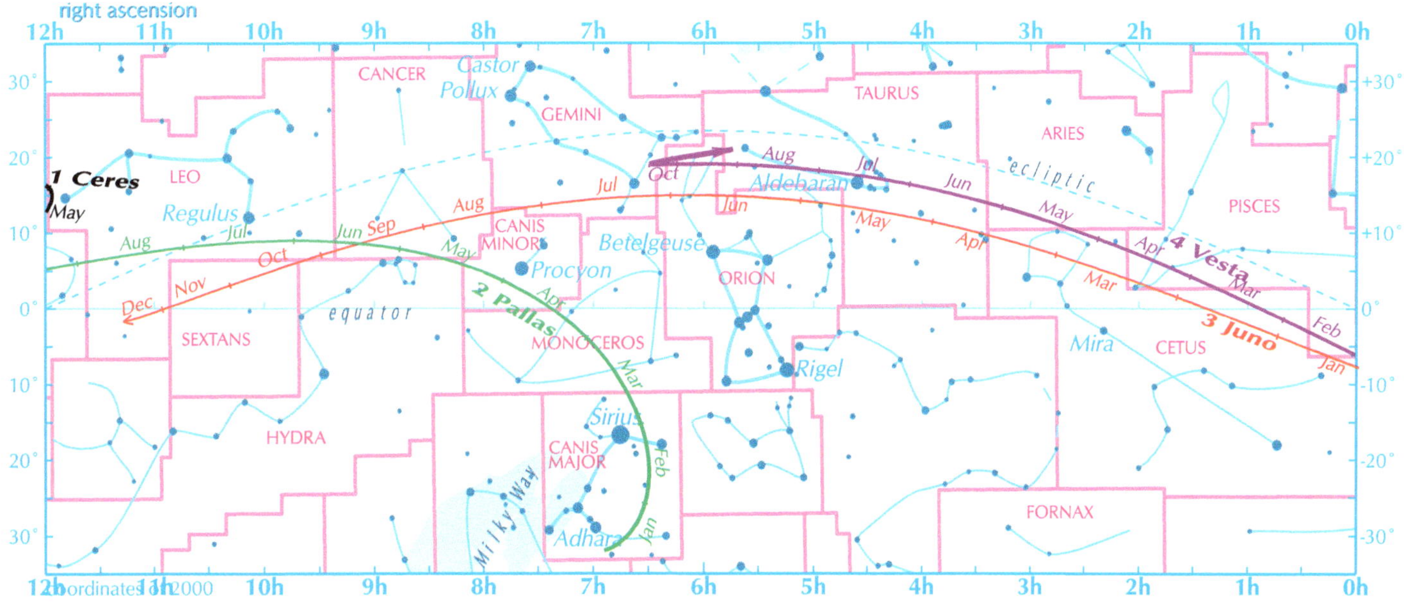

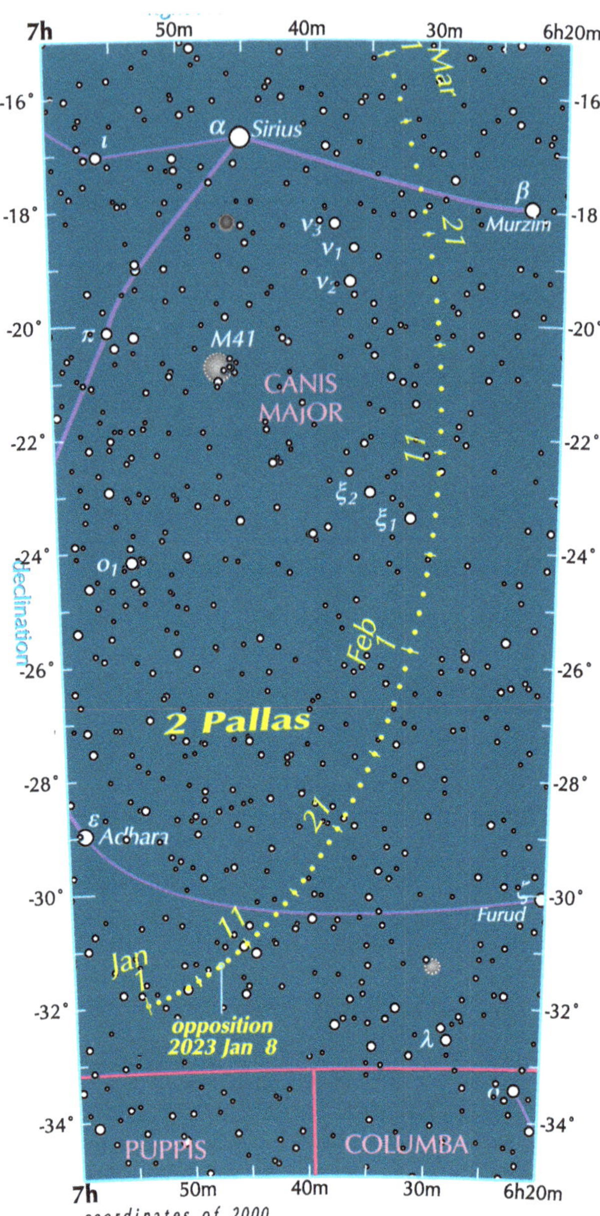

outside of Juno. At first it is in the evening sky; behind the Sun in April. In July it crosses the Hyades star cluster, passing close north of Aldebaran. As we overtake it, its retrograde path takes it from Gemini back into Taurus. At opposition on Dec. 21, it should reach magnitude 6.4: findable by the naked eye in good conditions. It can be as bright as 5.3, as in 2018 June and 2029 July, but is close to its dimmest opposition magnitude of 6.5 as in 2019 Nov. and 2030 Nov., because it was at aphelion in October.

Opposition is the center of the couple of months when an asteroid is nearest and brightest (and appears to be retrograding in the sky as we overtake it). So its exact date does not greatly matter. Opposition dates given here are calculated it in longitude, as for the planets; in other sources they may be slightly different because calculated in right ascension.

Phenomena. Columns: right ascension (hours, minutes) and declination (degrees, minutes), for epoch 2000; distance from Sun and Earth, in astronomical units; elongation from Sun (degrees; negative = westward); magnitude.

	RA(2000)	decl	hedis	gedis	elo	mag
1 Ceres						
Mar 20 21 opposition	12 28	15 29	2.569	1.600	-163	7.0
Nov 20 6 conjunc.with Sun	15 44	-16 41	2.725	3.711	3	8.6
2 Pallas						
Jan 8 8 opposition	6 48	-31 18	2.158	1.428	126	7.6
Mar 6 13 perihelion	6 39	-13 4	2.134	1.571	111	7.9
Oct 1 3 conjunc.with Sun	12 40	3 31	2.409	3.400	7	8.9
3 Juno						
Apr 2 13 perihelion	2 50	7 26	1.983	2.772	31	9.7
Jun 20 0 conjunc.with Sun	5 54	14 53	2.041	3.040	9	9.6
4 Vesta						
Apr 23 22 conjunc.with Sun	2 12	8 10	2.509	3.509	5	8.1
Oct 19 19 aphelion	6 28	18 59	2.570	2.059	-110	7.7
Dec 21 13 opposition	5 57	20 32	2.562	1.579	-177	6.4

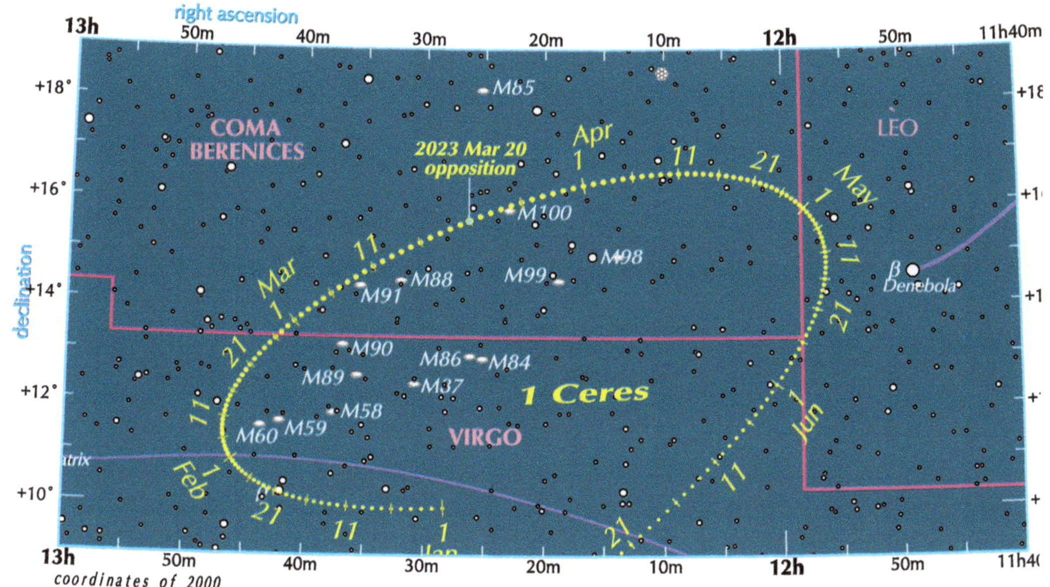

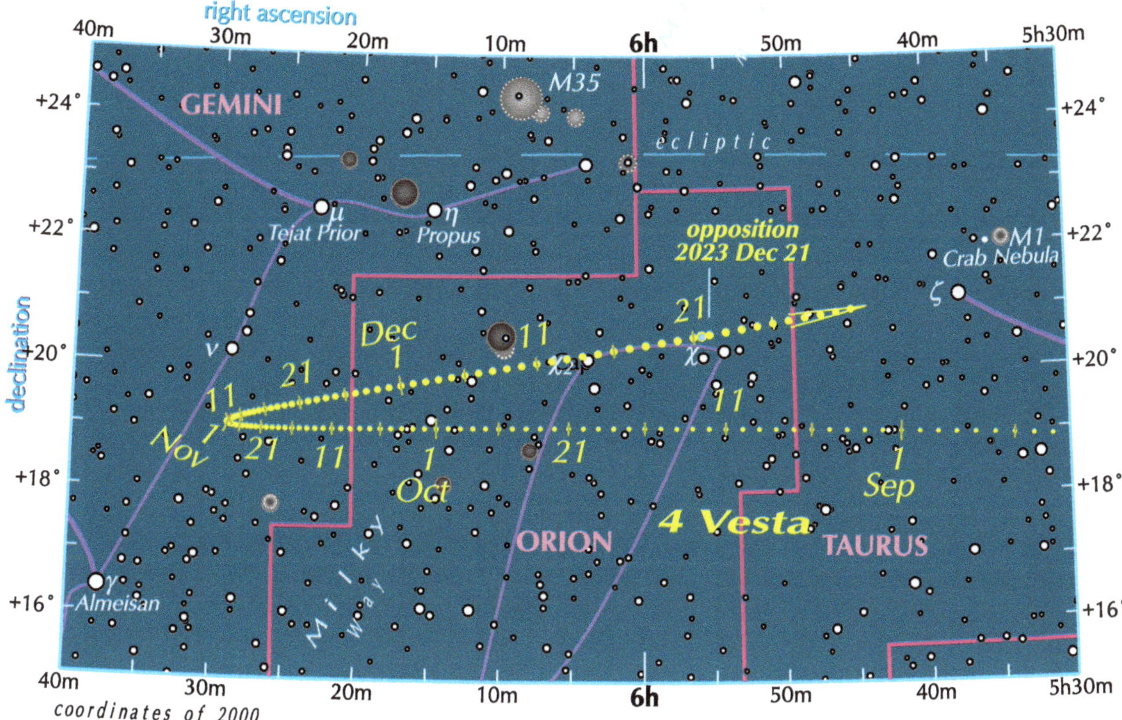

Favorability graph, as for the planets. The curve for each asteroid is blue when it is in the morning sky, black when in the evening sky, thicker when the asteroid is brighter.

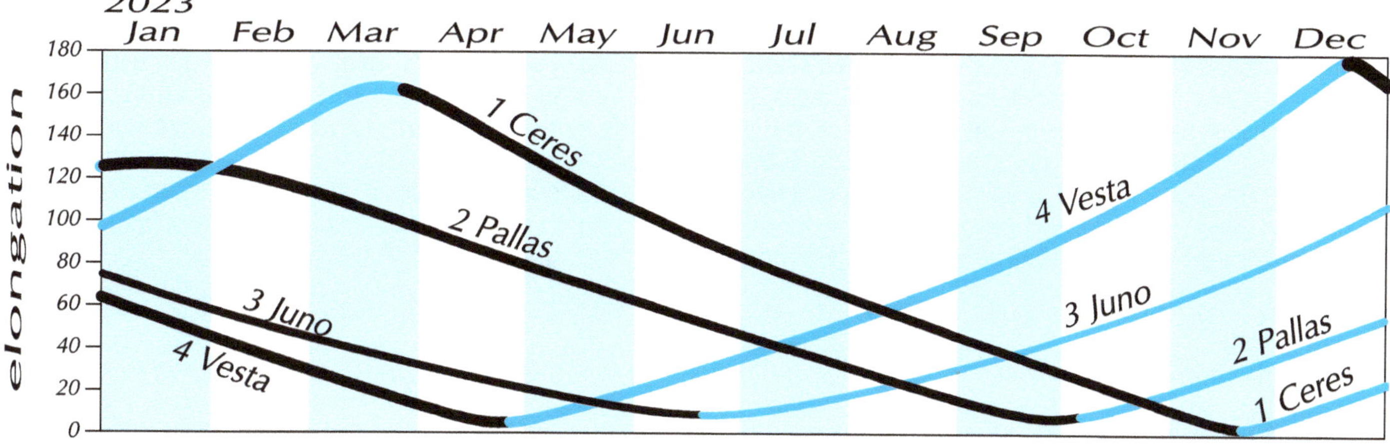

METEORS

These meteor showers are mentioned in our calendar. Here are more data about them.

	longit.	UT			active	radiant		drift	ZHR	vel	r	Moon
QUA	283.15	Jan 4	4	Quadrantids	Dec28-Jan12	230	49	2.4-0.2	110	41	2.1	full-3.0
LYR	32.32	Apr 23	1	Lyrids	Apr14-Apr30	271	34	4.4 0.0	18	49	2.1	new +2.8
ETA	45.50	May 6	15	Eta Aquarids	Apr19-May28	338	-1	3.5 0.4	50	66	2.4	full+0.3
ARI	76.60	Jun 7	9	Daytime Arietids	May14-Jun24	44	24	0.0 0.0	30	38	2.8	
SDA	127.00	Jul 30	18	Southern Delta Aquarids	Jul12-Aug23	340	-16	3.0 0.2	25	41	2.5	full-2.8
PER	140.00	Aug 13	7	Perseids	Jul17-Aug24	48	58	5.6 0.2	100	59	2.2	new -3.4
DRA	195.40	Oct 9	7	Draconids	Oct 6-Oct10	263	56	0.0 0.0	5	21	2.6	LQ +2.4
ORI	208.00	Oct 22	0	Orionids	Oct 2-Nov 7	95	16	2.6 0.1	20	66	2.5	FQ -0.1
STA	223.00	Nov 5	0	Southern Taurids	Sep20-Nov20	52	15	2.9 0.2	5	27	2.3	LQ -0.4
NTA	230.00	Nov 12	0	Northern Taurids	Oct20-Dec10	58	22	3.6 0.2	5	29	2.3	new -1.4
LEO	235.27	Nov 18	5	Leonids	Nov 6-Nov30	152	22	2.4-0.3	10	71	2.5	FQ -2.5
GEM	262.20	Dec 14	19	Geminids	Dec 4-Dec17	112	33	4.0-0.1	150	35	2.6	new +1.0
URS	270.70	Dec 23	4	Ursids	Dec17-Dec26	217	76	0.0-0.4	10	33	3.0	FQ +3.2

First column: standard abbreviation. Longit.: solar longitude. Peak times can be uncertain. Radiant: in degrees of right ascension and declination. Drift: in degrees per day. ZHR: zenithal hourly rate. vel: velocity entering atmosphere (km/sec). r: population index, the proportion of bright and faint meteors. r 2.0 or less: more meteors brighter than average; r 3.0 or higher: more faint. For sporadic meteors, r is typically 2.9-3.1. Moon: days from new, first quarter, full, last quarter.

About 40 are included in the **International Meteor Organization**'s yearly description, www.imo.net/files/meteor-shower/cal2023.pdf. But some are sparse or of debatable existence. The usual "big three" are the Quadrantids, Perseids, and Geminids. This year they respectively are near to full, new, and new Moon.

Meteoroids are bits of solid matter out in space. Encountering Earth's atmosphere, they are vaporized by friction, emitting light that is seen as **meteors** or "shooting stars"; sometimes leaving luminous **trains** for some minutes. **Fireballs** are of magnitude -3 or brighter; **bolides** cause sounds. Remnants large enough to reach the ground are **meteorites**. Most meteoroids are small — pebbles or dust — and have separated from comets (a few from asteroids), typically centuries ago. They orbit around the Sun in **streams**, appearing as **showers** when Earth passes through them at about the same dates each year.

Meteors of a shower can appear anywhere in the sky, but their apparent paths radiate from a **radiant** point or small area. For instance, a meteor of late November whose path can be traced back to Leo is a Leonid; otherwise, it is a **sporadic** meteor. Radiants drift slightly from day to day, because the direction from which the meteors appear to come changes as Earth proceeds around its orbit.

Particles are ejected from comets with differing directions and speeds, so streams become vastly wider than the Earth. So a shower may be active over weeks, though for part of this time it may have been detected only by camera or radar. A shower's **peak** may be broad and indefinite or as sharp as hours or minutes. Particles' orbits can be perturbed by the gravity of the planets, and the stream may contain clumps, causing meteor **storms**, and sub-streams, causing subsidiary peaks. The times we give, predictions by experts based on past evidence, may be only best guesses. The calendar dates can vary by a day because of leap years. So a more constant way to express the point in Earth's orbit at which a stream crosses it is the **longitude of the Sun**. Over centuries, the peak dates drift later through the year because of the precession of Earth's axis.

Counts of meteors per hour are the raw data. A shower's estimated **zenithal hourly rate (ZHR)** is the average number that one obaerver could see at the peak time if the radiant were overhead and conditions perfect. Actual counts are liable to be lower. Observations help to define a stream's radiant, peak date, composition, orbit, and origin.

Meteors hitting Earth's front — because traveling in retrograde orbits — are seen after midnight and enter the atmosphere at higher speeds. Meteor observing tends to be best before morning twilight!

Through most of the year, there are meteors coming at low rates from a large area roughly opposite to the Sun (culminating about 1 AM local time). Formerly they were treated as many minor showers, but since 2006 they have been treated as a general **Antihelion Source**.

A radiant at or below the horizon can produce long bright trails across the upper atmosphere. However, the lower the radiant, the fewer meteors from it aim above the horizon, so more meteors are seen when the radiant is at least 25° up.

The other factors are atmospheric conditions and the **Moon**: its glare drowns the dimmer meteors. A first quarter Moon sets around midnight, so does not spoil morning observation; last quarter rises around midnight.

Wrap warmly; be comfortable, as on a reclining chair; perhaps face east and gaze up at about 45°. Count the shower members seen in an hour. Record sporadics separately. If there's more than one of you, keep separate counts.

The clearinghouse for meteor information is the International Meteor Organization. If seriously interested, and if you'd like to report your counts, explore www.imo.net.

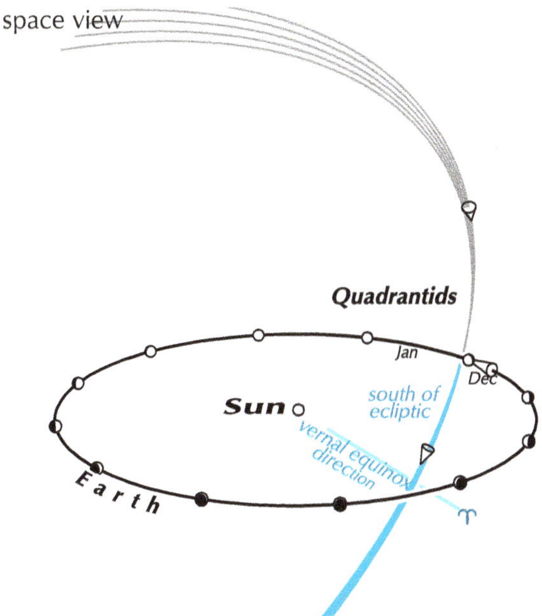

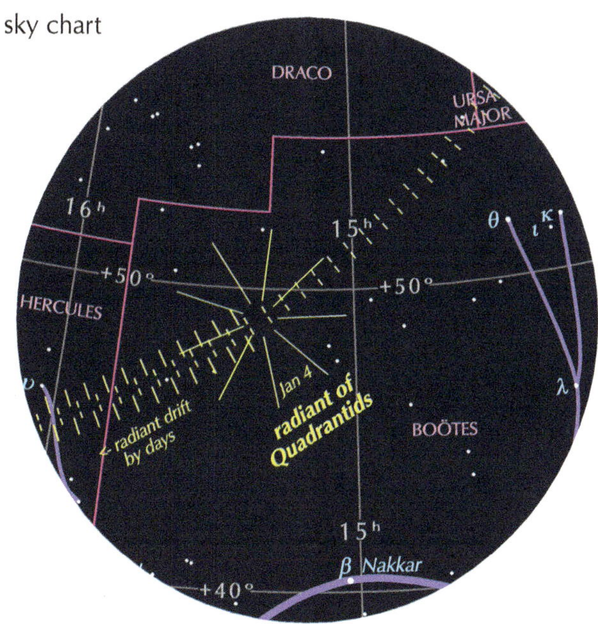

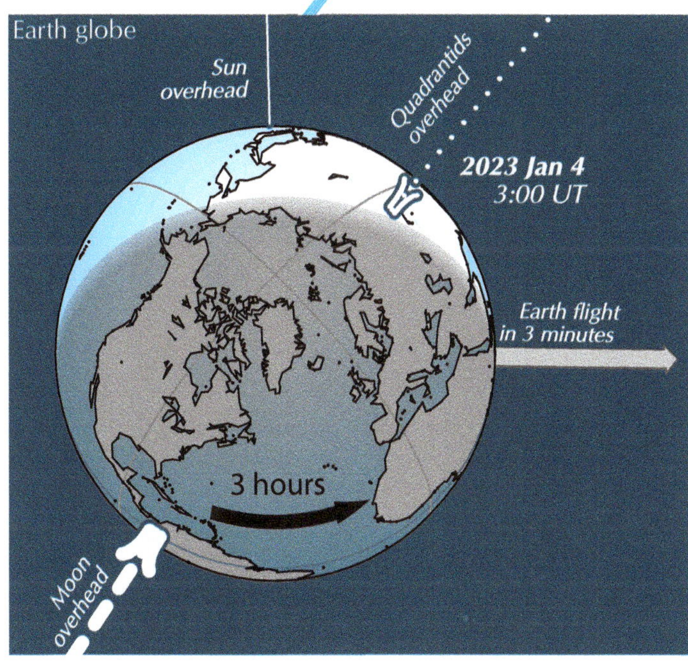

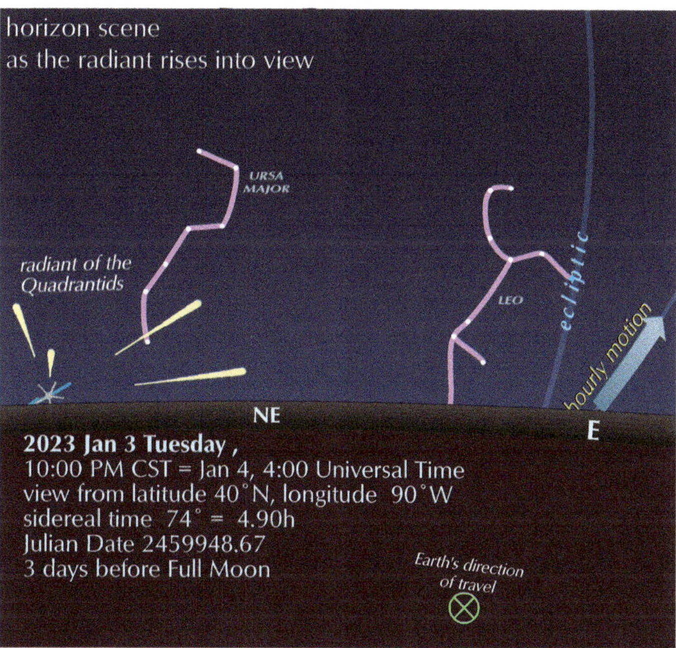

Jan 4: **Quadrantids**, named from a defunct constellation, Quadrans Muralis, the "wall quadrant", in a star-poor region east of the Big Dipper. The meteors come to us from the north (inclination 72° to the ecliptic) and slightly behind, so their relative speed is medium. For people north of latitude 41°, the radiant is in the sky all the time. It is low to the northern horizon in the early night, swings up in the northeast, is nearly overhead toward dawn. Though the ZHR is 120, the rate varies from 60 to 200. The main peak is fairly sharp. Faint Quadrantids caused by small particles may peak half a day earlier; there may be another peak some hours later, detected partly by radio observations. This year, **1 day before full Moon**.

Apr. 23: **Lyrids**. Derived from Comet C/1861 G1 Thatcher, which was seen only in 1861, having a period of 415 years. Coming down at about 80° into Earth's orbit, they are medium-slow; some are spectacularly bright; 20-25% leave persistent trains. The radiant, on the Lyra-Hercules border not far from Vega, is above the northeast horizon by 10 PM and overhead by 4 AM. The ZHR is usually around 15-20, but there were outbursts in 1803, 1922 (96/hour), 1982 (250/hour for a few minutes). The shower has the longest history, beginning with a Chinese chronicle of 687 BC (when the peak was on March 25) and summarized in Gary Kronk's invaluable book *Meteor Showers*. The shower has a narrow peak. **2 days after new Moon**.

May 6: **Eta Aquarids** are debris from the most famous comet, 1P Halley, which last came by in 1986 and will return in 2061. Its retrograde orbit crosses over the October part of Earth's orbit and back out just under the May part; so we see sister showers, the Orionids of October (inward) and the Eta Aquarids (outward). In both, the meteors are very swift (nearly head-on to Earth), often leaving trains; and there seem to be sub-streams spread over several days, with different average sizes of particles. The main radiant, near the Urn or Water-Jar or Y of Aquarius, is just south of the celestial equator. For latitude 40° north it rises about 2 AM and is highest toward 8 AM. For the southern hemisphere, now in autumn, there are more hours of viewing before dawn twilight, and Australians have said this is the best shower of their year. Hourly rates can be as low as 10 for northerners, as high as 85 for southerners. There is thought to be a 12-year periodicity caused by Jupiter, with one of the low-rate times in 2014-2016. Yet the 2013 May 6 Eta Aquarids were exceptionally strong, up to 140 an hour; meteor scientists think this extra dust separated from the comet three or four thousand years ago. **Full Moon**

June 7: **Daytime Arietids**. This shower is worth mentioning as the strongest example of a meteor stream that comes at us from the general direction of the Sun, therefore is barely observable except by radar. Yet some meteors from the radiant can be above the horizon in morning twilight, and the IMO has a project to collect and combine counts. The parent body may be the near-Earth asteroid 1566 Icarus.

July 30: **Southern Delta Aquarids.** Formerly there was thought to be a diffuse group, the Southern peaking around July 30 and the Northern Aug. 7. But the Northern stream has been found to be merely part of the Antihelion Source. The Aquarids are better for southerly observers, though for latitude 40° north the radiant is in the sky most of the night, highest around 2 AM. The meteors appear sparse, because they are spread widely, but may add up to one of the most massive of streams. Mostly faint, a few bright; 5-10% leave persistent trains; they move medium-slowly, because coming in sideways across Earth's orbit. **3 days before full Moon**.

Aug. 13: **Perseids**. Their morbid nickname "St. Lawrence's Tears" — he was martyred on a hot gridiron 258 Aug. 10 — may date back only to 1839. Long regarded as our most reliably great shower, now rivaled by the Geminids. Derived from Comet 109P Swift-Tuttle, which, with period around 130 years, appeared in 69 BC, AD 188, 1737, 1862, and 1992. The radiant, in the region where Perseus meets Cassiopeia and Camelopardalis, is in the sky all night (for northern latitudes), at first low in the northeast, overhead toward 6 AM. The orbit is steeply inclined to Earth's (113°, technically retrograde), hence passes near no other planet and is little perturbed. Records of the shower go back to China in AD 36 (when it was in July), Europe in 811. The 1866 occurrence was the first for which the link with a comet

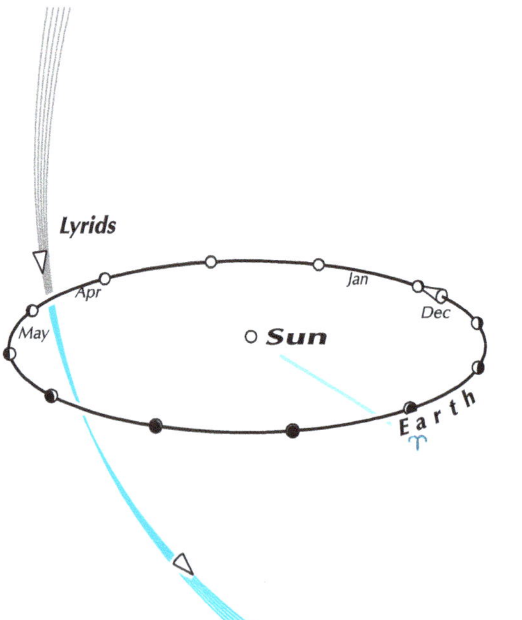

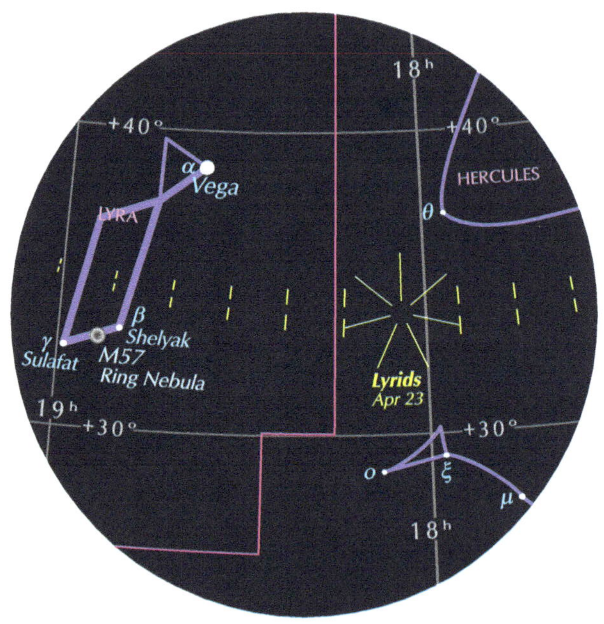

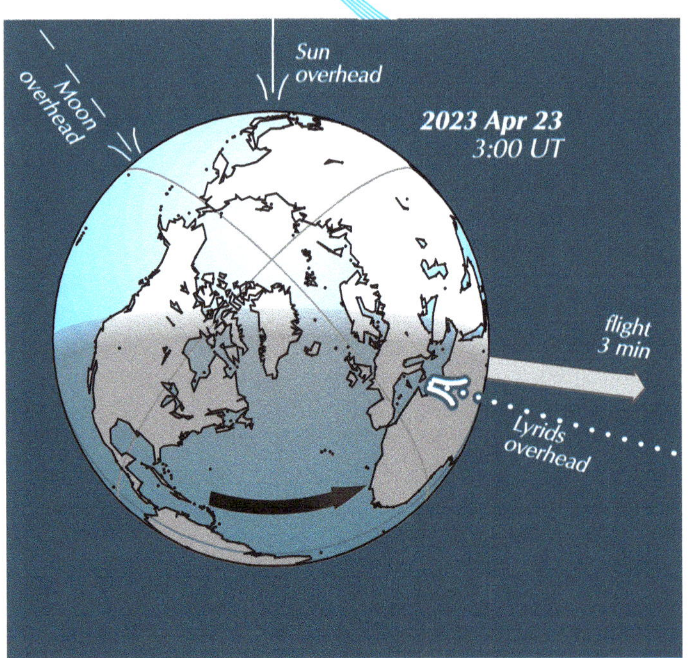

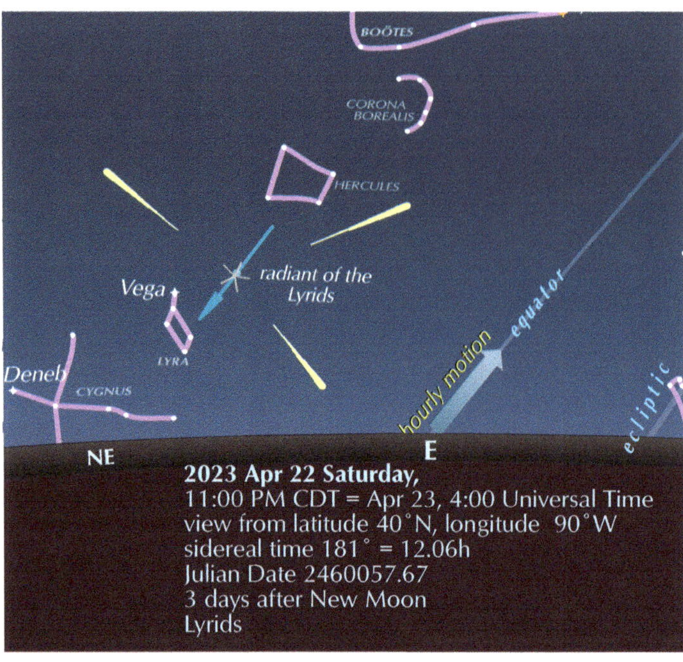

was made, by Schiaparelli. There were some amazing Perseid outbursts in 1980, the 1990s, and 2004. Numbers tend to rise slowly to the peak of 50 or more per hour, then drop faster. Sometimes two peaks have been noticed, or more. The meteors are swift, which helps to distinguish them from the far less numerous Aquarids and Capricornids of the same time. Many are bright; white, yellow, green, red, orange; leave spectacular long-lasting trains; end in flares. **3 days before new Moon**.

Oct. 9: **Draconids**, also called Giacobinids, because derived from Comet 21P Giacobini-Zinner, which in its 6.6-year orbit passes close to Earth's, last doing so in 2018. The radiant is in the Lozenge or head of Draco, only 13° from the north ecliptic pole, so that unlike other radiants it scarcely shifts from day to day. Descending steeply into the plane of the ecliptic, and from not far out, the meteors are slow-moving. Most are faint, some brilliant, some fragment easily. The radiant is high in the early night, low to the northern horizon 3-6 AM. In many years no Draconids are seen; in others the ZHR reaches 20 or 400. There have been storms near the comet's perihelion, as in 1926 (a Draconid fireball "lit up the sky"), 1933, 1946 (15 days after the comet passed; up to 10,000 an hour seen in the southwestern USA in full moonlight), 1985, 2005. **2 days after last quarter Moon**.

Oct. 22: **Orionids** are part of the stream coming inward along the approximate orbit of Halley's Comet, to be seen on the way out as the Eta Aquarids of May. The Orionid radiant, in the club of giant Orion near the feet of the Gemini twins, rises around 9-10 PM for mid-northern latitudes and is low till after midnight. Orionids are, like the Eta Aquarids, swift; they are sometimes bright, and more than half leave persistent trains. The ZHR can rise to 70, and sometimes there is more than one peak; there could be a strong sub-peak around Oct. 17. There were strong showers 2006-2009, but there may be a 12-year cycle (caused by Jupiter), 2014-16 being a "trough". **Near first quarter Moon**.

Nov. 5, **Southern Taurids**, and Nov. 12, **Northern Taurids**: a complex of streams derived from 2P Encke, the comet with the shortest period (3.3 years) and most frequent visits. The meteors radiate from a large area that moves

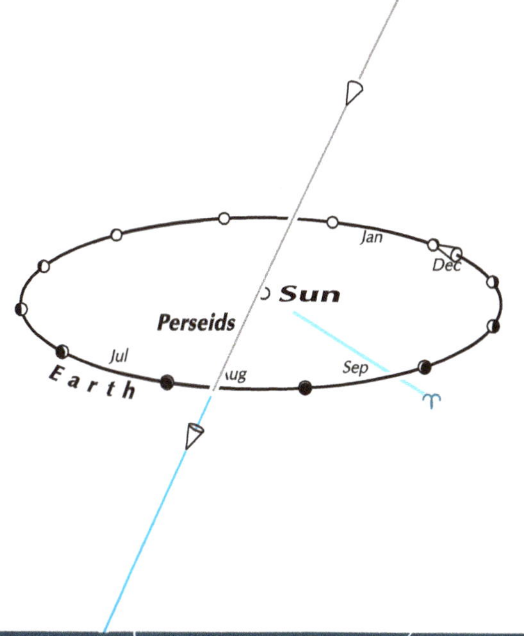

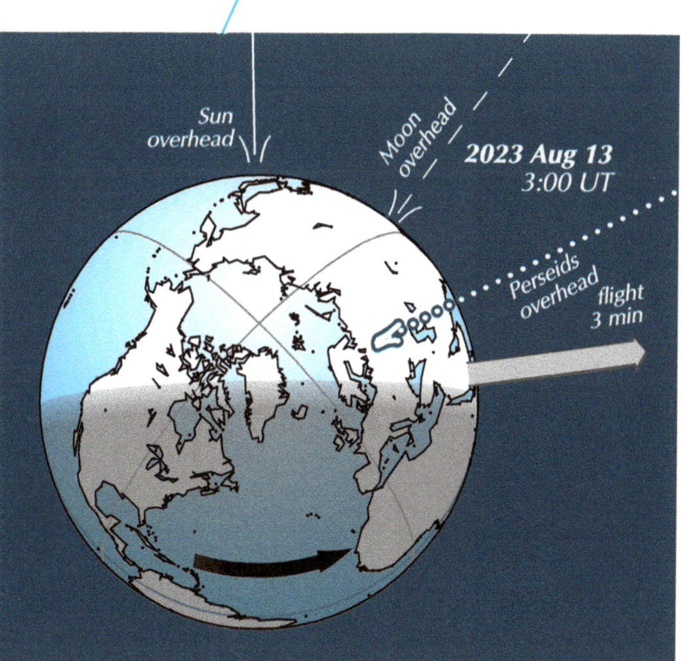

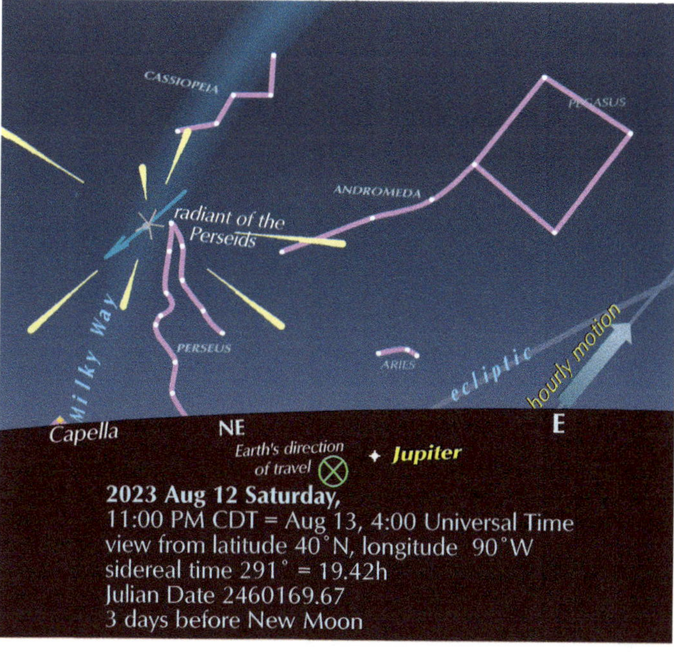

east along the ecliptic from Pisces through Aries into Taurus and is in view most of the long autumn night, highest about midnight. Spread over this time, they appear sparse on most nights. Because the general orbit lies in the inner solar system, with outer end near Jupiter, the stream has become perturbed into branches, which can scarcely be distinguished by visual observers. The most abundant component, the Southern, has an additional peak around Oct. 13, which used to be considered the main one. In some years we pass through a Taurid "swarm" with bright meteors from large particles; in 2005 they were popularly dubbed "Halloween fireballs", and a flash seen on the Moon was thought to be a Taurid impact. Taurids appear slow, because they are coming in across our orbit from behind. As the stream goes back out, it encounters Earth's daytime side and thus produces meteors detectable only as radio showers, the Zeta Perseids and Beta Taurids of June.

Nov. 18: **Leonids**, the most dramatically variable of all. Following the path of Comet 55P Tempel-Tuttle, they strike Earth's atmosphere slightly north of head-on (the inclination is 162°) and pierce it at almost the highest theoretical speed for meteors belonging to the solar system. So the shower is a morning one: the radiant, in the head of Leo (also called the "Sickle"), rises about 11 PM and is highest about 6 AM. Leonids are often bright, bluish; most leave persistent trains. Often only 5 to 20 per hour are seen at the maximum, but fantastic storms happen, usually but not always near the times of the comet's perihelion in its 33-year orbit. Many were vaguely recorded in early annals (902, Arabic "Year of the Stars"). In 1833, thousands per hour, woke people from their beds in eastern North America. This inspired Denison Olmsted to understand radiants and the periodic orbiting of the particles, thus founding meteor science. 1966 Nov. 17 was the most intense meteor storm known. At Kitt Peak in Arizona, Dennis Milon recorded 160,000 per hour for about 20 minutes; that is, over 40 per second! The comet's last visit was in 1998, the last storm in 2002. **1 day after last quarter Moon**.

Dec. 14: **Geminids** have in recent decades rivaled the Perseids as most reliable among the annual showers. Instead of a sharp peak they have a "plateau" of about 22 hours, during which from 50 to 130 an hour may be seen.

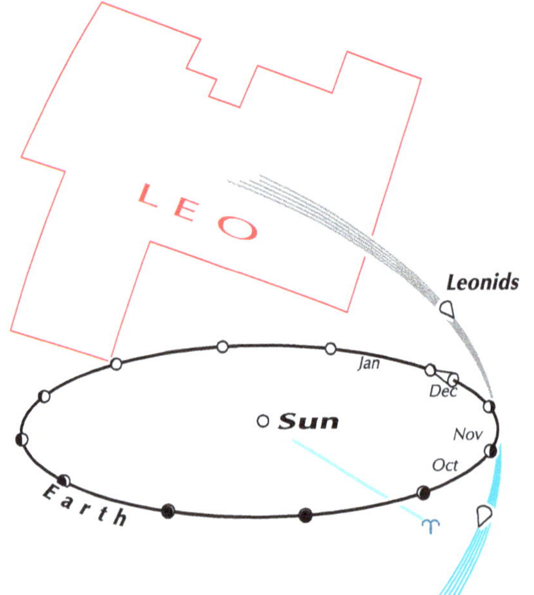

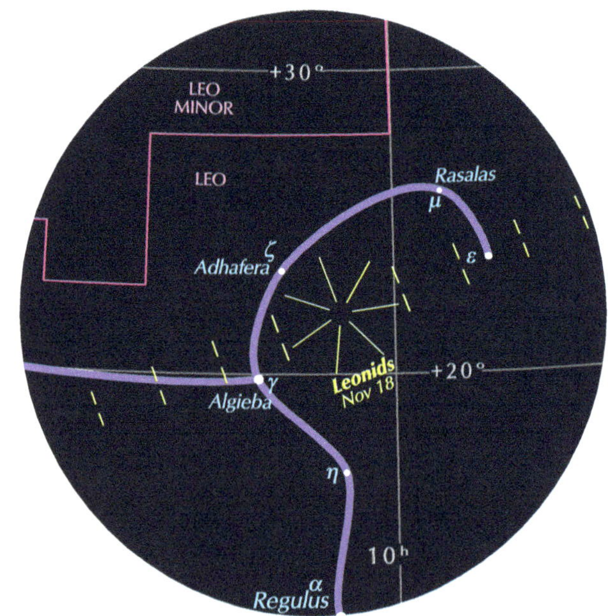

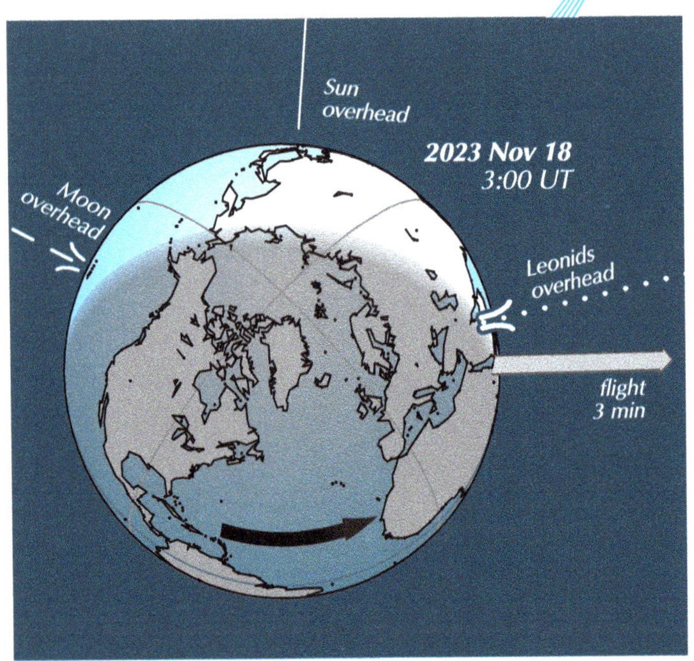

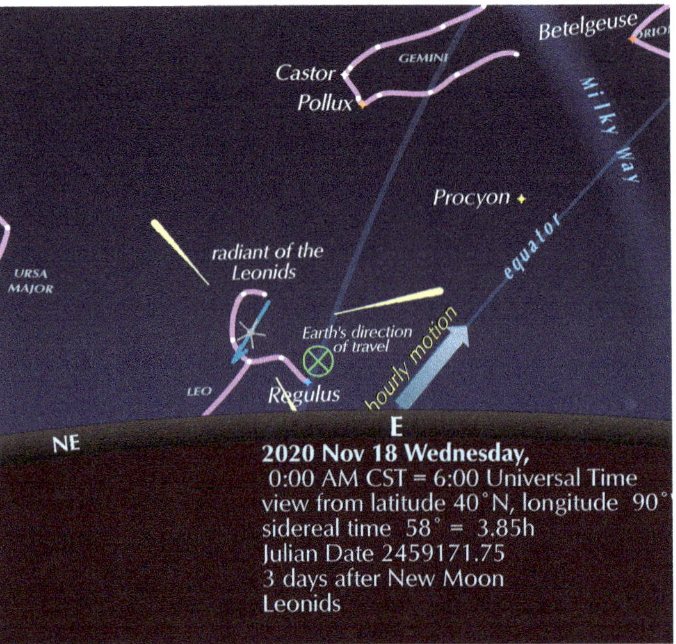

Their radiant, near Castor, is up for almost all of the long (northern) winter night, highest at 2 AM (so I once did a "star vigil", logging Geminids and waiting for the constellations that first set to come back around into view). Geminids are medium-slow because coming sideways into Earth's orbit. Their orbit is short, so it was puzzling that no parent comet was known, till in 1983 Fred Whipple pointed out that 1983 TB, discovered with the IRAS satellite, was in a "virtually coincident" orbit. Later named 3200 Phaethon, it is an asteroid (or possibly the denuded remains of a comet), 5 km wide, with a 1.52-year orbit, shorter than any comet's, passing over Earth's orbit by less than 1/10 of the Moon's distance (its last close approach was in 2017), then dipping 3 times nearer than Mercury to the Sun. This rocky origin may explain the nature of the Geminids: mostly bright, very few leaving trains. **1 day after new Moon**.

Dec. 22: **Ursids** radiate from near Kokab, the β star of Ursa Minor, at the other end of the Little Dipper from Polaris. They were also called (before names were regulated by the International Astronomical Union) Ursa-Minorids or Umids (would a shower from Ursa Major be Umads?). This radiant is (for latitude 40° north) in the sky all night, fairly low over the north horizon in the early night, almost overhead by dawn. These interesting and under-observed meteors fill a long cold winter-solstice night; it could be an even better pretext for a Star Vigil than the night of the Geminids! The parent comet is 8P Tuttle, which at intervals of 13.5 years drops steeply from the north through a perihelion close to Earth's orbit; its last perihelion was in August 2021. (Comet and meteors revolve almost in the plane of the Milky Way, though in the opposite direction to that of the stars.) The meteors are of medium speed, mostly faint but with a few fireballs; during the shower's brief peak 9 or 10 per hour may be seen, up to 50 especially when the comet is near; in 1945 and 1986 the rate was over 100. **3 days after first quarter Moon**.

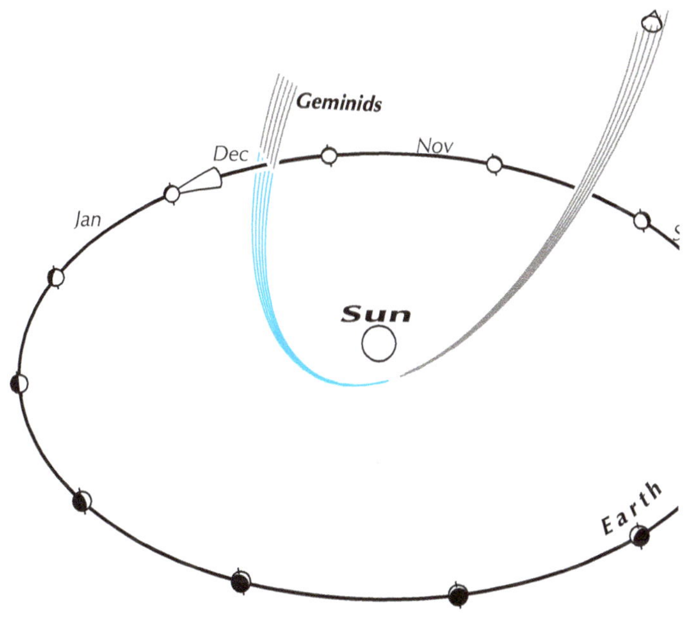

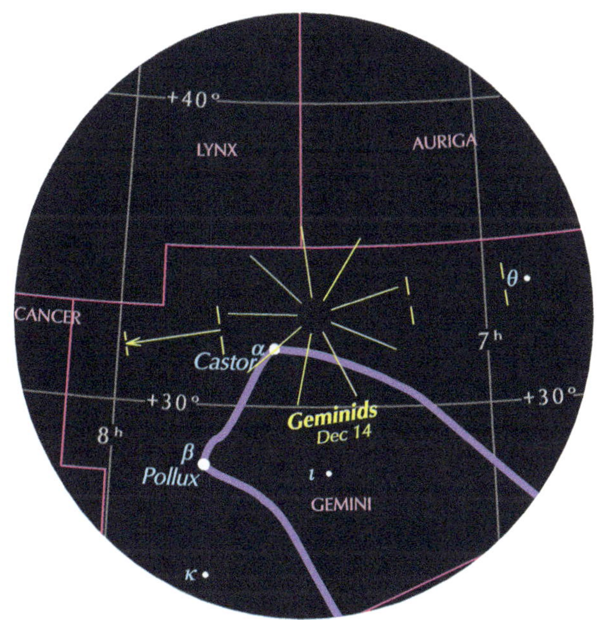

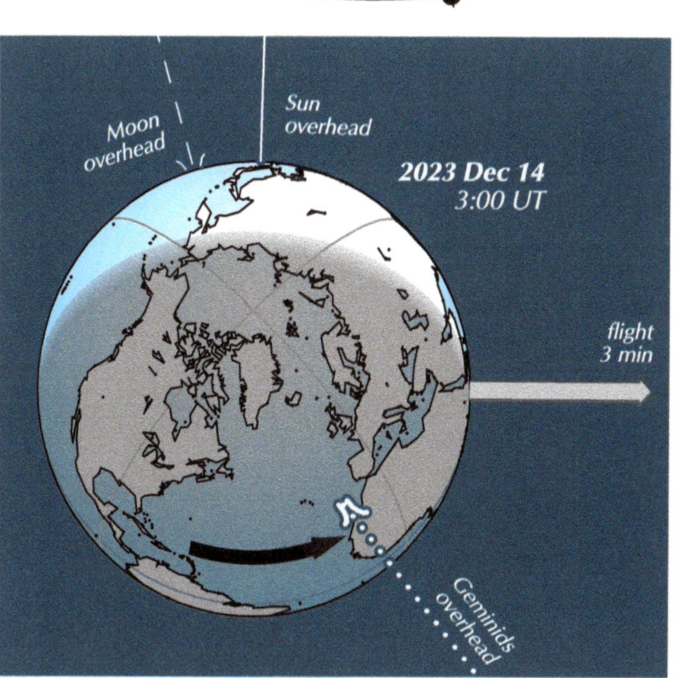

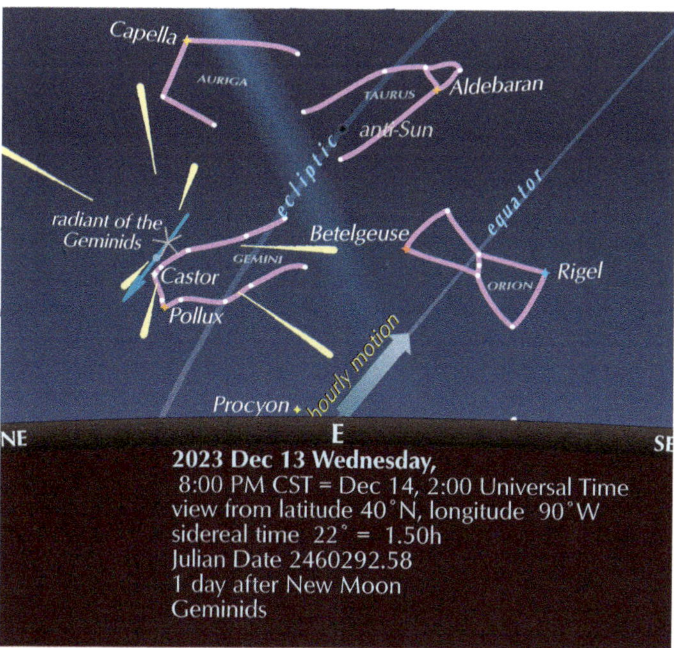

MAGNITUDE

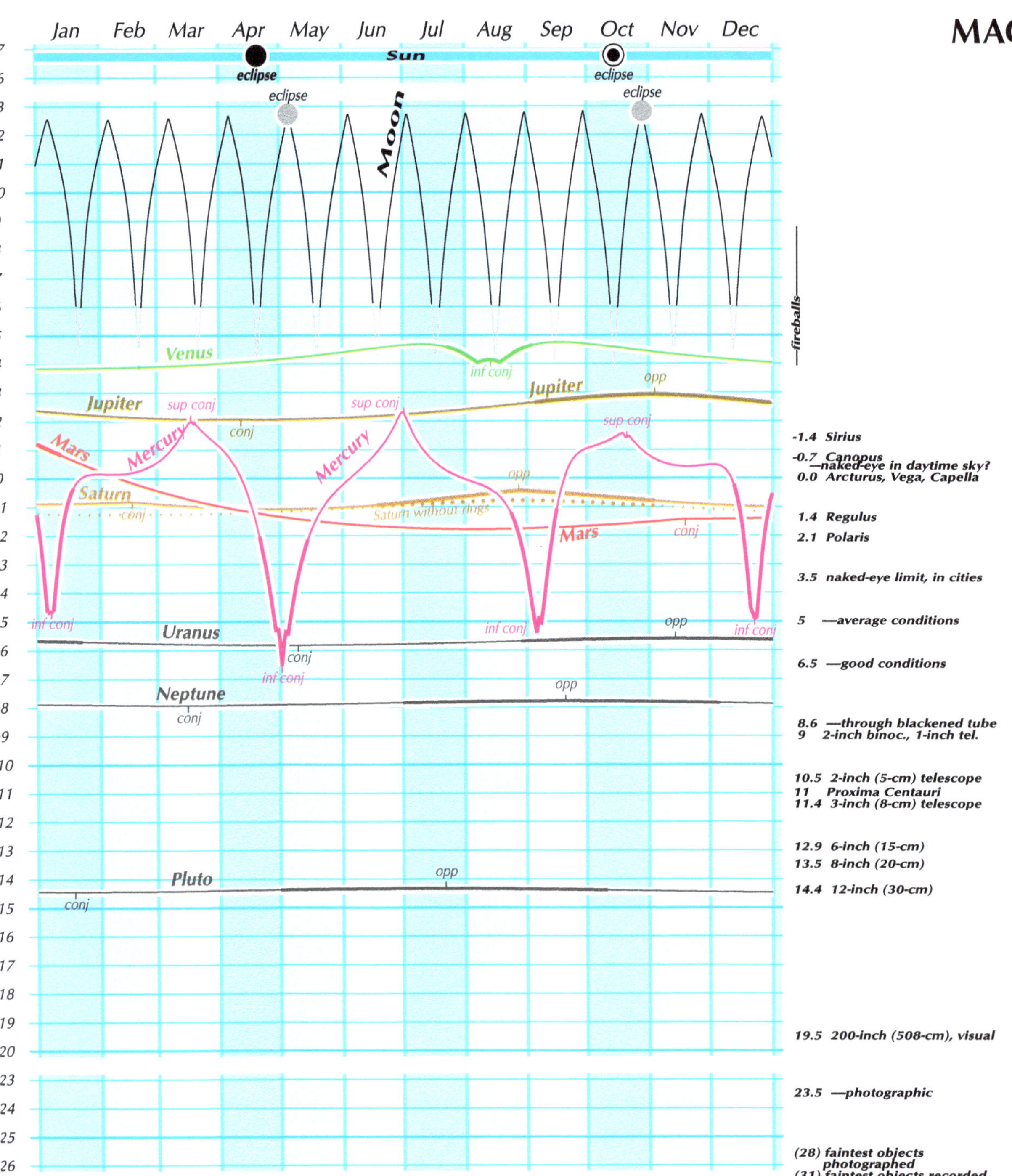

Magnitude is the astronomical way of measuring brightness. Each magnitude is roughly 2.5 times brighter than the one below it. (Magnitude 5 is exactly 100 times brighter than magnitude 10.) This graph shows the apparent magnitude (that is, as seen from Earth) of solar-system bodies.

Ticks on the Mercury and Venus curves mark superior conjunctions with the Sun (upward ticks) and inferior conjunctions (downward). For other bodies they mark opposition (upward) and conjunction with the Sun (downward).

Superior planets (Mars outward) are brightest near opposition. At superior conjunction Venus brightens slightly, Mercury greatly, because presenting full faces toward us, though we can't see Mercury past the Sun. Venus is brightest near maxima of elongation. Planets other than Venus are on an upward slope of brightness when seen in the morning sky, downward in the evening sky.

The curves are drawn slightly thicker when the bodies are retrograding, this being the time when Earth is closer to them. A dotted curve shows the brightness of the ball of Saturn alone, without its brilliant rings. Gradually closing since the northern face was most open to us in 2017, they contribute between 0.4 and 0.1 of a magnitude.

The Moon at first and third quarter is not, as one might suppose, half as bright as full (which would put it only 0.75 magnitude lower on the graph) but only about 1/11 as bright (2.6 magnitudes lower). See *Ast. Companion*, MOONLIGHT.

Magnitudes given are visual. Photographic magnitudes are about 0.7 to 0.9 greater (fainter); the peak sensitivity for traditional astronomical film is slightly blue-ward from that for the human eye. Magnitudes are calculated taking into account phase-angle (the angle Sun-body-Earth; hence, the part of the body that is not in shadow).

ELONGATION

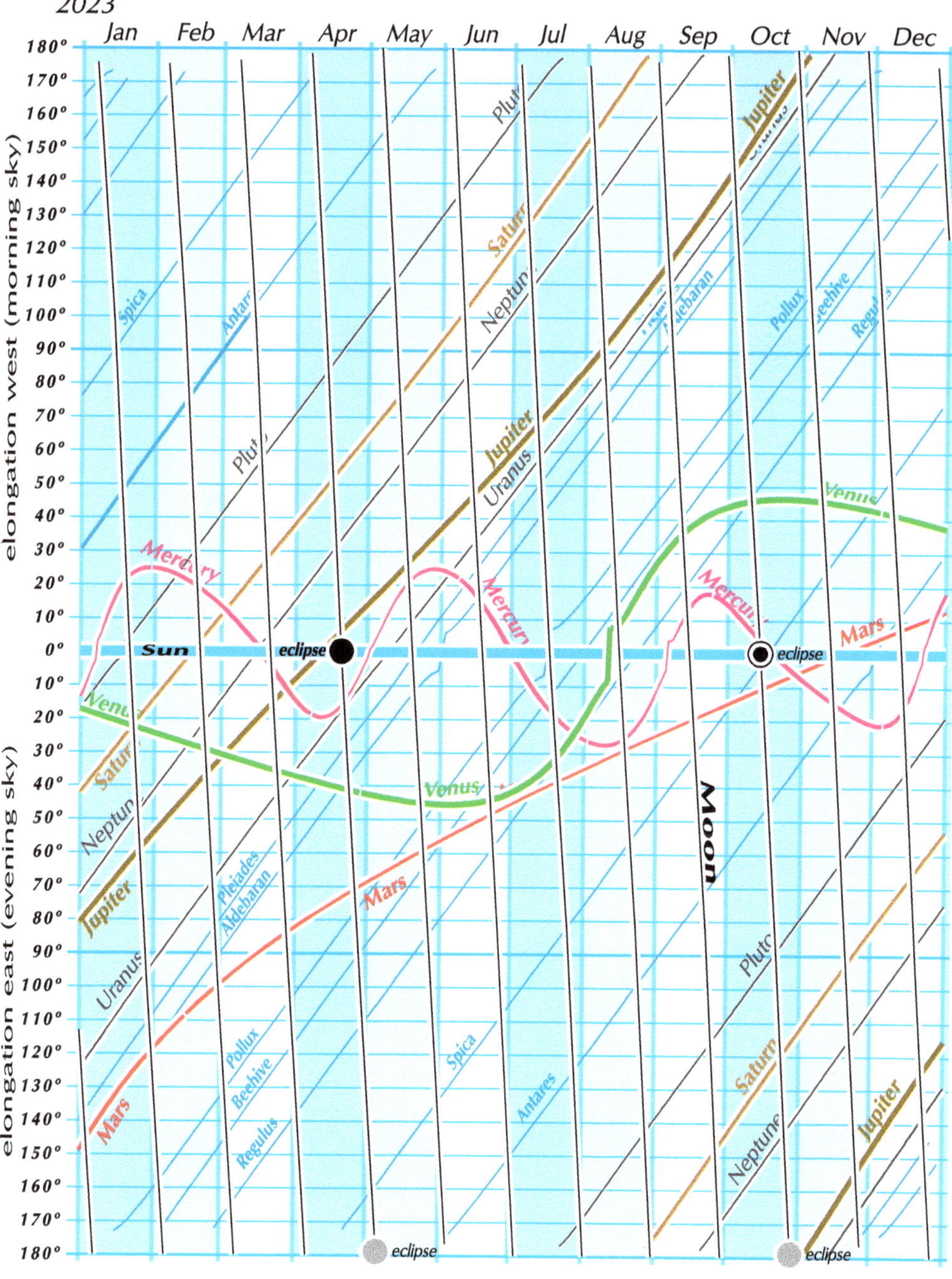

Try rotating the page so that January is at the top and the planets moving downward. There are two ways to look at it:

—Planets revolving around a star, as seen from one of those planets. Mercury and Venus spiral around the Sun. The others slip always backward (Mars slowest), because, being farther out than us, they are losing the race with us around the Sun. The Moon, blazing repeatedly across the foreground, describes a kind of time-cylinder around us.

—Imagine the diagram cut in half along the Sun-line, and put back the other way around: 0° (the Sun-line) at left and right, and 180° down the center. The Sun-line now represents the dawn horizon (on the left) and the sunset horizon (on the right). The line down the new middle (180°) is the meridian at midnight; the graph has become a graph of the night, from sunset to sunrise. On the new right, the superior planets sink into the sunset horizon; Mercury bobs out and drops back; the young Moon leaps repeatedly out. In the new middle, the superior planets cross the midnight meridian at their oppositions. On the new left, the dawn horizon, the superior planets emerge from their conjunctions with the Sun; Mercury keeps bobbing out; Venus heaves out for most of the year; the waning Moon dives repeatedly. Mars is the exception, failing to cross the 0° line in some years and the 180° line in others.

Crossings of lines represents a conjunction. A crossing of the 0° line is a conjunction with the Sun; of the 180° line, an opposition. At 90° east or west of the Sun, a planet is said to be at east or west "quadrature." The Moon is new when it crosses 0°; full at 180°; at first and last quarter when it is 90° east or west. The graph reveals the times when the Moon joins groupings of planets and stars.

The greatest elongation for Venus can vary from about 45.4° to 47.3°. This year it reaches 46.4° west. The greatest for Mercury varies between 17.87° and 27.83°; this year it varies between 17.9° and 27.4°.

Elongation really means angular distance measured from the Sun in any direction, not just along the ecliptic (difference in longitude). Even at conjunction, a planet is usually a little north or south of the Sun; at opposition it is north or south of the anti-Sun point. So elongation usually doesn't exactly reach 0° or 180°; instead, the lines on the graph curl away a little before reaching these limits, then after a jump resume on the other side.

Stars, too, have elongation from the moving Sun. Shown are the five bright stars and two clusters (Pleiades and Beehive or Praesepe) near the ecliptic, often seen in conjunctions with the Moon and planets. They serve to relate the rest of this time-diagram to the spatial background of the sky. Pollux, lying more than 7° north of the ecliptic, cannot have elongation less than that. Stars farther from the ecliptic would curl away even more; a star at the ecliptic pole always has elongation 90°.

The graph strictly shows the bodies' angular relations to the Sun, not to each other. Yet it serves to reveal times when they move side by side or in contrary directions, leave large parts of the sky bare, or gather in knots (visited fleetingly by the Moon).

RISING AND SETTING

This graph shows times when the Sun, Moon, and planets rise and set, for latitude 40° north, longitude 0°. (The times would differ little for other longitudes, much more for other latitudes.) These are mean solar times, roughly the same as Standard Time for your time zone.

Midnight is in the middle because we choose to show night rather than day undivided. Each day-line ends at the midnight point where the next day starts, so there is really just one time-line, a cut and flattened helix.

The dark hourglass-shaped area is night, between the curves of sunset and sunrise. The three gray bands are civil, nautical, and astronomical twilight, defined as being when the Sun is less than 6°, 12°, and 18° below the horizon.

The Moon is drawn at the moments each day when it rises and sets. Thus you can tell whether the Moon is in the sky at a time when you wish to observe. Dimmer objects are drowned out by moonlight, especially full moonlight. Full Moons are marked with a red cross. The Moon when full rises near sunset and sets near sunrise; vice versa when it is new.

Major meteor showers, such as the Quadrantids in January, are symbolized by bursts of radiating lines, at 1 AM, rather than at their predicted peak times (which can be uncertain) or the times when their radiants are highest (which can be in daylight). In general, meteor showers are strongest after midnight. See the meteors section for more on these showers and others.

Vertical red lines represent clock 6 PM and 6 AM, displaced by government rule an hour earlier in summer, when sunrise is earlier. The dates of displacement are for the USA; other countries have different rules. For more information on, and my opinion of, this interference with time, see www.universalworkshop.com/clock-shifting-times/

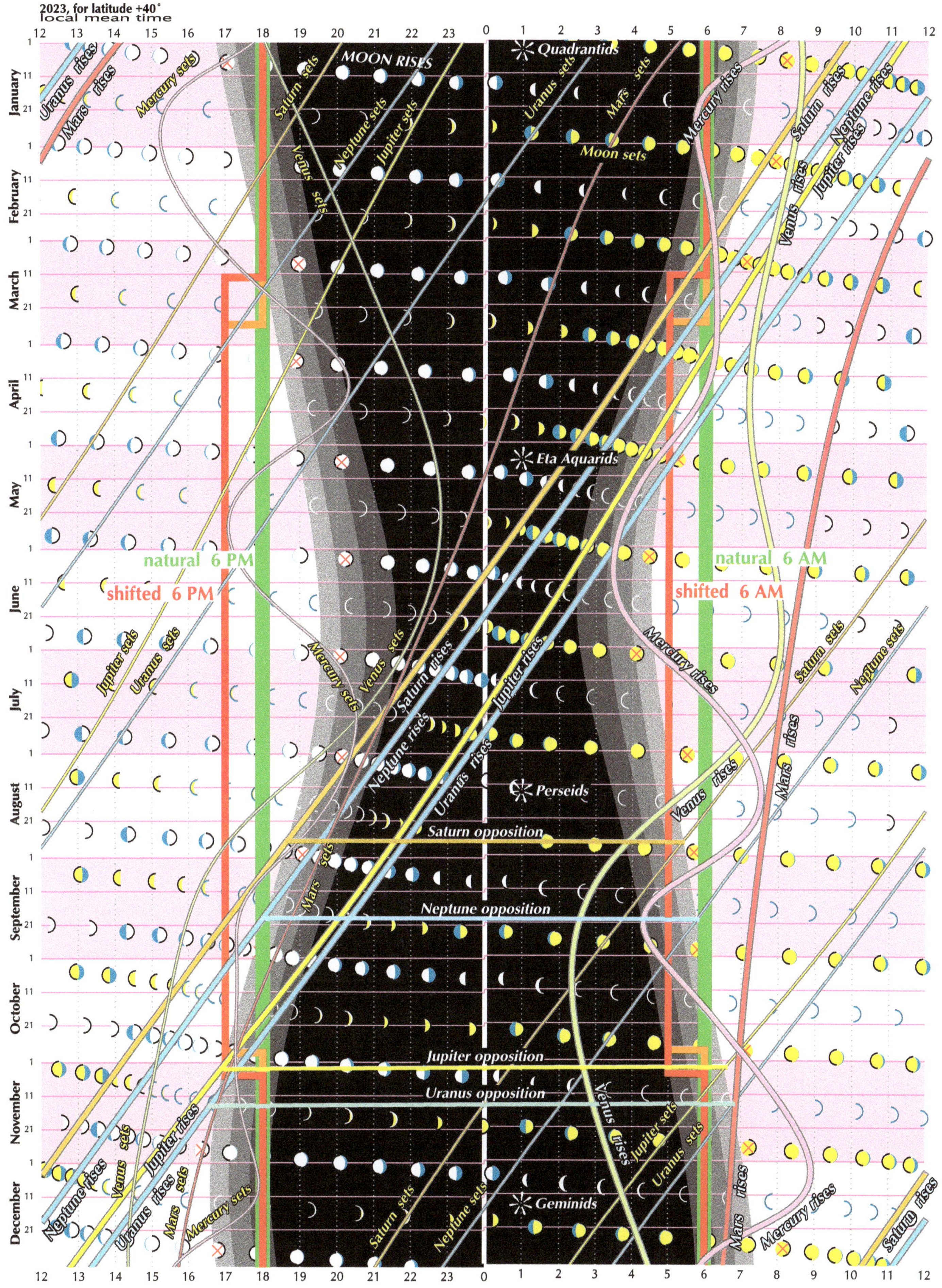
2023, for latitude +40°
local mean time
MOON RISES
Moon sets
Quadrantids
Eta Aquarids
Perseids
Geminids
natural 6 PM
shifted 6 PM
natural 6 AM
shifted 6 AM
Saturn opposition
Neptune opposition
Jupiter opposition
Uranus opposition
January
February
March
April
May
June
July
August
September
October
November
December

www.ingramcontent.com/pod-product-compliance
Lightning Source LLC
Chambersburg PA
CBHW051556030426
42334CB00034B/3455
9780934546904